NEUROMETHODS

For further volumes:
http://www.springer.com/series/7657

Neuromethods publishes cutting-edge methods and protocols in all areas of neuroscience as well as translational neurological and mental research. Each volume in the series offers tested laboratory protocols, step-by-step methods for reproducible lab experiments and addresses methodological controversies and pitfalls in order to aid neuroscientists in experimentation. *Neuromethods* focuses on traditional and emerging topics with wide-ranging implications to brain function, such as electrophysiology, neuroimaging, behavioral analysis, genomics, neurodegeneration, translational research, and clinical trials. *Neuromethods* provides investigators and trainees with highly useful compendiums of key strategies and approaches for successful research in animal and human brain function, including translational "bench to bedside" approaches to mental and neurological diseases.

Translational Research Methods for Tauopathies

Edited by

Jessica S. Fortin

College of Veterinary Medicine, Purdue University, West Lafayette, IN, USA

Humana Press

Editor
Jessica S. Fortin
College of Veterinary Medicine
Purdue University
West Lafayette, IN, USA

ISSN 0893-2336 ISSN 1940-6045 (electronic)
Neuromethods
ISBN 978-1-0716-5233-6 ISBN 978-1-0716-5234-3 (eBook)
https://doi.org/10.1007/978-1-0716-5234-3

This Humana imprint is published by the registered company Springer Science+Business Media, LLC, part of Springer Nature.
The registered company address is: 1 New York Plaza, New York, NY 10004, U.S.A.

Preface to the Series

Experimental life sciences have two basic foundations: concepts and tools. The *Neuromethods* series focuses on the tools and techniques unique to the investigation of the nervous system and excitable cells. It will not, however, shortchange the concept side of things as care has been taken to integrate these tools within the context of the concepts and questions under investigation. In this way, the series is unique in that it not only collects protocols but also includes theoretical background information and critiques which led to the methods and their development. Thus it gives the reader a better understanding of the origin of the techniques and their potential future development. The *Neuromethods* publishing program strikes a balance between recent and exciting developments like those concerning new animal models of disease, imaging, in vivo methods, and more established techniques, including, for example, immunocytochemistry and electrophysiological technologies. New trainees in neurosciences still need a sound footing in these older methods in order to apply a critical approach to their results.

Under the guidance of its founders, Alan Boulton and Glen Baker, the *Neuromethods* series has been a success since its first volume published through Humana Press in 1985. The series continues to flourish through many changes over the years. It is now published under the umbrella of Springer Protocols. While methods involving brain research have changed a lot since the series started, the publishing environment and technology have changed even more radically. Neuromethods has the distinct layout and style of the Springer Protocols program, designed specifically for readability and ease of reference in a laboratory setting.

The careful application of methods is potentially the most important step in the process of scientific inquiry. In the past, new methodologies led the way in developing new disciplines in the biological and medical sciences. For example, Physiology emerged out of Anatomy in the 19th century by harnessing new methods based on the newly discovered phenomenon of electricity. Nowadays, the relationships between disciplines and methods are more complex. Methods are now widely shared between disciplines and research areas. New developments in electronic publishing make it possible for scientists that encounter new methods to quickly find sources of information electronically. The design of individual volumes and chapters in this series takes this new access technology into account. Springer Protocols makes it possible to download single protocols separately. In addition, Springer makes its print-on-demand technology available globally. A print copy can therefore be acquired quickly and for a competitive price anywhere in the world.

Saskatoon, Canada *Wolfgang Walz*

Preface

Tauopathies encompass a group of neurodegenerative diseases defined by the misfolding and aggregation of one of the six tau protein isoforms. These neurological conditions are under the umbrella of protein folding disorders, which are characterized by the accumulation of proteins prone to misfold inside and outside the cells. In the context of Alzheimer's disease (AD), the accumulation of amyloid-beta plaques and neurofibrillary tangles (NFTs) in the brain was first reported by German psychiatrist and neuropathologist Alois Alzheimer. These amyloid-beta plaques and NFTs in the AD brains are the two pathological hallmarks of AD.

The structure of these tau fibrils found in AD brains was first discovered in 1963 by Dr. M. Kidd and Dr. Robert D. Terry. Much later, in 1986, the main nature of the filaments found in AD brain was determined to consist of tau by Dr. Inge Grundke-Iqbal et al. The tau protein is expressed predominantly in the neurons and is associated with the microtubules. One of its roles is to aid in the proper functioning of the cytoskeleton network by promoting microtubule assembly. Regarding its structure, the amino acid composition of tau varies between 352 and 441 residues due to alternative mRNA splicing. These isoforms, specifically 0N3R, 0N4R, 1N3R, 1N4R, 2N3R, and 2N4R, differ in their number of N-terminal inserts (N) and microtubule-binding repeat domains (R). Each isoform exhibits a unique signature in different neurologic diseases.

In 1986, several groups reported that a major component of NFTs, one of the main pathological hallmarks of AD, consisted of hyperphosphorylated tau. Since then, studies of this protein have expanded to examine the mechanisms of tau phosphorylation, aggregation, and degradation. It was in this era that individuals who believed tau more heavily contributes to AD progression, so-called "tauists," emerged. The alternative group, known as the "baptists," believes that amyloid-beta is the main driver of the disease. While therapeutic approaches targeting amyloid-beta have failed to show efficacy in slowing the progression of dementia, the past 20 years have seen a growing interest in targeting tau for AD treatment. This shift provides a major stimulus for present-day research on tau and related biological targets.

Introduced in 1991 and applied to AD diagnosis by 1997, the Braak staging system placed an emphasis on the importance of tau in the pathophysiology of AD. Braak's staging demonstrated that NFTs first appear in the transentorhinal region (stage I) and then spread to the entorhinal cortex (stage II), hippocampus (stage III), superior temporal neocortex (stage IV), and finally to the entire neocortex (stages V and VI). In contrast to amyloid-beta plaques, the spatiotemporal distribution of misfolded tau correlates with the loss of cognition. All six tau isoforms are present in NFTs, and the 1 N isoform is the most abundant in the AD brain. The 4R isoform is capable of spreading very extensively in rodent brains. Moreover, one of the most studied post-translational modifications is hyperphosphorylation. In a normal brain, tau is phosphorylated. However, the hyperphosphorylation of tau found in AD leads to the loss of microtubule banding, disruption of axonal transport, and neurodegeneration. This hyperphosphorylated state increases tau's propensity to misfold, causing faster aggregation, and exhibits higher toxicity to cultured neuroblastoma cells.

Although tau mutations do not naturally occur in AD, they are the primary genetic drivers of a distinct group of neurodegenerative conditions known as primary tauopathies. For example, mutations in the tau gene on chromosome 17 were found in frontotemporal dementia with parkinsonism-17 (FTDP-17) in 1998. These mutations include coding mutations (missense and two deletions) and splicing-related mutations affecting exon 10. The coding mutations impact tau's ability to bind to microtubules and may speed up its aggregation, while the splicing mutations often lead to an increased ratio of 4R:3R tau isoforms. Apart from its etiology, FTDP-17 is a familial brain disorder marked by a loss of neurons in the frontal and temporal lobes. Over time, an affected individual can exhibit changes in personality, behavior, cognition, and movement. No curative treatments are available, and efforts should be focused on the development of novel disease-modifying therapies.

Another notable tauopathy is chronic traumatic encephalopathy (CTE), which is diagnosed by the neuropathological examination of brain tissue. American neurologist and neuropathologist Dr. Ann McKee has studied and characterized the long-term cognitive impact of repetitive head injuries on the brains of National Football League and National Hockey League players. In CTE, the misfolding and spread of phosphorylated tau, accompanied by a shift from predominant 4R to 3R isoforms, advance as the disease progresses. This hyperphosphorylated tau misfolds and generates NFTs, abnormal neurites, and inclusions in astrocytes clustered around small blood vessels at the sulcal depths of the cortex. Moving forward, it is important to push research efforts toward specific neuropathological characterization of CTE and identifying biomarkers for diagnosis, monitoring disease progression, and establishing new treatments.

Furthermore, Pick's disease (PiD) is a form of frontotemporal dementia characterized by the abnormal accumulation of misfolded tau with a shift toward increased prevalence of the 3-repeat (3R) tau inclusions. This tauopathy affects the frontal and temporal lobes of the brain, typically causing progressive aphasia, dementia, and behavioral changes. Patients affected are often younger than 60 years of age, with a life expectancy varying from 2 to 20 years. The disease is officially diagnosed with a postmortem examination of the brain. The brain presents with severe atrophy of the frontal and temporal lobes with gliosis, aggregated tau structures ("Pick bodies"), and enlarged neurons ("Pick cells"). Improvements in diagnostics and treatment modalities of PiD are needed as well.

This volume is centered on translational research methods for tauopathies. It provides the novice tauist with detailed protocols for establishing in vitro and in vivo laboratory techniques designed to provide novel insights into the pathophysiology of tau in neurodegenerative diseases. The protocols included in this book were intentionally selected to serve as a basis for research laboratories focusing on tau-related neurodegenerative drug discovery.

Part I of this methodology book describes the production of recombinant tau protein. Chapter **1** focuses on the generation of tau with insect cells, which may produce some level of phosphorylation. The generation of the different tau isoforms can be expressed using specific plasmids and bacterial cell culture, as explained in Chap. **2**. Chapters **3** and **4** allow for the generation and evaluation of tau phosphorylated at ≥16 sites. In vitro functional assays with the hyperphosphorylated tau are depicted in Chap. **4**. The monitoring of tau fibril formation kinetics using the thioflavin S assay and the direct visualization of fibrils by transmission electron microscopy are detailed in Chap. **5**. Protocols for the generation of tau oligomers by simple incubation or cross-linking reactions are explained in Chap. **6**. Chapter **6** also includes methods for detecting oligomers non-specifically by gel staining or specifically by commercially available monoclonal antibodies.

Part II describes techniques to obtain insoluble tau and soluble tau from disease-affected and aged normal brains collected and archived in biobanks. The isolation of paired helical filaments alongside amyloid-beta plaques is detailed in Chap. **7**. Chapter **8** includes a protocol to specifically capture tau-seeds and outlines techniques for their proteomic analysis.

Part III focuses on conditions to capture interactions or mimic certain processes, such as the spreading of tau oligomerization. Chapter **9** covers specific methods for the detection, isolation, amplification, and characterization of tau oligomers from diseased brain tissue. Tau interacts with RNA, and these specific interactions and their downstream effects can be monitored, as explained in Chap. **10**. Several cell-based assays are included in Chaps. **11** and **12**. Chapter **11** contains protocols to evaluate the effects of tau oligomers isolated from brains or generated in vitro on the induction of normal intracellular tau. These so-called biosensor cells contain tau marked with fluorochrome and utilize fluorescence resonance energy transfer (FRET) technology. They are used to evaluate the effect of tau oligomers inducing the aggregation of tau. Upon aggregation, a fluorescent signal emerges. The protocol included in Chap. **12** involves specific split constructs with green fluorescent protein (GFP) fluorochrome. This specific split construct allows for the generation of tau polymers, and the polymerization process can be detected by the generation of the fluorescence signal. Drug candidates can be tested as well for their anti-seeding effects (Chap. **11**) or anti-oligomer effects (Chap. **12**).

Part IV provides protocols for robust mechanistic and proof-of-concept studies using nematode, insect, and animal models. Chapters **13–16** cover a variety of in vivo protocols to perform physiological studies or evaluate the effects of drug candidates using diverse tauopathy models, including *Caenorhabditis elegans* (Chap. **13**), *Drosophila* (Chap. **14**), zebrafish (Chap. **15**), and tau transgenic mice (Chap. **16**). These chapters allow for the evaluation of drug candidates for the prevention of neurodegeneration using these advanced in vivo models.

The editor is delighted to share her enthusiasm for the tau translational field with the readers and looks forward to seeing new developments in understanding the biology of this complex protein, with the ultimate goal of discovery and advancing different therapeutic approaches for these devastating, progressive, and heterogeneous neurodegenerative diseases known as tauopathies.

West Lafayette, IN, USA *Jessica S. Fortin*

Contents

Preface to the Series v

Preface vii

Contributors xiii

PART I RECOMBINANT TAU PROCURATION

1 Production of Hyperphosphorylated Tau Using a Baculovirus-Insect Cell System 3
Sehong Min, Natalie G. Horgan and Jean-Christophe Rochet

2 Expression and Purification of Tau in *E. coli* Cells 11
Natalie G. Horgan, Taiwo A. Ademoye, Jessica S. Fortin and Raluca Ostafe

3 Preparation and Examination of Hyperphosphorylated Tau Produced by PIMAX System 23
Hsiao-Tien Hagar, Kuang-Wei Wang and Min-Hao Kuo

4 Structural and Functional Assays for the Characterization of Recombinant Hyperphosphorylated Tau Protein 35
Kuang-Wei Wang, Hsiao-Tien Hagar and Min-Hao Kuo

5 Evaluation of the Kinetics of Tau Fibril Formation 45
Natalie G. Horgan, Taiwo A. Ademoye and Jessica S. Fortin

6 Tracking Tau Oligomer Formation In Vitro 57
Natalie G. Horgan, Taiwo A. Ademoye and Jessica S. Fortin

PART II ISOLATION OF TAU FROM DISEASED PATIENTS

7 Isolation and Characterization of Amyloid-β Plaque Cores and Neurofibrillary Tangles from Alzheimer's Disease Brain 69
Germán Plascencia-Villa and George Perry

8 Isolation and Proteomic Analysis of Pathological Tau-Seeds 87
Henika Patel, Pablo Martinez, Armando Amaro and Cristian A. Lasagna-Reeves

PART III MIMICKING NEURODEGENERATIVE PROCESSES IN VITRO

9 Neurotoxic Tau Oligomers: Isolation, Preparation, and Characterization 109
Sheeza Khan, Md. Anzarul Haque and Rakez Kayed

10 Capturing Tau and RNA Interactions 149
Sophie J. F. van der Spek and Benjamin L. Wolozin

11 Generation of Stable Tau Biosensor Cell Lines 159
Jaime Vaquer-Alicea and Vaibhav Bommareddy

12 Cell-Based Tau Polymerization Assays 177
Natalie G. Horgan, Omnia M.H. Ibrahium, Taiwo A. Ademoye, Jessica S. Fortin and Raluca Ostafe

PART IV IN VIVO MODELS

13 Modeling Functions and Pathology of Tau Protein in *C. elegans*. 193
Ghulam Jeelani Pir, Bikash Choudhary and Eckhard Mandelkow

14 Tracking Tau Multimerization in the Adult *Drosophila* Brain. 215
Meghna N. Chinchankar and Bess Frost

15 Testing Small Molecules for Activity Against Neurological Phenotypes in a Zebrafish 4-Repeat Tauopathy Model 251
Oluwaseyi Pearce, Qing Bai, Hugh S. Hammer and Edward A. Burton

16 Tau Transgenic Mice 281
Natalie G. Horgan, Emma N. Poston, Jessica S. Fortin and David G. Morgan

Subject Index. *297*

Contributors

TAIWO A. ADEMOYE • *Department of Basic Medical Sciences, College of Veterinary Medicine, Purdue University, West Lafayette, IN, USA*
ARMANDO AMARO • *Department of Neurology, Baylor College of Medicine, Houston, TX, USA*
QING BAI • *Department of Neurology, University of Pittsburgh, Pittsburgh, PA, USA; Pittsburgh Institute for Neurodegenerative Diseases, Pittsburgh, PA, USA*
VAIBHAV BOMMAREDDY • *Center for Alzheimer's and Neurodegenerative Diseases, Peter O'Donnell Jr. Brain Institute, University of Texas Southwestern Medical Center, Dallas, TX, USA*
EDWARD A. BURTON • *Department of Neurology, University of Pittsburgh, Pittsburgh, PA, USA; Pittsburgh Institute for Neurodegenerative Diseases, Pittsburgh, PA, USA; Geriatrics Research, Education and Clinical Center, Pittsburgh VA Healthcare System, Pittsburgh, PA, USA*
MEGHNA N. CHINCHANKAR • *Barshop Institute for Longevity and Aging Studies, University of Texas Health San Antonio, San Antonio, TX, USA; Department of Cell Systems and Anatomy, University of Texas Health San Antonio, San Antonio, TX, USA*
BIKASH CHOUDHARY • *Department of Biology, The University of the South, Sewanee, TN, USA*
JESSICA S. FORTIN • *Department of Basic Medical Sciences, College of Veterinary Medicine, Purdue University, West Lafayette, IN, USA*
BESS FROST • *Center for Alzheimer's Disease Research, Brown University, Providence, RI, USA; Department of Molecular Biology, Cell Biology and Biochemistry, Brown University, Providence, RI, USA*
HSIAO-TIEN HAGAR • *Department of Biochemistry and Molecular Biology, Michigan State University, East Lansing, MI, USA; Institute of Biochemistry and Molecular Biology, National Yang-Ming University, Taipei, Taiwan*
HUGH S. HAMMER • *Department of Laboratory Animal Resources, University of Pittsburgh, Pittsburgh, PA, USA*
MD. ANZARUL HAQUE • *Mitchell Center for Neurodegenerative Diseases, University of Texas Medical Branch, Galveston, TX, USA; Departments of Neurology, Neuroscience and Cell Biology, University of Texas Medical Branch, Galveston, TX, USA*
NATALIE G. HORGAN • *Department of Basic Medical Sciences, College of Veterinary Medicine, Purdue University, West Lafayette, IN, USA*
OMNIA M. H. IBRAHIUM • *Molecular Evolution Protein Engineering and Production, Purdue University, West Lafayette, IN, USA*
RAKEZ KAYED • *Mitchell Center for Neurodegenerative Diseases, University of Texas Medical Branch, Galveston, TX, USA; Departments of Neurology, Neuroscience and Cell Biology, University of Texas Medical Branch, Galveston, TX, USA*

SHEEZA KHAN • *Mitchell Center for Neurodegenerative Diseases, University of Texas Medical Branch, Galveston, TX, USA; Departments of Neurology, Neuroscience and Cell Biology, University of Texas Medical Branch, Galveston, TX, USA; School of Life Sciences, B. S. Abdur Rahman Crescent Institute of Science and Technology, Vandalur, Chennai, Tamil Nadu, India*

MIN-HAO KUO • *Department of Biochemistry and Molecular Biology, Michigan State University, East Lansing, MI, USA*

CRISTIAN A. LASAGNA-REEVES • *Department of Neurology, Baylor College of Medicine, Houston, TX, USA*

ECKHARD MANDELKOW • *German Center for Neurodegenerative Diseases (DZNE), Bonn, Germany; UKB, University of Bonn Medical School, Department Gerontopsychiatry and Neurodegenerative Diseases, Bonn, Germany*

PABLO MARTINEZ • *Department of Neurology, Baylor College of Medicine, Houston, TX, USA*

SEHONG MIN • *Biological Sciences Division, Pacific Northwest National Laboratory, Richland, WA, USA; Borch Department of Medicinal Chemistry and Molecular Pharmacology, College of Pharmacy, Purdue University, West Lafayette, IN, USA*

DAVID G. MORGAN • *Department of Translational Neuroscience, College of Human Medicine, Grand Rapids Research Center, Michigan State University, Grand Rapids, MI, USA*

RALUCA OSTAFE • *Molecular Evolution Protein Engineering and Production, Purdue University, West Lafayette, IN, USA; Purdue Institute for Inflammation, Immunology, and Infectious Disease, Purdue University, West Lafayette, IN, USA*

HENIKA PATEL • *Department of Neurology, Baylor College of Medicine, Houston, TX, USA*

OLUWASEYI PEARCE • *Department of Neurology, University of Pittsburgh, Pittsburgh, PA, USA; Pittsburgh Institute for Neurodegenerative Diseases, Pittsburgh, PA, USA*

GEORGE PERRY • *Department of Neuroscience, Developmental and Regenerative Biology, The University of Texas at San Antonio (UTSA), San Antonio, TX, USA*

GHULAM JEELANI PIR • *The Neuroscience Institute, Academic Health System, Hamad Medical Corporation, Doha, Qatar*

GERMÁN PLASCENCIA-VILLA • *Department of Neuroscience, Developmental and Regenerative Biology, The University of Texas at San Antonio (UTSA), San Antonio, TX, USA*

EMMA N. POSTON • *Department of Basic Medical Sciences, College of Veterinary Medicine, Purdue University, West Lafayette, IN, USA*

JEAN-CHRISTOPHE ROCHET • *Borch Department of Medicinal Chemistry and Molecular Pharmacology, College of Pharmacy, Purdue University, West Lafayette, IN, USA; Purdue Institute for Integrative Neuroscience, Purdue University, West Lafayette, IN, USA*

SOPHIE J. F. VAN DER SPEK • *Department of Anatomy and Neurobiology, Boston University Chobanian & Avedisian School of Medicine, Boston, MA, USA*

JAIME VAQUER-ALICEA • *Center for Alzheimer's and Neurodegenerative Diseases, Peter O'Donnell Jr. Brain Institute, University of Texas Southwestern Medical Center, Dallas, TX, USA*

KUANG-WEI WANG • *Department of Biochemistry and Molecular Biology, Michigan State University, East Lansing, MI, USA*

BENJAMIN L. WOLOZIN • *Department of Anatomy and Neurobiology, Boston University Chobanian & Avedisian School of Medicine, Boston, MA, USA*

Part I

Recombinant Tau Procuration

Chapter 1

Production of Hyperphosphorylated Tau Using a Baculovirus-Insect Cell System

Sehong Min, Natalie G. Horgan and Jean-Christophe Rochet

Abstract

Hyperphosphorylation of the microtubule-associated protein tau is a hallmark of tauopathies, including Alzheimer's disease and related neurodegenerative disorders. Detailed biochemical and biophysical analyses of recombinant hyperphosphorylated tau are vital for understanding tau pathobiology. The protocol described herein outlines an efficient method for producing hyperphosphorylated tau using a baculovirus-insect cell expression system. The approach involves the preparation of a baculovirus vector, viral amplification, infection of Sf9 insect cells, tau expression in Sf9 cells, and tau purification via anion exchange and size exclusion chromatography. This protocol enables the consistent production of tau with a pathologically relevant phosphorylation profile, advancing research into the role of hyperphosphorylated tau in disease mechanisms.

Keywords Alzheimer's disease, Baculovirus, Chromatography, Hyperphosphorylation, Protein expression, Protein purification, Sf9 insect cells, Tau

1 Introduction

Tau is a microtubule-associated protein that plays a critical role in regulating axonal transport, an essential process for neuronal health and function [1]. Abnormal posttranslational modifications of tau, especially hyperphosphorylation, cause the protein's dissociation from microtubules and aggregation into neurofibrillary tangles, which are pathological hallmarks of a group of neurodegenerative diseases collectively termed tauopathies [1–3]. To date, over 26 distinct tauopathies have been identified, including Alzheimer's disease, frontotemporal dementia, progressive supranuclear palsy, and corticobasal degeneration [4, 5]. To understand the role of hyperphosphorylated tau in these disorders, biochemical and biophysical analyses of recombinant tau with disease-relevant phosphorylation patterns are needed. The baculovirus-insect cell expression system (Fig. 1A) provides a powerful platform for producing recombinant tau with a pathologically relevant phosphorylation profile. Since

Jessica S. Fortin (ed.), *Translational Research Methods for Tauopathies*, Neuromethods, vol. 233,
https://doi.org/10.1007/978-1-0716-5234-3_1,

the first report of tau expression in Sf9 insect cells in 1991 [6], refined protocols have enabled the production of recombinant tau with a large number of phosphorylated residues per tau molecule (Fig. 1B, C) and yields up to 50 μg/10[6] cells [7–9]. Several groups (including ours) have leveraged this baculovirus-insect cell system to characterize recombinant hyperphosphorylated tau variants in terms of their biochemical properties, aggregation propensities, and antibody affinities [7–10]. The following protocol outlines the efficient production of hyperphosphorylated tau in a baculovirus-insect cell system, providing a valuable biochemical tool for investigating the molecular underpinnings of tauopathy diseases [9].

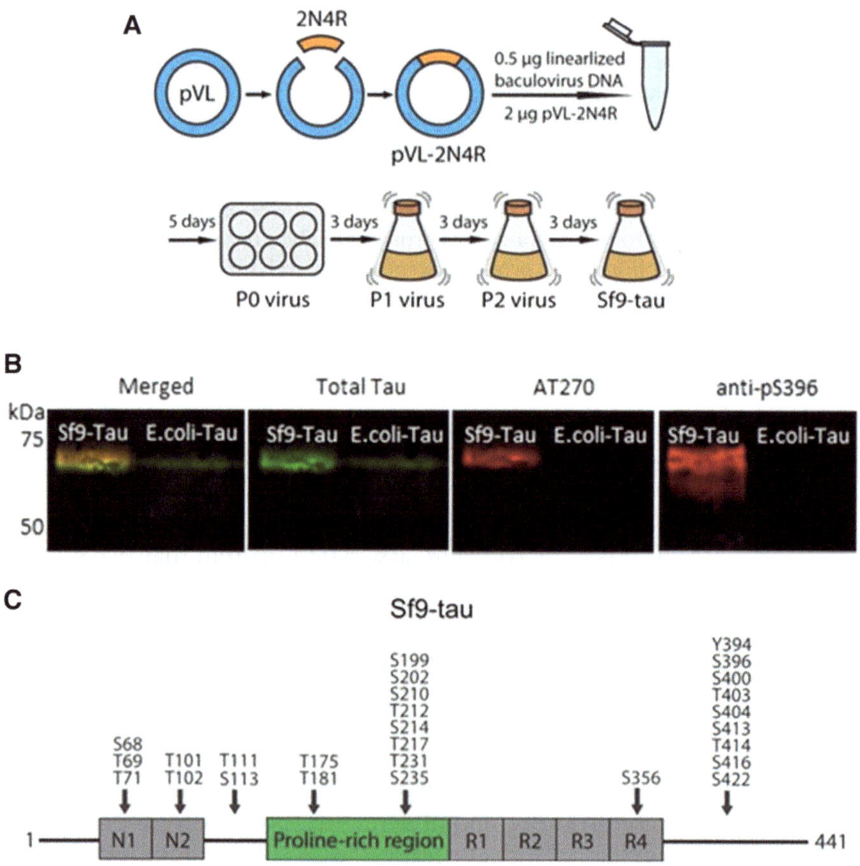

Fig. 1 Tau 2N4R is expressed in a hyperphosphorylated form in Sf9 insect cells, but not in *E. coli* cells. (**A**) Schematic representation of tau 2N4R expression using a baculovirus/Sf9 cell system. (**B**) Images obtained via Western blot analysis of purified tau 2N4R expressed in Sf9 or *E. coli* cells (left or right band in each image, respectively). The membranes were probed with antibodies specific for total tau, pT181-tau (AT270), or pS396-tau. The merged image on the left consists of an overlay of the total tau and pT181-tau bands. Tau 2N4R generated in Sf9 cells, but not in *E. coli* cells, is phosphorylated at both T181 and S396. (**C**) Diagram illustrating phosphorylated residues of tau 2N4R expressed in the baculovirus/Sf9 cell system, mapped onto the tau domain structure. The phosphorylated residues were identified by liquid chromatography tandem mass spectrometry [9] and are consistent with phosphosites identified in other studies [7, 8]. (Adapted from [15])

2 Materials

2.1 Reagents and Kits

1. Baculovirus transfer vector pVL1393 (Expression Systems #91-013).
2. Human tau cDNA (encoding the 2N4R variant).
3. NEBuilder HiFi DNA assembly (New England Biolabs #E5520).
4. BestBac 2.0, Baculovirus Cotransfection Kit (Expression Systems #91-200) containing linearized baculovirus DNA, $Expres^2$ TR transfection reagent, and transfection medium.

2.2 Cell Lines and Cell Culture Media

1. Sf9 insect cells (Expression Systems #94-001 S).
2. ESF 921 Insect Cell Culture Medium (Expression Systems #96-001-01).

2.3 Buffers and Solutions

1. Lysis buffer: 50 mM Tris-HCl, 500 mM NaCl, 5 mM DTT, 10 mM EGTA, 20 mM NaF, 1 mM Na_3VO_4, 5 μM microcystin, protease inhibitor cocktail (Sigma-Aldrich #P8340), pH 7.4.
2. Streptomycin sulfate (Thermo Fisher Scientific #S9136-100G): 1% (w/v) solution, prepared fresh; used to precipitate DNA during lysate clarification.
3. Anion exchange buffer A: 100 mM MES, 2 mM DTT, 1 mM EGTA, 1 mM $MgSO_4$, 0.1 mM PMSF, pH 6.8.
4. Anion exchange buffer B: 100 mM MES, 2 mM DTT, 1 mM EGTA, 1 mM $MgSO_4$, 0.1 mM PMSF, 1 M NaCl, pH 6.8.
5. Dialysis buffer (PBS with 1 mM DTT): 10 mM phosphate buffer, 2.7 mM KCl, 137 mM NaCl, 1 mM DTT, pH 7.4.

2.4 Equipment

1. Biosafety cabinet for insect cell culture.
2. Shaker-equipped incubator with a nonhumidified, ambient-air atmosphere (27°C).
3. Luna FL dual fluorescence cell counter (Logos Biosystems) for Sf9 cell counting.
4. Erlenmeyer shaker flasks with vented cap.
5. Centrifuge capable of 30,000 × *g*.
6. Dounce homogenizer (e.g., Fisherbrand #06435B).
7. HiPrep Q HP 16/10 column (Cytiva #29018182).
8. HiLoad 16/600 Superdex 200 pg column (Cytiva #28989335).

2.5 Additional Consumables

1. Tissue culture-treated 6-well plates (Fisherbrand FB012927).
2. Sterile polystyrene tubes, 15 or 50 mL (Corning 14-959-70C or 14-959-49A).

3. Plastic bag and paper towel.
4. Micropipettes, serological pipettes, and sterile tips.
5. SnakeSkin dialysis tubing MWCO 3.5 kDa (Thermo Scientific #68035).
6. Sterile 0.2 μm syringe filters (Fisherbrand 09-720-004) and syringes (BD 309653).

3 Methods

3.1 Production of Recombinant Baculovirus

Several methods exist for generating recombinant baculovirus, including the Bac-to-Bac and BaculoGold systems. In this protocol, we employ the BestBac system, which involves cotransfecting insect cells with linearized baculovirus DNA and a transfer plasmid encoding the gene of interest to produce baculovirus expressing 2N4R tau (Fig. 1A). The BestBac system uses a highly efficient linearized baculovirus DNA backbone derived from *Autographa californica* multiple nucleopolyhedrovirus (AcMNPV). This backbone is engineered for superior recombination efficiency with polyhedrin gene locus-based transfer vectors, resulting in enhanced viral stability and purity.

BestBac 2.0 lacks the v-*cath/chiA* protease genes, reducing protease-mediated degradation of expressed proteins, thereby increasing yield and functional integrity. Following virus generation, viral stocks can be stored and used to infect insect cells for tau expression. This protocol is adapted from the BestBac manual provided by Expression Systems.

Construction of Recombinant Expression Vector

BestBac DNA is compatible with widely used baculovirus transfer vector families, including pBAC, pAc, and pVL vectors. Here, human tau cDNA is subcloned into the pVL1393 transfer vector (yielding pVL-tau) using the NEB HiFi DNA assembly system (Fig. 1A).

Sf9 Cell Culture

Optimal Sf9 cell health and transfection efficiency are achieved when cells are split during mid-log phase growth. This protocol employs monolayer Sf9 cultures for transfection and suspension cultures for viral amplification and protein expression (Fig. 1A). Although antibiotics (e.g., penicillin/streptomycin) and serum (e.g., fetal bovine serum) can be used, this protocol is optimized for antibiotic-free and serum-free conditions. All culturing steps must adhere strictly to best practices for sterile technique in a biosafety cabinet.

Transfection of Sf9 Cells

1. Seed 0.9×10^6 Sf9 cells per well in a 6-well plate with 1 mL ESF 921 cell culture medium (*see* **Note 1**).
2. Allow cells to attach at room temperature for 30 min.
3. Prepare cotransfection mixtures:

a. Tube A: 100 μL transfection medium, 5 μL (0.5 μg) BestBac linearized DNA, 2 μg pVL-tau plasmid.
b. Tube B: 100 μL transfection medium, 6 μL Expres^2 TR transfection reagent.
c. Incubate solutions A and B separately for 5 min at room temperature.
d. Combine solutions A and B; mix gently by pipetting up and down twice with a large bore pipette tip.
e. Incubate the combined A and B mixture for 20 min at room temperature.
f. Add 800 μL transfection medium to reach 1 mL total volume.

4. Aspirate medium from the well, add transfection solution.
5. Incubate at 27°C for 4 h in a plastic bag with a damp paper towel to prevent drying.
6. Aspirate transfection solution and add fresh cell culture medium.
7. Incubate at 27°C for 4–5 days in a plastic bag with a damp paper towel to maintain a humid environment.

Harvest of P0 Virus Stock

Cytopathic effects, including cell growth inhibition and cell detachment, indicate virus production.

1. Collect conditioned medium in a sterile microcentrifuge tube.
2. Centrifuge at 2000 × *g* for 5 min to remove cell debris.
3. Transfer supernatant containing the virus to a new sterile tube.
4. Store the virus at 4°C in the dark. This is the P0 virus stock.

Generation of P1 Virus Stock

P0 virus stocks typically exhibit low titers ($<1 \times 10^8$ pfu/mL). P1 virus stocks are amplified in Sf9 suspension cultures to increase viral titer (*see* **Note 2**). When cells are infected, they usually swell, have darker nuclei, and appear granular.

1. Seed Sf9 suspension culture at 1×10^6 cells/mL and incubate overnight at 27°C with shaking at 120 rpm.
2. The following day, infect cells at 2×10^6 cells/mL with P0 virus at a multiplicity of infection (MOI) of 0.1.
3. Incubate for 72 h at 27°C with shaking at 120 rpm.
4. Pellet cells by centrifugation at 2000 × *g* for 10 min and filter supernatant through a 0.2 μm sterile filter.
5. Store supernatant at 4°C in the dark. This is the P1 virus stock.

Generation of P2 Virus Stock

The P2 virus stock, used for protein expression, is generated as follows.

1. Seed Sf9 suspension culture at 1×10^6 cells/mL and incubate overnight at 27°C with shaking at 120 rpm.
2. The following day, infect cells at 2×10^6 cells/mL with P1 virus at an MOI of 0.1.
3. Incubate for 72 h at 27°C with shaking at 120 rpm.
4. Pellet cells by centrifugation at $2000 \times g$ for 10 min.
5. Store supernatant at 4°C in the dark. This is the P2 virus stock.

3.2 Tau Protein Expression

When Sf9 cells express high levels of tau protein, they develop cellular processes [6, 10].

1. Seed Sf9 suspension culture at 1×10^6 cells/mL and incubate at 27°C with shaking at 120 rpm.
2. When the culture reaches 2×10^6 cells/mL, infect the culture with P2 virus at an MOI of 1.
3. Incubate for 72 h at 27°C with shaking at 120 rpm.

3.3 Tau Protein Purification

1. Harvest cells by centrifugation at $2000 \times g$ for 10 min.
2. Resuspend pellet with lysis buffer at 10 mL per gram of cell pellet.
3. Homogenize using a Dounce homogenizer (~20 strokes).
4. Add 1% (w/v) of streptomycin sulfate (predissolved) to precipitate the DNA. Incubate with rotation at 4°C for 30 min.
5. Centrifuge at $30{,}000 \times g$ for 30 min at 4°C.
6. Heat supernatant by boiling for 10 min.
7. Centrifuge at $30{,}000 \times g$ for 30 min at 4°C.
8. Dialyze supernatant overnight against anion exchange buffer A at 4°C, using a dialysis membrane with a MWCO of 3.5 kDa.
9. Filter dialysate through 0.22 μm membrane.
10. Apply the filtered supernatant from Step 9 to an anion-exchange column (Hiprep Q HP 16/10, Cytiva) following the manufacturer's instructions.
11. Elute tau protein using a linear gradient ranging from 0 to 1 M NaCl.
12. Pool fractions containing eluted tau.
13. Apply the protein solution from Step 12 to a size exclusion column (HiLoad 16/600 Superdex 200 pg, Cytiva) following the manufacturer's instructions.

14. Analyze the fractions by SDS-PAGE with Coomassie blue staining to assess tau purity (*see* **Note 3**). Pool fractions exhibiting high protein yield with >95% purity.
15. Divide the pooled protein solution from Step 15 into aliquots, with volumes chosen for single-thaw use in future experiments; snap freeze, and store at –80°C.

4 Notes

1. One additional well can be included for a control reaction, which serves as a negative control for transfection.
2. This protocol is based on the assumption that baculovirus titers are within the average range. However, it is highly recommended that accurate titers be determined. Plaque assays, quantitative PCR, end point dilution, and flow cytometry-based approaches are commonly used for baculovirus titration. Detailed protocols for these methods can be found in standard manuals and in the manufacturers' documentation [11–14].
3. Characterization of the purified Sf9-tau by SDS-PAGE, bicinchoninic acid (BCA) assay, Western blotting, and mass spectrometry analysis is strongly recommended before using the protein in experiments.

Conflict of Interest

The authors have no conflicts of interest to declare that are relevant to the content of this chapter.

References

1. Wang Y, Mandelkow E (2016) Tau in physiology and pathology. Nat Rev Neurosci 17(1):5–21. https://doi.org/10.1038/nrn.2015.1
2. Das V, Miller JH (2012) Microtubule stabilization by peloruside A and paclitaxel rescues degenerating neurons from okadaic acid-induced tau phosphorylation. Eur J Neurosci 35(11):1705–1717. https://doi.org/10.1111/j.1460-9568.2012.08084.x
3. Arendt T, Stieler JT, Holzer M (2016) Tau and tauopathies. Brain Res Bull 126 (Pt 3):238–292. https://doi.org/10.1016/j.brainresbull.2016.08.018
4. Zhang Y et al (2022) Tauopathies: new perspectives and challenges. Mol Neurodegener 17(1):28. https://doi.org/10.1186/s13024-022-00533-z
5. Gotz J, Halliday G, Nisbet RM (2019) Molecular pathogenesis of the tauopathies. Annu Rev Pathol 14:239–261. https://doi.org/10.1146/annurev-pathmechdis-012418-012936
6. Knops J et al (1991) Overexpression of tau in a nonneuronal cell induces long cellular processes. J Cell Biol 114(4):725–733. https://doi.org/10.1083/jcb.114.4.725
7. Tepper K et al (2014) Oligomer formation of tau protein hyperphosphorylated in cells. J Biol Chem 289(49):34389–34407. https://doi.org/10.1074/jbc.M114.611368

8. Mair W et al (2016) FLEXITau: quantifying post-translational modifications of tau protein in vitro and in human disease. Anal Chem 88(7):3704–3714. https://doi.org/10.1021/acs.analchem.5b04509
9. Min S et al (2023) Effects of neighboringphosphorylation events on the affinities of pT181-tau antibodies. Anal Chem 95(49):18241–18248. https://doi.org/10.1021/acs.analchem.3c04081
10. Baas PW, Pienkowski TP, Kosik KS (1991) Processes induced by tau expression in Sf9 cells have an axon-like microtubule organization. J Cell Biol 115(5):1333–1344. https://doi.org/10.1083/jcb.115.5.1333
11. Yang JP (2016) Small-scale production of recombinant proteins using the baculovirus expression vector system. Methods Mol Biol 1350:225–239. https://doi.org/10.1007/978-1-4939-3043-2_10
12. Wu HC, Cha HJ, Bentley WE (2016) Evaluating baculovirus infection using green fluorescent protein and variants. Methods Mol Biol 1350:447–459. https://doi.org/10.1007/978-1-4939-3043-2_22
13. Thompson CM, Montes J, Aucoin MG, Kamen AA (2016) Recombinant protein production in large-scale agitated bioreactors using the baculovirus expression vector system. Methods Mol Biol 1350:241–261. https://doi.org/10.1007/978-1-4939-3043-2_11
14. King LA, Hitchman R, Possee RD (2016) Recombinant baculovirus isolation. Methods Mol Biol 1350:73–94. https://doi.org/10.1007/978-1-4939-3043-2_4
15. Min S (2022) Diversity of TAU proteoforms in tauopathies: relevance to biomarker analysis and preclinical modeling. Purdue University Graduate School

Check for updates

Chapter 2

Expression and Purification of Tau in *E. coli* Cells

Natalie G. Horgan, Taiwo A. Ademoye, Jessica S. Fortin and Raluca Ostafe

Abstract

Tau aggregation is a defining pathological feature of Alzheimer's disease and related tauopathies, making reliable access to recombinant tau essential for mechanistic and translational studies. Here, we describe a robust protocol for the bacterial production and purification of recombinant human 0N4R tau, the shortest isoform expressed in the adult brain. Expression in *Escherichia coli* enables high yields of unmodified tau suitable for controlled in vitro aggregation assays. The protocol includes transformation, culture in either IPTG-induced or autoinduction media, cell lysis, and a heat-clarification step that exploits tau's intrinsic heat stability. Subsequent purification is achieved by cation-exchange chromatography, producing tau of >90% purity with typical yields of 10–20 mg/L of culture. The purified protein can be directly applied to biochemical and biophysical assays of tau aggregation, including thioflavin T fluorescence, transmission electron microscopy, and photoinduced crosslinking (PICUP). Representative results demonstrate reproducible fibril formation under defined conditions and clear visualization of filament morphology. Notes are provided to guide optimization, including choice of induction method, buffer systems, and storage conditions, as well as troubleshooting low yields or protein instability.

Keywords Tau 0N4R, Recombinant protein, Bacterial expression, Protein aggregation, Cation-exchange chromatography, Autoinduction, Fibril formation, Neurodegeneration

1 Introduction

Tau is a microtubule-associated protein predominantly expressed in neuronal axons and, to a lesser extent, in oligodendrocytes. It plays a critical role in microtubule assembly and stabilization, as well as in facilitating axonal transport [1, 2]. In the adult human brain, tau exists in six isoforms generated by alternative splicing of exons 2, 3, and 10 of the MAPT gene. These isoforms differ in the number of N-terminal inserts (0 N, 1 N, 2 N) and in whether they contain three or four microtubule-binding repeats (3 R, 4 R) [3] (Fig. 1). Under pathological conditions such as oxidative stress, inflammation, or genetic mutations, tau becomes hyperphosphorylated and prone to self-aggregation. This process leads to the formation of paired helical filaments (PHFs) and neurofibrillary tangles (NFTs) [4–11].

Jessica S. Fortin (ed.), *Translational Research Methods for Tauopathies*, Neuromethods, vol. 233,
https://doi.org/10.1007/978-1-0716-5234-3_2,

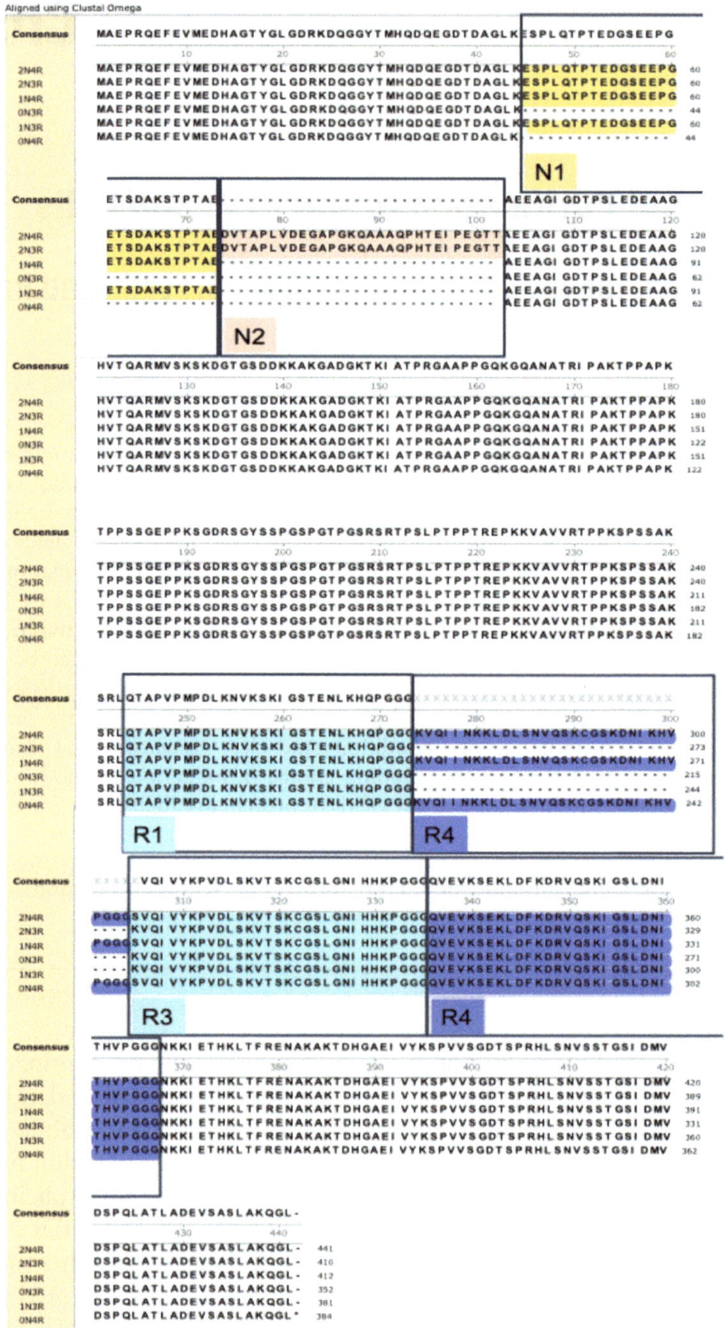

Fig. 1 Sequence alignment of the six major human tau isoforms. The isoforms 0N3R, 1N3R, 2N3R, 0N4R, 1N4R, and 2N4R are shown, highlighting conserved and variable regions. The N-terminal inserts (N1, N2) are indicated in yellow and orange, respectively. The four microtubule-binding repeats (R1–R4) are marked in light to dark blue. The alignment illustrates the modular structure of tau, including the N-terminal inserts, proline-rich regions, and repeat domains critical for microtubule binding and aggregation. The 0N4R isoform was used in this protocol

The accumulation of these aggregates contributes to synaptic dysfunction, neuronal loss, and cognitive decline, hallmarks of Alzheimer's disease and related tauopathies [12, 13].

To investigate the mechanisms underlying tau aggregation and to identify potential therapeutic strategies, reliable access to recombinant tau protein is essential. Among available expression systems, *Escherichia coli* remains the most widely used due to its rapid growth, cost-effectiveness, and ability to produce high yields of recombinant protein [14]. Bacterial expression has proven especially valuable in tau research, as it allows for scalable production of unmodified tau suitable for controlled in vitro aggregation assays [15, 16]. However, tau is an intrinsically disordered protein that is difficult to purify stably, and standard purification efforts often rely on adding affinity tags followed by multistep chromatography, including size-exclusion and anion-exchange, to achieve sufficient purity [17, 18]. Heterologous expression of tau in *E. coli* presents further challenges due to its proline- and arginine-rich sequence and strong codon bias, which often result in poor expression levels. To overcome these limitations, specialized *E. coli* strains have been employed. In particular, Rosetta (DE3) cells provide additional tRNAs for rare codons, thereby alleviating codon bias and enhancing expression. KrishnaKumar and Gupta (2017) demonstrated that using Rosetta (DE3) cells, together with codon optimization, improved tau expression yields by up to eight-fold [19]. Their method also simplified downstream processing: instead of conventional mechanical or enzymatic lysis, expressed cells were directly boiled. Because tau remains soluble after heat treatment while most bacterial proteins precipitate, this "boiling extraction" step enriched tau in the soluble fraction without additional lysis steps.

However, a key limitation of *E. coli* is the absence of posttranslational modifications (PTMs), which strongly influence tau biology and aggregation [20]. Thus, while bacterial expression provides an efficient means of producing tau, complementary systems are needed when PTMs are under investigation.

This chapter describes a protocol for the expression and purification of the human 0N4R tau isoform in *E. coli*. The 0N4R variant is the shortest isoform, which facilitates expression and purification, while the presence of four microtubule-binding repeats makes it particularly relevant for studying aggregation, as 4R tau is enriched in the hippocampus of Alzheimer's disease brains compared to 3R tau [21, 22]. Following transformation and colony selection on kanamycin LB-agar plates, tau expression is induced and purified using an ÄKTA chromatography system with cation-exchange. The resulting protein is suitable for downstream analyses including transmission electron microscopy (TEM), photoinduced crosslinking of unmodified proteins (PICUP), and Thioflavin T (ThT) fluorescence assays, enabling detailed studies of tau's aggregation potential and fibril formation.

2 Materials

2.1 DNA, Strains, and Antibiotics

1. **Expression plasmid**. Human Tau 0N4R in pET-30a (+) (Kan^R) (Fig. 2). Gift of B. Wolozin (*see* **Note 1**). The sequence alignment and plasmid map are presented in Fig. 1 and Fig. 2, respectively.
2. **Host cells**. *E. coli* Rosetta (DE3) pLysS (Novagen)(*see* **Note 2**).
3. **Selection antibiotics** (use in all stages: agar plates, starter cultures, expressions) (*see* **Note 3**).

   a. Kanamycin: 50 μg/mL (for pET-30a).

   b. Chloramphenicol: 34 μg/mL (for Rosetta host plasmids).

2.2 Media and Solutions

1. LB (1 L): 10 g tryptone, 5 g yeast extract, 10 g NaCl per liter; adjust pH 7.4.
2. LB-agar: LB + 1.5% agar; supplement with Kan (and Cam if using Rosetta) after cooling to ~60°C.

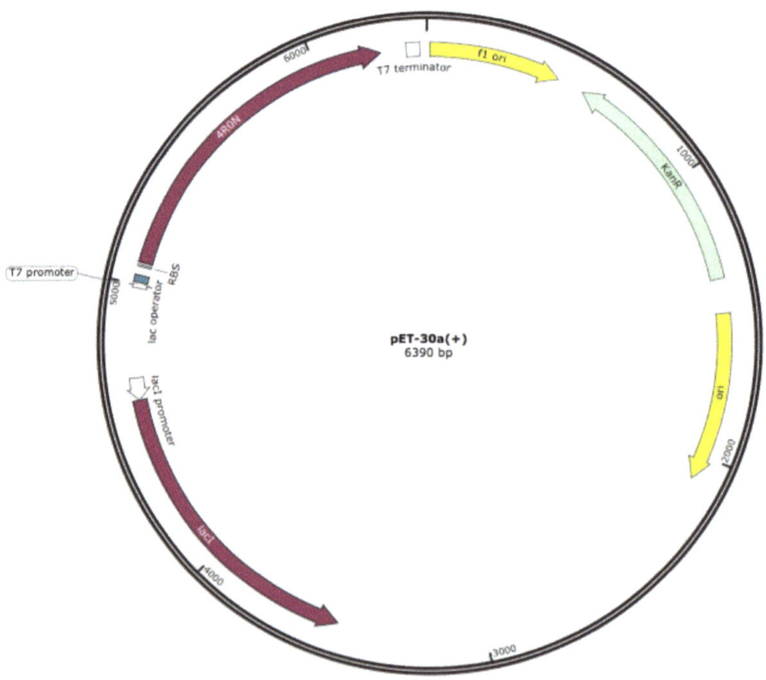

Fig. 2 Plasmid map of the expression construct used for recombinant tau production. The vector contains a T7 promoter driving expression of the 0N4R tau insert. Key features include the kanamycin resistance gene (KanR) for selection in *E. coli*, the lacI gene for induction control, and the multiple cloning site (MCS), where the tau sequence was inserted. The plasmid backbone is pET-based, optimized for high-level expression in *E. coli* under IPTG induction or autoinduction conditions

3. S.O.C. medium, normally commercial competent cells are delivered with S.O.C media. (Tryptone 20 g/L; yeast extract 5 g/L; NaCl 10 mM; KCl 2.5 mM; $MgCl_2$ 10 mM; $MgSO_4$ 10 mM; glucose 20 mM; pH 7.0).
4. Autoinduction medium: Overnight Express Instant TB Medium—Novagen—prepare according to manufacture's instructions (e.g., 54 g/L) with glycerol (0.5%–1% v/v) after autoclave.
5. Lysis buffer (pH 7.4): 10 mM HEPES, 50 mM NaCl, 0.5 mM DTT, PMSF, complete protease inhibitor cocktail (Roche).
6. Cation exchange buffer Buffer A: 50 mM MES, pH 6.0, 50 mM NaCl, 1 mM DTT.
7. Cation exchange buffer B: 20 mM MES, 1 M NaCl, 1 mM EGTA, 2 mM DTT (add fresh), 0.1 mM PMSF (add fresh), pH 6.0.
8. Gel filtration buffer: PBS (137 mM NaCl, 3 mM KCl, 10 mM Na2HPO4, 2 mM KH2PO4, pH 7.4) with 1 mM DTT (freshly added), pH 7.4.
9. Other: PBS, TBS, 100% ethanol, bleach.

2.3 Equipment and Consumables

1. Baffled Erlenmeyer flasks (250 mL, 1 L, 4 L) with breathable caps; incubator shakers (25–37°C, 200–250 rpm).
2. Cell disruption: probe sonicator (3–6 mm tip) or French press/ microfluidizer (~20,000 psi).
3. Refrigerated centrifuges: microcentrifuge; tabletop rotor for 50 mL tubes; bottle rotor for 1 L bottles.
4. Water bath or dry heat block capable of 80°C; ice buckets.
5. Syringe/vacuum filters: 0.45 μm PES.
6. Dialysis tubing: 3–7 kDa MWCO with clamps.
7. Cation-exchange column: HiPrep SP HP 16/10 (or SP FF/ HP equivalent).
8. ÄKTA or other chromatography system with fraction collector.
9. Ultrafiltration concentrators: 10 kDa MWCO.
10. SDS-PAGE system, gel documentation imager.
11. Spectrophotometer for A_{600}/A_{800} (cuvettes or NanoDrop).
12. Optional analytics: CD spectropolarimeter, TEM access (negative stain), microplate reader for ThT, SEC column(e.g., Superdex 200 Increase 10/300).

3 Methods

3.1 E. coli Transformation(see Note 4)

1. Thaw Rosetta (DE3) pLysS competent cells on ice (20–30 min).
2. Add 1–5 μL plasmid DNA (10 pg–100 ng) to 20–50 μL cells; flick to mix; incubate on ice 30 min.
3. Heat-shock 42°C, 30–60 s; return to ice 2–5 min.
4. Add 950 μL S.O.C.; shake 1 h, 37°C, 225 rpm.
5. Plate on LB + kanamycin (50 μg/mL) + chloramphenicol (34 μg/mL).
6. Incubate overnight at 37°C.

3.2 Starter Cultures (Day 0 → Day 1)

1. Inoculate a single colony into 10 mL LB + Kan (50 μg/mL) + Cam (34 μg/mL).
2. Shake overnight, 37°C, 225 rpm.

3.3 Expression Options (choose A or B) (see Note 5)

A. **IPTG Induction**

1. Inoculate 10 mL overnight into 1 L LB + Kan + Cam in a 4 L baffled flask (1:100).
2. Grow 37°C, 225 rpm to OD_{600} ≈ 0.6–0.8 (≈2–4 h).
3. Induce with IPTG 0.1–1 mM (Rosetta pLysS benefits from the lower range).
4. Shift to 18–25°C; shake 16–20 h. *(Alternative: 30°C for 12–18 h, or 37°C for 3–6 h if short induction is desired.)*

B. **Autoinduction media**

1. Preculture 50–100 mL LB + antibiotics, then inoculate 1 L at 1:10.
2. Grow in a 4 L baffled flask at 30°C for 16–24 h (gentler; often higher solubility with pLysS). *(Alternative: 37 °C, ~14–20 h.)*
3. Regardless of the methods used for induction, check culture for protein expression using SDS-PAGE (Fig. 3) (*see* **Note 6**).

3.4 Harvest (Day 2)

1. Pellet at 10,000 × *g*, 10 min, 4°C.
2. Discard supernatant; freeze pellets (−80°C) or proceed.

3.5 Lysis and Heat Clarification (Day 3)

Use either a microfluidizer or sonication.

1. Resuspend pellet in 10–30 mL cold lysis buffer per liter of culture (*see* **Note 7**).

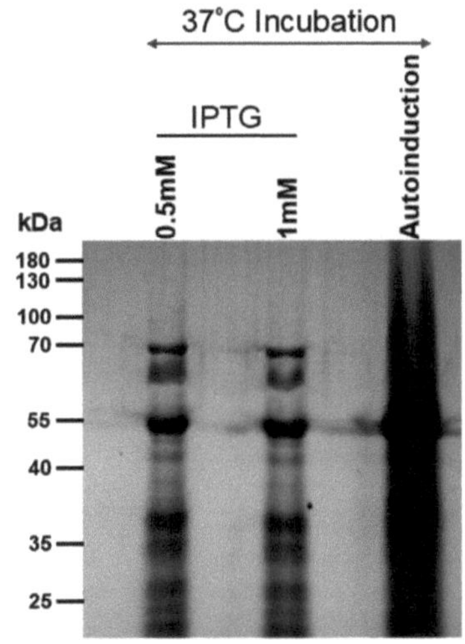

Fig. 3 The tau 0N4R isoform is more highly expressed using autoinduction compared to IPTG induction at both tested concentrations. Autoinduction is more efficient and produces more tau protein than IPTG induction. The IPTG inductions, at 0.5 mM and 1 mM, were incubated at 37°C for 4 h, while the autoinduction was incubated at 37°C overnight

2. Lyse (choose one):
   - Microfluidizer: ~20,000 psi, three passes, at 4°C. Add protease inhibitors (e.g., PMSF 0.1 mM, fresh after lysis).
   - Probe sonication: total 3–5 min (30 s on/30 s off, 30%–45% amplitude) on ice.
3. Clarify 30 min, 10,000 × *g*, 4°C; collect supernatant.
4. Raise NaCl to 0.5 M; cap loosely/vented and heat 80°C, 10 min; cool on ice 10 min (*see* **Note 8**).
5. Add DTT to 2 mM; clarify again 30 min, 15,000 × *g*, 4°C.
6. Proceed to dialysis.

3.6 Dialysis

1. Prepare 3 L of 50 mM MES, pH 6.0 (add 2 mM DTT fresh).
2. Load clarified supernatant into 3–7 kDa MWCO dialysis tubing, rinse tube with ddH_2O prior to use.
3. Dialyze sequentially at 4°C with stirring: 1 h in 0.5 L, replace with 1 L for 3–4 h, and then 2 L overnight.

3.7 Cation-Exchange Purification (Day 4)

1. Filter dialysate through a 0.45 μm PES syringe filter.
2. Equilibrate HiPrep SP HP 16/10 with Buffer A.
3. Load sample at ≤1 mL/min; wash 2–3 CV with Buffer A.
4. Elute with a linear gradient from 0% to 100% buffer B (50 mM to 1 M NaCl) over ~20 CV; monitor A_{280} (Fig. 4, top).

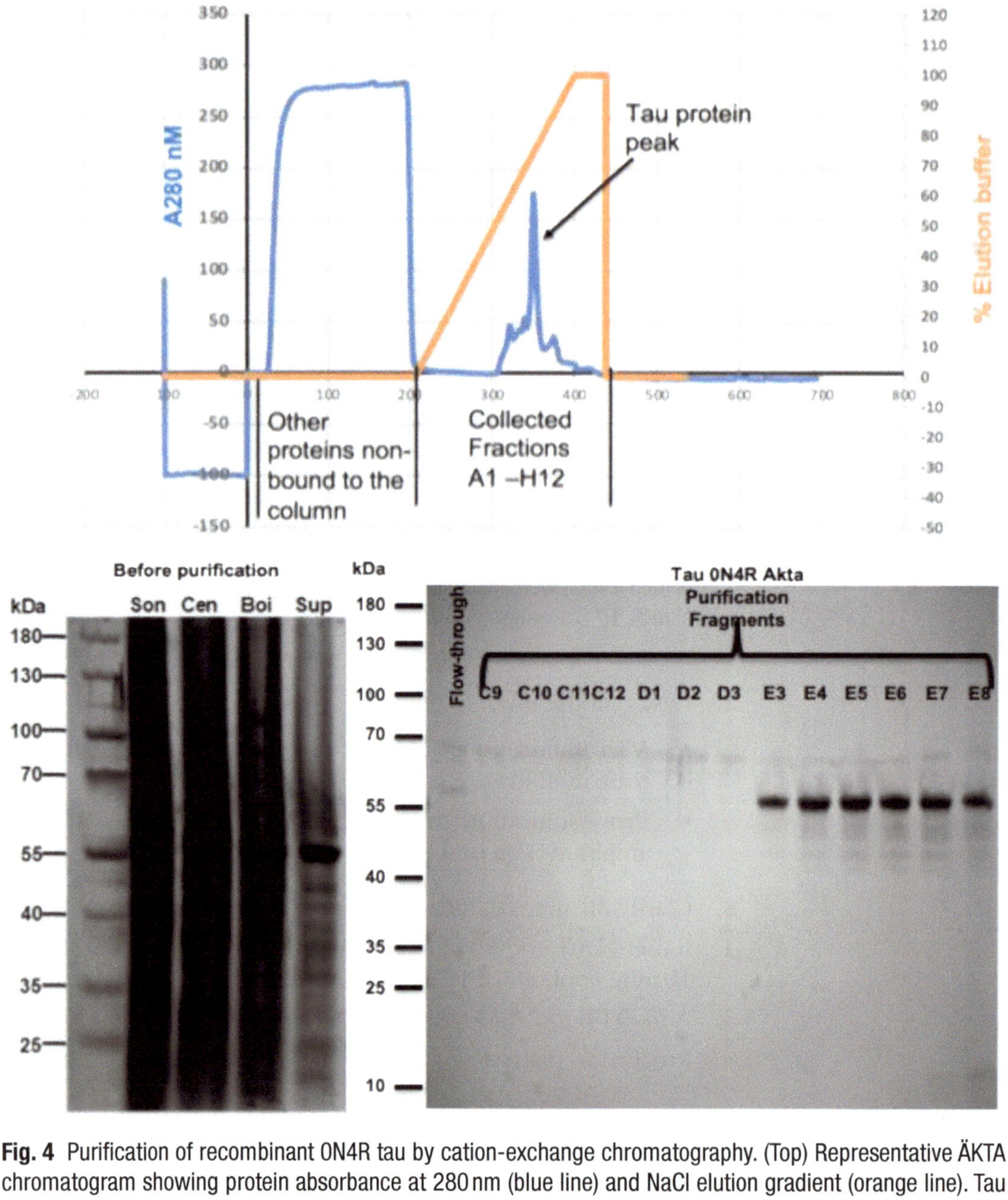

Fig. 4 Purification of recombinant 0N4R tau by cation-exchange chromatography. (Top) Representative ÄKTA chromatogram showing protein absorbance at 280 nm (blue line) and NaCl elution gradient (orange line). Tau eluted as a distinct peak during the salt gradient. (Bottom) SDS-PAGE analysis of purification fractions. Tau appears as a prominent band at ~55 kDa in elution fractions E3–E8, while earlier fractions (C9–D3) contain little or no detectable protein. The gel demonstrates enrichment of tau to high purity after cation-exchange chromatography

5. Analyze fractions by SDS-PAGE/Coomassie; pool tau-rich fractions (Fig. 4, bottom).
6. Concentrate pooled protein using 10 kDa MWCO concentrators at 4000 × g, 10°C to the desired volume.

3.8 Optional Polishing and Buffer Exchange

1. If impurities persist, perform SEC into PBS or assay-specific buffer.
2. For storage, exchange into PBS + 1 mM DTT (optional 10% glycerol for freeze-thaw stability); aliquot; snap-freeze in liquid N_2; store at −80°C.

3.9 Quality Control and Downstream Assays

1. SDS-PAGE: Expect a prominent band near ~55 kDa after CE (see Fig. 4) (*see* **Note 9**).
2. CD spectroscopy: 6 μM tau in low-salt buffer typically shows a random coil spectrum (Fig. 5).
3. Aggregation assays: ThT fluorescence, PICUP, and TEM after incubation (e.g., 6 μM tau with DTT, heparin, arachidonic acid; 37°C, 3 days) reveal mature fibrils (Fig. 5).
4. Once the protein is identified, use Tau 0N4R for circular dichroism (Fig. 5B), PICUP, TEM (Fig. 5C), and ThT experiments (Fig. 5A),

   - Alternatively, use impure Tau 0N4R (prior to dialysis/ÄKTA) for preliminary experiments.

4 Notes

1. Any T7 promoter–based expression vector carrying Tau (other isoforms or mutants) is suitable. Adjust antibiotic selection to the vector (e.g., AmpR vectors require ampicillin, pET-30a is KanR).
2. Other BL21(DE3)-type strain that expresses T7 RNA polymerase can be used. Rosetta(DE3) strains carry a rare-tRNA plasmid (pRARE) that improves expression of eukaryotic genes containing codons infrequently used in *E. coli*. The pLysS element expresses T7 lysozyme, which inhibits T7 RNA polymerase and reduces basal (leaky) expression—useful for burdensome/toxic proteins—but typically slows induction kinetics, so plan for slightly longer or cooler inductions. Rosetta strains require chloramphenicol (34 μg/mL) to maintain the host plasmid(s).
3. Maintain both kanamycin (50 μg/mL) and chloramphenicol (34 μg/mL) in all media when using Rosetta (DE3) pLysS with a KanR vector. Add antibiotics after media cools to ≤ 60°C. Refresh antibiotic stocks regularly (light/heat degrade antibiotics). If you switch vectors/hosts, match the selection markers accordingly and document the final concentrations.

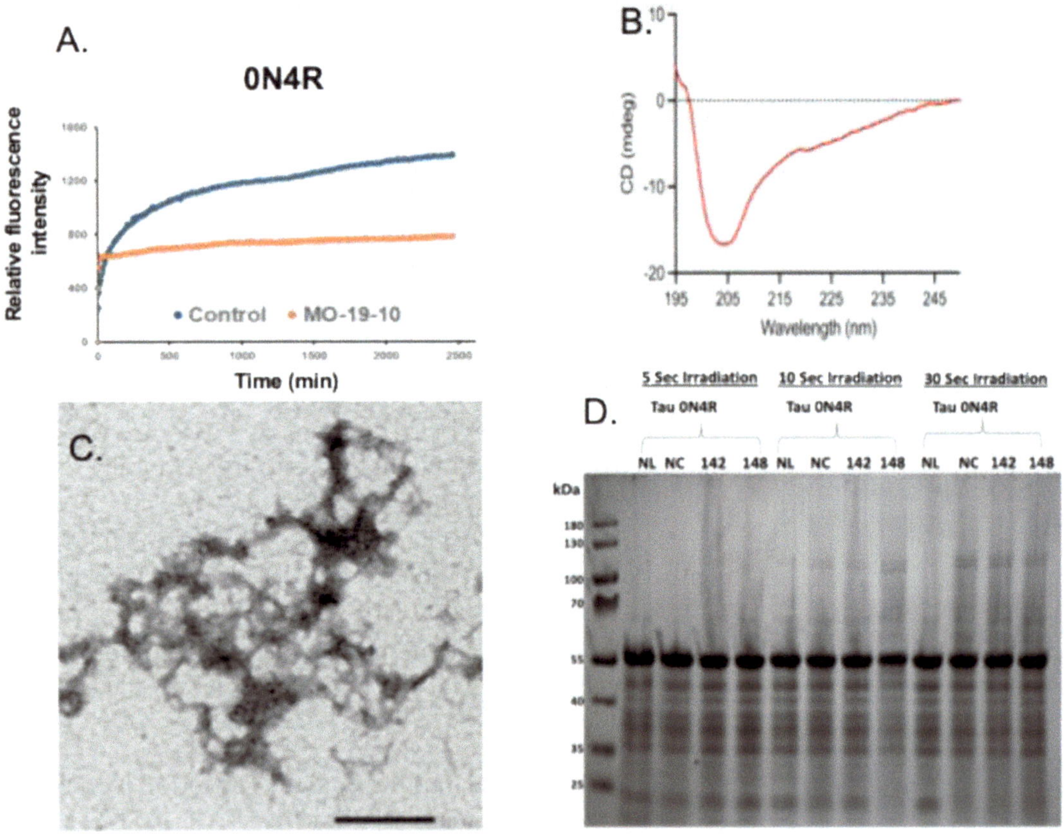

Fig. 5 (**A**) Thioflavin T (ThT) fluorescence assay of tau aggregation with and without inhibitor. Purified 0N4R tau was incubated under fibrillization conditions in the presence of ThT dye. In control reactions (blue), a time-dependent increase in fluorescence was observed, consistent with β-sheet–rich fibril formation. In contrast, samples treated with an inhibitory compound (orange) showed markedly reduced fluorescence, indicating suppression of tau fibril formation. (**B**) Circular dichroism (CD) spectrum of purified 0N4R tau at 6 μM, demonstrating a minimum near 200 nm consistent with a random coil conformation. (**C**) Transmission electron microscopy (TEM) image of mature fibrils formed by 0N4R tau. Aggregation was induced by incubating tau at 6 μM in Chelex-treated PBS containing 0.25% DMSO, 5 mM DTT, 150 μM heparin, and 10 mg/mL arachidonic acid at 37°C for 3 days. Scale bar: 200 nm. (**D**) Photoinduced crosslinking of unmodified proteins (PICUP) assay showing tau oligomerization. Crosslinking reactions were performed with purified 0N4R tau followed by SDS-PAGE analysis, confirming the presence of oligomeric species in addition to the monomeric band at ~55 kDa. Compounds 142 and 148 failed to reduce the presence of oligomers (bands between 100 and 130 kDa). The oligomers are generated from the light exposure for a duration of 10–30 s. NL no light exposure, NC no compound treatment

4. Tau expresses best in strains that tolerate repetitive sequences (e.g., BL21(DE3), Rosetta). Some strains may show instability due to the repeat-rich tau sequence.
5. Autoinduction typically yields higher amounts of tau than IPTG induction and requires less handling.
6. It is generally advisable to perform small-scale expression trials (4–5 mL cultures) before scaling up. Test a range of induction

conditions, including different temperatures (37, 30, and 25°C), IPTG concentrations (if using LB medium), and multiple postinduction time points (e.g., 4, 6, 8, 12, and 24 h), and monitor expression by SDS-PAGE. If tau expression is low or the protein appears largely insoluble (i.e., present mainly in the pellet after lysis), consider lowering the induction temperature or IPTG concentration. Alternatively, switching to autoinduction medium often improves yield and solubility. Finally, confirm that the plasmid sequence is intact, as the repeat-rich tau gene can undergo recombination in certain *E. coli* strains, leading to reduced or failed expression.

7. DNA release during lysis can increase viscosity; adding DNase I improves clarification.
8. Tau remains soluble at 80°C while most bacterial proteins precipitate, making the heat step an efficient enrichment method.
9. On SDS-PAGE, 0N4R tau migrates at ~55 kDa despite its calculated mass (~45 kDa) due to charge distribution.

Acknowledgments

The authors would like to thank Eduardo Ramirez for technical contributions. The construct for tau 0N4R was a generous gift from Dr. Benjamin Wolozin (Boston University).

Conflict of Interest The authors have no conflicts of interest to declare that are relevant to the content of this chapter.

References

1. Kahlson MA, Colodner KJ (2015) Glial tau pathology in tauopathies: functional consequences. J Exp Neurosci 9s2:JEN.S25515
2. Weingarten MD, Lockwood AH, Hwo SY et al (1975) A protein factor essential for microtubule assembly. Proc Natl Acad Sci USA 72:1858–1862
3. Goedert M, Spillantini MG, Jakes R et al (1989) Multiple isoforms of human microtubule-associated protein tau: sequences and localization in neurofibrillary tangles of Alzheimer's disease. Neuron 3:519–526
4. Alonso ADC, Grundke-Iqbal I, Iqbal K (1996) Alzheimer's disease hyperphosphorylated tau sequesters normal tau into tangles of filaments and disassembles microtubules. Nat Med 2:783–787
5. Gamblin TC, King ME, Dawson H et al (2000) *In vitro* polymerization of tau protein monitored by laser light scattering: method and application to the study of FTDP-17 mutants. Biochemistry 39:6136–6144
6. Gamblin TC, King ME, Kuret J et al (2000) Oxidative regulation of fatty acid-induced tau polymerization. Biochemistry 39:14203–14210
7. Goedert M, Jakes R (2005) Mutations causing neurodegenerative tauopathies. Biochim Biophys Acta (BBA)–Mol Basis Dis 1739:240–250
8. Hong M (1998) Mutation-specific functional impairments in distinct tau isoforms of hereditary FTDP-17. Science 282:1914–1917
9. Kidd M (1963) Paired helical filaments in electron microscopy of Alzheimer's disease. Nature 197:192–193
10. Kitazawa M, Oddo S, Yamasaki TR et al (2005) Lipopolysaccharide-induced inflammation exacerbates tau pathology by a cyclin-dependent kinase 5-mediated pathway in a transgenic model of Alzheimer's disease. J Neurosci 25:8843–8853
11. Sy M, Kitazawa M, Medeiros R et al (2011) Inflammation induced by infection potentiates tau pathological features in transgenic mice. Am J Pathol 178:2811–2822

12. Honson NS, Kuret J (2008) Tau aggregation and toxicity in tauopathic neurodegenerative diseases. JAD 14:417–422
13. Tracy TE, Gan L (2018) Tau-mediated synaptic and neuronal dysfunction in neurodegenerative disease. Curr Opin Neurobiol 51:134–138
14. Rosano GL, Ceccarelli EA (2014) Recombinant protein expression in *Escherichia coli*: advances and challenges. Front Microbiol 5:172
15. Elbatrawy AA, Ademoye TA, Alnakhala H et al (2024) Discovery of small molecule benzothiazole and indole derivatives tackling tau 2N4R and α -synuclein fibrils. Bioorg Med Chem 100:117613
16. Ademoye TA, Ganegamage SK, Masoudi B et al (2025) *In vitro* evaluation of amide-linked coumarin scaffolds for the inhibition of α -synuclein and tau aggregation. ACS Omega 10:38498–38514
17. Combs B, Tiernan CT, Hamel C et al (2017) Production of recombinant tau oligomers *in vitro*. In: Methods in cell biology. Elsevier, Amsterdam, The Netherlands, pp 45–64
18. Ferrari L and Rüdiger SGD (2018) Recombinantproduction and purification of the humanprotein Tau. Protein Eng Des Sel 31:447–455
19. KrishnaKumar VG, Gupta S (2017) Simplified method to obtain enhanced expression of tau protein from *E. coli* and one-step purification by direct boiling. Prep Biochem Biotechnol 47:530–538
20. Combs B, Voss K, Gamblin TC (2011) Pseudohyperphosphorylation has differential effects on polymerization and function of tau isoforms. Biochemistry 50:9446–9456
21. Cherry JD, Esnault CD, Baucom ZH et al (2021) Tau isoforms are differentially expressed across the hippocampus in chronic traumatic encephalopathy and Alzheimer's disease. Acta Neuropathol Commun 9:86
22. Buée L, Bussière T, Buée-Scherrer V et al (2000) Tau protein isoforms, phosphorylation and role in neurodegenerative disorders. Brain Res Rev 33:95–130

Check for updates

Chapter 3

Preparation and Examination of Hyperphosphorylated Tau Produced by PIMAX System

Hsiao-Tien Hagar, Kuang-Wei Wang and Min-Hao Kuo

Abstract

Abnormally phosphorylated tau is increasingly recognized as a critical pathogenic factor in Alzheimer's disease and other neurodegenerative tauopathies. However, studies that rely on unmodified tau species may impede our understanding of tau-mediated pathology. Given the limitations of nondisease-relevant tau species, obtaining disease-relevant tau species is crucial for advancing molecular studies and drug development. The hyperphosphorylated tau (p-tau) produced by the PIMAX system represents one of the most convenient and pathophysiologically relevant tools for basic and translational research in tauopathies. This chapter details the preparation of hyperphosphorylated tau from *E. coli* expression, while the subsequent chapter (Chap. 4–Wang et al.) presents protocols for the biochemical and functional characterization of PIMAX p-tau.

Keywords Alzheimer's disease, Hyperphosphorylated tau, Neurodegeneration, Tauopathy, PIMAX

1 Introduction

Tau is a microtubule-binding protein that is encoded by the single MAPT gene in the human genome [1]. The central nervous system (CNS) tau exists as six isoforms resulting from alternative splicing, including 0N3R, 0N4R, 1N3R, 1N4R, 2N3R, and 2N4R [2]. The repeat domains of tau create relatively structured core that is thought to be responsible for microtubule binding and the formation of higher-ordered structures [3]. Except the repeat domains, the rest of tau is intrinsically disordered [4]. The brain deposition of tau due to misfolding is a key feature shared by numerous tauopathies including Alzheimer's disease (AD), Pick's disease, Frontotemporal lobar degeneration, chronic traumatic encephalopathy, and so on [5]. Moreover, tau is distinctive in that, out of the 441 residues in its longest isoform, 84 are serine, threonine, or tyrosine residues. Of these, around 50 sites have been identified phosphorylated in AD [6]. While physiological tau is phosphorylated at two to three sites per

Jessica S. Fortin (ed.), *Translational Research Methods for Tauopathies*, Neuromethods, vol. 233,
https://doi.org/10.1007/978-1-0716-5234-3_3,

molecule in heathy brains, the number of phosphorylation increases three-fold to four-fold in AD brains [7]. Hyperphosphorylation of tau is pivotal to the formation of neurofibrillary tangles (NFTs), one of the major hallmarks of AD [8]. Aberrant activation of glycogen synthase kinase-3-beta (GSK-3β) and other kinases has been associated with tau hyperphosphorylation [9, 10]. Because of the strong spatiotemporal association of tau precipitates and the cognitive impairments, hyperphosphorylated tau has become an important therapeutic target [11, 12]. Initially, NFTs consisting of tau were considered as pathogenic elements. However, accumulating evidence shows that NFT formation is separable from cognitive dysfunction and neural loss [13]. Instead, injection of tau oligomers into C57BL/6 mice led to cognitive, synaptic and mitochondrial abnormalities whereas NFT and tau monomers had no effect [14], suggesting soluble oligomeric form is more likely as the toxic species, which causes neurodegeneration.

Tau oligomers are frequently obtained via the use of an inducer, such as heparin and arachidonic acid [15]. However, tau aggregates triggered by these inducers are structurally different from those associated with the disease and lack the pathological properties [16]. Given the limitations associated with nondisease-relevant tau species, it is essential to obtain disease-relevant tau species for molecular studies and drug development. To this aim, the Protein Interaction Module–assisted Function X (PIMAX) system was developed to efficiently generate hyperphosphorylated tau [17], which exhibits characteristics relevant to neurodegeneration. In PIMAX, tau and the kinase (GSK-3β in this work) are brought together via the heterodimerization of Fos and Jun leucine zipper domains, each fused to GSK-3β and tau, respectively. The proximity facilitates highly efficient yet specific interaction between the enzyme and the substrate protein, resulting in tau hyperphosphorylation. The phosphorylation pattern of hyperphosphorylated tau (p-tau) synthesized by PIMAX is confirmed to be tightly associated with the disease. Hyperphosphorylation also facilitates autonomous tau aggregation without an inducer. Moreover, p-tau causes cell death under a physiological concentration [18]. A high-throughput drug screening on the basis of the cytotoxicity and aggregation of p-tau identified several promising candidates that inform AD drug development [19]. Mechanistically, PIMAX p-tau helped identify the pathobiological basis underlying tauopathy [20, 21], including the activation of the Unfolded Protein Response, endoplasmic reticulum (ER) stress, inflammation, apoptosis, and acute alterations in calcium and proteomic homeostasis. These findings attest to the value of using PIMAX p-tau for both disease mechanism and drug discovery studies.

The purification method employed in the PIMAX system is well designed. The heat-resistant nature of tau protein allows for efficient separation from the most other bacterial proteins through a boiling step. Additionally, the incorporation of a TEV protease cleavage site between the Jun and tau provides the removal of the Jun fragment, yielding purified p-tau or tau protein following TEV protease digestion. If the kinase is omitted from the plasmid, an unmodified tau protein can be produced and used as a control. The purpose of this chapter is to detail the procedure for PIMAX p-tau induction and purification, and how we examine the whole process by SDS-PAGE. For further studies, the biochemical and functional assays of p-tau are demonstrated in the next chapter.

2 Equipment and Materials

1. Transformation and Induction.
   a. PIMAX-based p-tau plasmid (available upon request; MH Kuo, kuom@msu.edu).
      Note: pMK1013-GSK-tau(1N4R) plasmid is as representative here. 0N3R, 1N3R, and 2N4R tau isoforms are available upon request.
   b. *E. coli* BL21(DE3)-CodonPlus competent cells (Agilent Technologies).
   c. LB agar plate (100 μg/mL ampicillin, 25 μg/mL chloramphenicol).
   d. Temperature-controlled shaking incubator (LabLine 3526 or equivalent).
   e. UV/visible spectrophotometer (Ultrospec™ 2100 pro or equivalent).
   f. Temperature-controlled centrifuge (Thermo Scientific Sorvall RC 6 Plus Superspeed Centrifuge or equivalent).
   g. Ampicillin (Sigma-Aldrich). 100 mg/mL stock dissolved in deionized water.
   h. Chloramphenicol (Sigma-Aldrich). 25 mg/mL stock dissolved in 70% ethanol.
   i. Magnesium sulfate ($MgSO_4$) (Sigma-Aldrich). 1 M stock dissolved in deionized water.
   j. Isopropyl β- d-1-thiogalactopyranoside (IPTG) (GoldBio). Fresh 1 M solution in deionized water.
   k. Purification.
   l. Lysis buffer: 20 mM Tris, pH 5.8, 100 mM NaCl.
   m. Purification buffer: 20 mM Tris, pH 7.4, 100 mM NaCl.

n. cOmplete™, Mini, EDTA-free Protease Inhibitor Cocktail (Roche #11836170001).

o. Home-made recombinant TEV protease (available upon request; MH Kuo, kuom@msu.edu).

p. Amicon® Ultra Centrifugal Filter, 30 kDa MWCO (Sigma-Aldrich #UFC9030).

q. Temperature-controlled centrifuge (Thermo Scientific Sorvall RC 6 Plus Superspeed Centrifuge or equivalent).

r. Sonicator (Branson 450 Digital Sonifier or equivalent).

s. Orthovanadate (Sigma-Aldrich #567540). 0.1 M stock dissolved in deionized water.

t. Lysozyme (Sigma-Aldrich #L6876). 20 mg/mL stock dissolved in deionized water.

u. Phenylmethylsulfonyl fluoride (PMSF) (Sigma-Aldrich). 0.1 M stock dissolved in 70% ethanol.

v. Dithiothreitol (DTT) (Sigma-Aldrich). 1 M stock dissolved in deionized water.

w. Ethylenediaminetetraacetic acid (EDTA) (Invitrogen). 200 mM, pH 8.0, stock dissolved in deionized water.

3 Methods

3.1 Transformation and Induction

1. Transform (see Appendix 1) pMK1013-GSK-tau(1N4R) plasmid into *E. coli* BL21(DE3) cells and grow the transformed cells on a fresh LB plate (100 µg/mL ampicillin, 25 µg/mL chloramphenicol) overnight at 37°C.
2. Pick a colony and place in a 250-mL flask containing 50 mL LB medium with 100 µg/mL ampicillin and 25 µg/mL chloramphenicol.
3. Grow the culture overnight, typically 12–18 h, at 37°C in a shaking incubator at 230 rpm.
   Note: From Step (4) till the end of purification, it is imperative to keep a small portion after each step for later SDS-PAGE examination.
4. Transfer the overnight culture (usually around 10 mL) to a 3-L flask containing 1 L LB medium with 100 µg/mL ampicillin and 2 mM $MgSO_4$ with a starting OD_{600} at 0.03–0.05.
   Note: When making the dilution at the day of induction, it is important to avoid the use of an overgrown overnight culture. We suspect that the secreted β-lactamase for ampicillin resist-

ance may increase the number of bacterial cells that have lost the plasmid for hyperphosphorylated tau production. This will inevitably reduce the yield of p-tau.

5. Grow the culture at 37°C in a shaking incubator at 230 rpm.
6. Measure the OD_{600} of cells after approximately 2 h. Monitor the OD_{600} until it reaches 0.3–0.4.
   Note: Overgrown cultures produce lower yields.
   Note: Collect 0.8 OD_{600} cells as the "before-induction" sample. Centrifuge cells at 14,000 rpm for 1 min and save the pellet for later SDS-PAGE analysis.
7. Add freshly prepared 1 M IPTG and bring the concentration to 0.2 mM to induce protein expression.
8. Incubate for no more than 2 h at 37°C, 230 rpm.
   Note: 1 h is typically sufficient. Do NOT go beyond 2 h, as prolonged induction results in lower yield.
   Note: Measure the OD_{600} of cells and collect 0.8 OD_{600} cells as the "after-induction" sample for later SDS-PAGE analysis.
9. Pour the culture to centrifuge bottles and spin at 4000 × g for 10 min at 4°C. Discard the supernatant and store the cell pellet at –80°C before purification.

3.2 Purification

Note: Before further processing the purification procedures, check induction efficiency especially in the first few trials. See details in the section of SDS-PAGE Examination.

Cell Lysis

1. Suspend the cell pellet with 7 mL of lysis buffer per liter of culture on ice. Make sure cells are homogeneously suspended.
2. Transfer cell suspension to a 50-mL conical tube.
3. Wash the bottle with another mL of lysis buffer and pool the suspensions. The final volume should be approximately 10 mL.
4. Add final concentration of 1 mg/mL lysozyme, 1 mM PMSF, 0.2 mM orthovanadate, and 1 tablet of Roche protease inhibitor and mix gently.
5. Incubate at 30°C for 30 min (with gentle inversion every 10 min) or 45 min at RT with gentle shaking.
6. Chill the cell solution on ice for 5 min.
7. Sonicate the cell solution by sonicator at 30% amplitude; 5 s ON + 5 s OFF; 30 cycles.
   Note: Cap and invert the tube up and down a few times in between each ten cycles.
   Note: Make sure that samples are always submerged in ice water. Avoid foaming. If foaming occurs, stop sonication and let samples sit for a few minutes to cool down and allow the bubble to dissipate before continuing. The solution should be less

viscous after good sonication. Do more cycles if cell lysis is incomplete. Avoid overheating.

8. Centrifuge cell lysate at 17,000 × g, 30 min at 4°C.
9. Transfer the supernatant to another 50-mL tube.

Note: Collect 1 μL of soluble fraction as the "lysis-supernatant" sample for SDS-PAGE analysis.
Note: For the lysis pellet, add the equivalent volume of lysis/purification buffer as original, approximately 10 mL, and sonicate with 40% amplitude until the pellet disperses. Save some pellet fraction as the "lysis-pellet" sample for later SDS-PAGE analysis. Make sure you record the volume so that the subsequent SDS-PAGE analysis uses materials that are equivalent to the number of cells (see below SDS-PAGE section).

Boiling

1. Boil the supernatant of the cell lysate in a boiling water bath for 30 min.
   Note: A rice cooker is a good and safe option.
   Note: Shake the tube occasionally and keep the lid loose to avoid explosion.
2. Cool down on ice for 30 min with occasional shaking.
3. Centrifuge the boiling solution at 17,000 × *g*, 30 min at 4°C.
4. Transfer the supernatant to another tube.

   Note: Collect 5 μL of supernatant as the "boil-supernatant" sample for later SDS-PAGE analysis.
   Note: For the boil pellet, collect the same way as the lysis pellet for the "boil-pellet" sample for later SDS-PAGE analysis.

TEV Digestion

1. To liberate tau and p-tau from Jun, add 1 OD_{280} of TEV protease for each 100 OD_{280} of the boil supernatant with 0.5 mM DTT and 1 mM EDTA.
2. Incubate the reaction at 4°C overnight with gentle shaking.
3. Centrifuge to remove protein aggregates during the proteolysis reaction at 17,000 × *g* for 30 min at 4°C.

   Note: Collect 5 μL of supernatant as the "after-TEV digestion" sample for later SDS-PAGE analysis.
   Note: TEV protease would precipitate after overnight incubation and could be spun down by centrifugation.

Concentration

1. To concentrate the sample, centrifuge it in an Amicon 30 kD spin column at 5000 × *g* for 20 min. Note: Repeat the centrifugation until the remnant volume is less than 2 mL.
2. To adjust the pH of the sample to 7.4 for storage, add 4 mL of purification buffer and centrifuge again. Repeat the buffer exchange step two more times to bring the pH up to 7.4.

3. Transfer the protein solution to another tube.
4. Wash the spin column with a small volume of purification buffer to collect all the proteins and pool to the same tube.
5. Add glycerol to a final concentration of 10% (w/w) for protein storage.
   Note: Collect 5 μL of final product for later SDS-PAGE analysis.
6. Snap freeze the protein in liquid nitrogen and store at –80°C.
   Note: Aliquot before freezing if desired.

3.3 SDS-PAGE Examination

Note: It is imperative to make sure that the induction and purification are efficient. SDS-PAGE is conducted to compare the samples from each step as well as to determine the final protein concentration.

Preparation of "Before- and After-Induction" Samples

1. Add 200 μL of 1× SDS-PAGE dye to the pellets of 0.8 OD_{600} cells.
2. Pipette to fully suspend cells.
3. Boil for 5 min in a water bath. Note: Use lid lock to cap the tubes can prevent inadvertent opening and evaporation.
4. Vortex vigorously for 30 s.
5. Centrifuge at 14,000 rpm for 1 min.

SDS-PAGE for Induction and Purification Efficiency

1. Run an 8% or 10% SDS-PAGE to examine the induction.
   Note: Typically, 0.04 OD_{600} gives good staining intensity (i.e., load 10 μL of the 200 μL induction samples). Tau and p-tau expression should be readily visible. If you have a hard time identifying either protein, it is better to start over again instead of proceeding to the purification procedures.
2. Run an 8% or 10% SDS-PAGE to examine the purification efficiency (i.e., 1 μL of lysis-supernatant, 1 μL of lysis-pellet, 1 μL of boil-supernatant, 1 μL of boil-pellet, 1 μL of after-TEV digest). See Fig. 1 for an example.

Protein Concentration Determination

1. Dilute 1 μL protein sample ten times with 9 μL purification buffer.
2. Prepare 100 ng/μL BSA standard solution for comparison.
3. Add SDS-PAGE dye and boil for 5 min in a water bath.
4. Run the samples and BSA standards on an 8% or 10% SDS-PAGE (i.e., 2 μL, 4 μL of diluted sample along with 200, 400, 600, 800, 1000 ng of BSA on a same gel). See Fig. 2 for an example.

   Note: In addition to visual inspection, an imaging program such as ImageJ can be used for quantification.

   Note: Ideally, the yield should be more than 2 mg/L of culture, and the final concentration should be around 2 mg/mL.

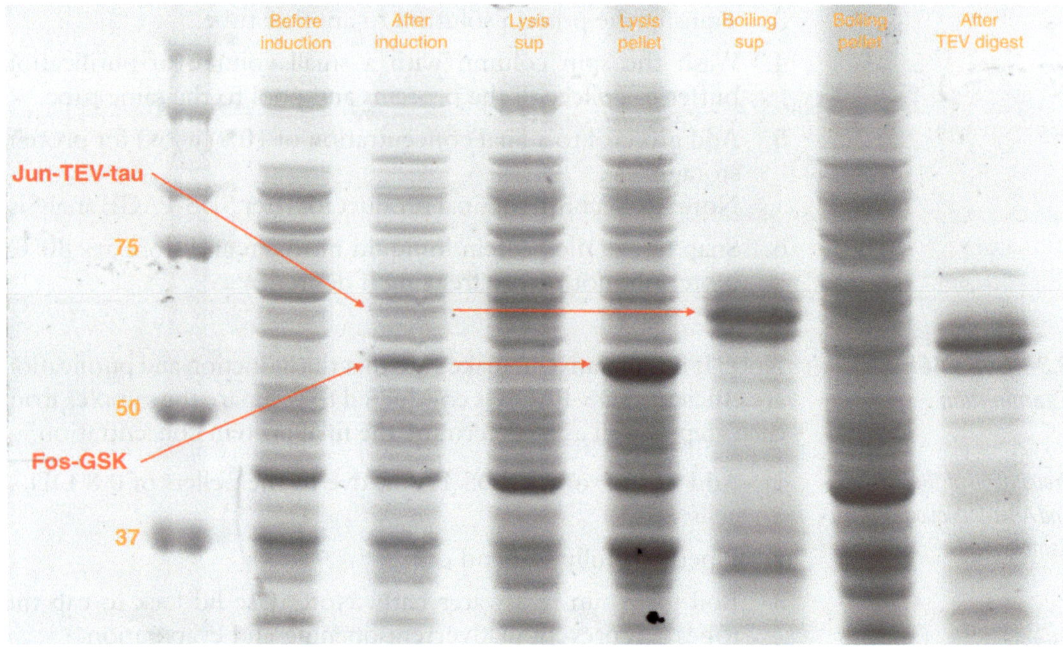

Fig. 1 SDS-PAGE examination of induction and purification. Notes: (1) Lysis pellet in the image shows the incompletion of lysis. This can be improved by incubating the lysis solution longer or sonicating more. (2) The impurity seen in the last lane might be due to overloading or incompletion of transferring solution during the steps

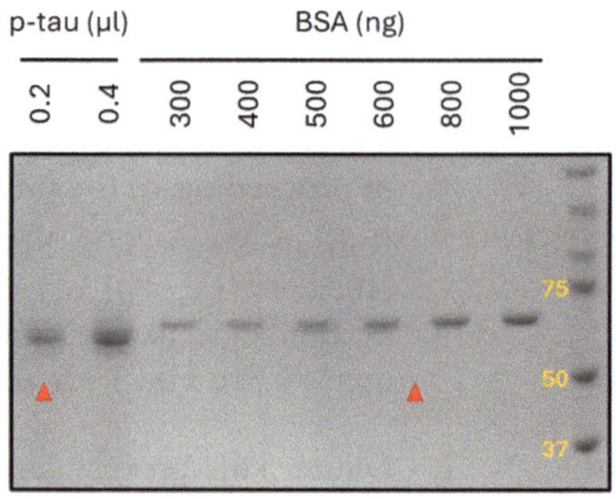

Fig. 2 Protein concentration determination. Red arrows indicate the comparable intensity of 0.2 μL p-tau and 700 ng of BSA by visual estimation. Note: The concentration of enriched p-tau can be analyzed using ImageJ compared with a series of diluted BSA

Troubleshooting

Common issues encountered during the procedure and their solutions are summarized in (Table 1).

Table 1
Common issues and their solutions

Issues	Possible reasons	Solution
Induction failure	1. IPTG is inactive 2. Incorrect plasmid 3. Overgrown overnight culture	1. Parallel testing with other stab culture as positive control. 2. Confirming vector by Sanger sequencing and template sequence can be found in [17]. Primers for checking such as T7 terminator primer CCGCTGAGCAATAACTAGCZAC AS primer GATTATGCGGCCGTGTACAA 3. When SDS-PAGE analysis reveals a low induction of p-tau, and at the same time, an increase of a 50-kDa band, repeating the induction with care given to the overnight seed culture is warranted.
Low protein yield in the purification	1. Incomplete sonication 2. Unevenly heating	1. There are still plenty of proteins in the sonication pellet fraction. Increase sonication time while keeping the temperature low and no bubble. 2. Making sure to swirl the tube in the heating step several times, around once per 10 min.
Protein loss in the concentration step	Protein is trapped or precipitated on the membrane surface	1. High concentrated protein becomes sticky and rinse the membrane well with a small volume of purification buffer after transferring proteins to the tube. 2. Reduce the rounds and time for concentration by using the right cutoff and the right volume of Amicon. 3. Mix the concentrated protein solution evenly before measuring the concentration.

Acknowledgments

The authors thank Stacy Hovde, Mengyu Liu, and all the people in the Kuo group for their technical support and discussion throughout the progression of this work. We also thank John Henige for sharing his advice of p-tau induction. This work was supported by NIH grants (R01NS135693, RF1AG077475, and R01AG062435) to MHK.

Conflict of Interest The authors have no conflicts of interest to declare that are relevant to the content of this chapter.

Appendix 1

Support protocol for transformation using rubidium chloride (RbCl) treated competent cells [22]:

1. Thaw competent cells on ice.
2. Add 50 ng DNA into 50 μL of competent cells. GENTLY mix by flicking the bottom of the tube.

3. Incubate the competent cell/DNA mixture on ice for 30 min.
4. Heat shock each transformation tube in 42°C water bath for 45 s.
5. Cool the tubes on ice for 2 min.
6. Add 800 μL LB (without antibiotics) to the mixture and put on the rotor at 230 rpm at 37°C for 45 min.
   *The outgrowth step allows the bacteria to have time to generate the antibiotic resistance proteins.
7. Evenly plate 50 μL of cells on 10 cm LB agar plates with 100 μg/mL ampicillin and 25 μg/mL chloramphenicol and incubate it at 37°C for 12–18 h.
8. Select a single colony from the LB agar plate using a sterile pipette tip or wooden stick and inoculate into the liquid LB with 100 μg/mL ampicillin and 25 μg/mL chloramphenicol. Incubate bacterial culture at 37°C for 12–18 h in a shaking incubator at 230 rpm.
9. Mix 600 μL of overnight culture with 300 μL 100% glycerol. Snap freeze the solution and store the stab culture at –80°C.

References

1. Goedert M et al (1988) Cloning and sequencing of the cDNA encoding a core protein of the paired helical filament of Alzheimer disease: identification as the microtubule-associated protein tau. Proc Natl Acad Sci U S A 85(11):4051–4055
2. Hernández F, Avila J (2007) Tauopathies. Cell Mol Life Sci 64(17):2219–2233
3. Shammas SL et al (2015) A mechanistic model of tau amyloid aggregation based on direct observation of oligomers. Nat Commun 6: 7025
4. Fitzpatrick AW et al. (2017) Cryo-EM structures of tau filaments from Alzheimer's disease. Nature 547(7662):185–190
5. Götz J, Halliday G, Nisbet RM (2019) Molecular pathogenesis of the tauopathies. Annu Rev Pathol: Mech Dis 14(1):239–261
6. Noble W et al (2013) The importance of tau phosphorylation for neurodegenerative diseases. Front Neurol 4:83
7. Köpke E et al (1993) Microtubule-associated protein tau. Abnormal phosphorylation of a non-paired helical filament pool in Alzheimer disease. J Biol Chem 268(32):24374–24384
8. Gong CX, Iqbal K (2008) Hyperphosphorylation of microtubule-associated protein tau: a promising therapeutic target for Alzheimer disease. Curr Med Chem 15(23):2321–2328
9. Rankin CA, Sun Q, Gamblin TC (2007) Tau phosphorylation by GSK-3β promotes tangle-like filament morphology. Mol Neurodegener 2:1–14
10. Patrick GN et al (1999) Conversion of p35 to p25 deregulates Cdk5 activity and promotes neurodegeneration. Nature 402(6762):615–622
11. Nelson PT et al (2012) Correlation of Alzheimer disease neuropathologic changes with cognitive status: a review of the literature. J Neuropathol Exper Neurol 71(5):362–381
12. Braak H et al (2006) Staging of Alzheimer disease-associated neurofibrillary pathology using paraffin sections and immunocytochemistry. Acta Neuropathol 112(4):389–404
13. Shafiei SS, Guerrero-Muñoz MJ, Castillo-Carranza DL (2017) Tau oligomers: cytotoxicity, propagation, and mitochondrial damage. Front Aging Neurosci 9:83
14. Lasagna-Reeves CA et al (2011) Tau oligomers impair memory and induce synaptic and mitochondrial dysfunction in wild-type mice. Mol Neurodegener 6:1–14
15. Goedert M et al (1996) Assembly of microtubule-associated protein tau into Alzheimer-like filaments induced by sulphated glycosaminoglycans. Nature 383(6600):550–553

16. Despres C et al (2019) Major differences between the self-assembly and seeding behavior of heparin-induced and *in vitro* phosphorylated tau and their modulation by potential inhibitors. ACS Chem Biol 14(6):1363–1379
17. Sui D et al (2015) Protein interaction module–assisted function X (PIMAX) approach to producing challenging proteins including hyperphosphorylated tau and active CDK5/p25 kinase complex. Mol Cell Proteomics 14(1):251–262
18. Liu M et al (2020) Hyperphosphorylation renders tau prone to aggregate and to cause cell death. Mol Neurobiol 57:4704–4719
19. Liu M et al (2020) Hyperphosphorylated tau aggregation and cytotoxicity modulators screen identified prescription drugs linked to Alzheimer's disease and cognitive functions. Sci Rep 10(1):16551
20. Song Z et al (2024) Hyperphosphorylated tau inflicts intracellular stress responses that are mitigated by apomorphine. Mol Neurobiol 61(5):2653–2671
21. Wang KW, Zhang G, Kuo MH (2023) Frontotemporal dementia P301L mutation potentiates but is not sufficient to cause the formation of cytotoxic fibrils of tau. Int J Mol Sci 24(19):14996
22. Madison (1996) Protocols and applications guide, 3rd edn, Vol. 3. Promega Corporation, Madison, WI, USA, pp 45–46

Structural and Functional Assays for the Characterization of Recombinant Hyperphosphorylated Tau Protein

Kuang-Wei Wang, Hsiao-Tien Hagar and Min-Hao Kuo

Abstract

Abnormally phosphorylated tau is increasingly accepted as a critical pathogenic factor of Alzheimer's disease and other neurodegenerative tauopathies. The molecular, biochemical, biophysical, and cell biological studies of hyperphosphorylated tau will facilitate our understanding of the disease mechanisms and promote the quest for efficacious therapeutics. To date, the hyperphosphorylated tau (p-tau) produced by the PIMAX system provides one of the most convenient and pathophysiologically relevant materials for the basic and translational research of tauopathies. In the accompanying chapter (Chap. 3—Hagar et al.), we describe how to prepare hyperphosphorylated tau from *E. coli*. In this chapter, protocols for the biochemical and functional characterization of this protein tau are described in detail.

Keywords Alzheimer's disease, Hyperphosphorylated tau, Neurodegeneration, Tauopathy, PIMAX, Apomorphine

1 Introduction

Tau is an intrinsically disordered protein encoded by the MAPT gene on chromosome 17 [1]. Although its in vitro microtubule stabilization activity is well-documented [2], the normal, biological role of tau remains to be fully delineated [3]. In Alzheimer's disease (AD), tau is found to possess a remarkable degree of post-translational modifications (PTMs), including acetylation, methylation, phosphorylation, N- and O-glycosylation, ubiquitination, truncation, glycation, oxidation, ubiquitylation, etc., amounting to 63 PTMs [4]. Despite the diversity in PTMs, a cohort of neurodegenerative tauopathies shares the trait of the deposition of hyperphosphorylated tau, whether wild-type or bearing a mutation [5, 6], suggesting that abnormal phosphorylation of tau plays one of the key roles in the disease [7]. Hyperphosphorylated tau can assemble and propagate in neuronsthrough a prion-like molecular process [8]. The spatiotemporal distribution of hyperphosphorylated tau in filaments or neurofibrillary tangles

Jessica S. Fortin (ed.), *Translational Research Methods for Tauopathies*, Neuromethods, vol. 233,
https://doi.org/10.1007/978-1-0716-5234-3_4,

closely correlates with the progression of cognitive impairments in human patients [8, 9]. Indeed, the aggregation of tau protein can be induced by the intracerebral infusion of postmortem human disease tau into transgenic mice expressing human tau or, to a lesser extent, into wild-type, nontransgenic mice [10–12], strongly suggesting a pathogenic role for abnormally phosphorylated tau.

The oligomeric form of tau is increasing thought to be the pathogenic species that drives the neurodegeneration [13]. In our previous work, we characterized recombinant hyperphosphorylated tau (p-tau) produced by the PIMAX system [14] with critical disease-relevant features including hyperphosphorylation patterns associated with AD, inducer-free aggregation, and activation of intracellular stresses and cell death [15, 16]. Notably, none of these characters is associated with the unphosphorylated tau counterpart. These observations showcase the invaluable advantages of p-tau for AD mechanism and drug discovery studies. The accompanying chapter (Hagar et al.) describes in detail the preparation of p-tau from the PIMAX system. This chapter shows methods that we developed for the functional characterization of p-tau, which include aggregation, Photoinduced crosslinking of unmodified proteins (PICUP assay), transmission electron microscopy and Fourier-transform infrared spectroscopy (FTIR) for measuring the aggregation level and morphology of p-tau. The ability of propagation and cell-killing is also considered as a critical property of pathological tau [17]. We detail the procedures of using SH-SY5Y neuroblastoma cells for cytotoxicity assay. Lastly, the uptake of p-tau and dysregulation of intracellular calcium can be examined with the same cell line.

Since its discovery in the eighties [18], numerous studies centering on tau have been published. However, the majority of these works focused on unmodified tau that deviates from the disease-associated protein [19]. One of the purposes of this chapter is to promote the use of a pathophysiologically relevant p-tau for basic and translational studies, hence, facilitating the design of efficacious therapeutics for Alzheimer's disease and other tauopathies.

2 Equipment and Materials

2.1 ThS Aggregation Assay

1. Microplate Readers (BioTek Synergy H1 plate reader or equivalent).
2. MicroAmp Clear Adhesive Film (Thermo Fisher # 4360954).
3. 20 mM Tris (pH 7.4) filtered with a 0.22 μm filter.
4. Dithiothreitol (DTT) (Sigma-Aldrich). 20 mM DTT stock dissolved in deionized water and filtered with a 0.22 μm filter.
5. Thioflavin S (ThS) (Sigma-Aldrich). 1 mM ThS stock dissolved in deionized water and filtered with a 0.22 μm filter.

2.2 Cytotoxicity Assay

1. SH-SY5Y neuroblastoma (ATCC CRL-2266).
2. DMEM/F12 medium (Gibco).
3. Fluorescein diacetate (FDA) (Sigma-Aldrich). 0.5 mg/mL FDA stock in DMSO. Diluted to 5 μg/mL FDA with sterile PBS when use.
4. Propidium iodide (PI) (Sigma-Aldrich). 1 mg/mL PI stock in deionized water. Diluted to 10 μg/mL FDA with sterile PBS when use.
5. Fluorescence microscope with filter sets for FITC and Texas Red to observe FDA-positive and PI-positive cells, respectively (Olympus BX51 or equivalent).
6. Cell Counting Kit-8 (CCK-8) (GlpBio).

2.3 Photoinduced Cross-Linking of Unmodified Proteins (PICUP Assay)

1. Tris(bipyridine)ruthenium (II) ($Ru(Bpy)_3$ (Sigma-Aldrich). 1 mM $Ru(Bpy)_3$ in 10 mM sodium phosphate, pH 7.4.
2. Ammonium persulfate (APS) (Sigma-Aldrich). 20 mM APS solution in 10 mM sodium phosphate, pH 7.4.
3. Dithiothreitol (DTT) (Sigma-Aldrich). 1 M DTT stock in deionized water.

2.4 LED Work Light (Porter Cable or Equivalent)

1. Transmission electron microscopy (TEM).
2. Transmission electron microscope (JEOL JEM-1400 or equivalent).
3. 150-mesh copper square grids (VWR).
4. 25% glutaraldehyde solution, EM grade (Polysicences #01909-100).
5. 4% uranyl acetate (EMS #22400-4).

2.5 Examination of p-tau Uptake by Cells

1. Alexa Fluor 594 protein labeling kit (Invitrogen #A10239).
2. Chambered coverglass (Lab-Tek #155380 or equivalent).
3. Confocal microscope (Nikon A1 or equivalent).
4. Intracellular calcium level imaging.

2.6 Fluo-4 AM (Invitrogen #F14201)

1. Chambered coverglass (Lab-Tek #155380 or equivalent).
2. Confocal microscope (Nikon A1 or equivalent).

2.7 Fourier-Transform Infrared Spectroscopy (FTIR)

1. Mattson Galaxy series GL5020 spectrometer (or equivalent).
2. Demountable FTIR Liquid Cells (CaF_2 window) (PIKE Technologies #162-1100 or equivalent).
3. Acetone (Sigma-Aldrich).
4. Deuterium oxide (D_2O) (Sigma-Aldrich #151890).

3 Methods

1. **Aggregation Assay Based on Thioflavin S (ThS) Fluorescence**

   (1) Thaw the p-tau and, if necessary, the unmodified tau control protein on ice before reaction assembly.
   Note: Proteins can be ultracentrifuged to remove large protein aggregates (100,000 x*g*, 60 min). However, this step typically did not result in a significant reduction of the initial ThS fluorescence.

   (2) Mix 6 μM tau protein with or without compounds of interest in 20 mM Tris (pH 7.4) with 1 mM DTT and 20 μM ThS on ice.
   Note: 1 mM DTT is included in the reaction to avoid disulfide bond formation.

   (3) Place 10 μL of the solution into a 384-well black plate. Triplicate can avoid technical errors.

   (4) Cover the plate with transparent adhesive film and place into a preheated 37° plate reader.

   (5) Measure the ThS fluorescence every 10 min at 440/490 nm Ex/Em for up to 16 h. See Fig. 1 for an example.

2. **Cytotoxicity Assay**

   (1) Seed 2000 SH-SY5Y cells per well in a 96-well plate with 100 μL DMEM/F12 media (10% FBS, 1% Pen/

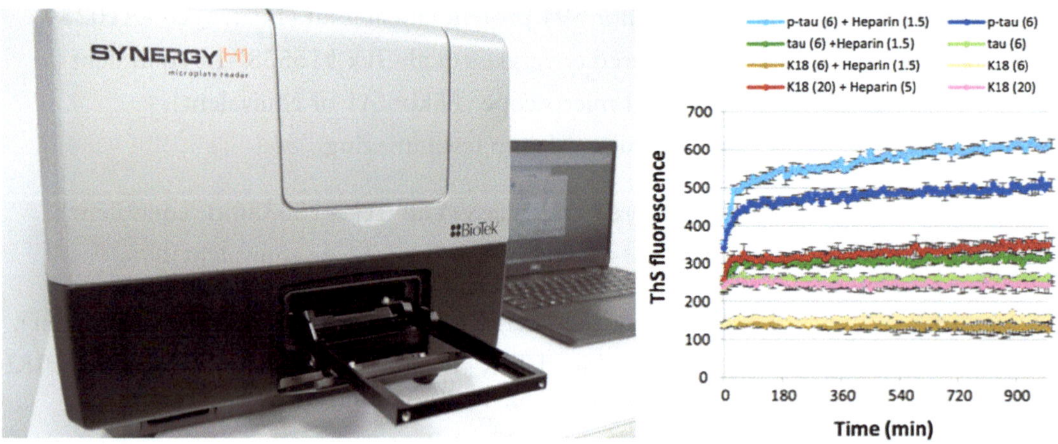

Fig. 1 (**A**) BioTek Synergy H1 plate reader (**B**) Inducer-free fibrillization of p-tau. Aggregation of p-tau, tau, and K18 with or without heparin was quantified by thioflavin S (ThS) fluorescence changes through the entire course of assays [16]

Strep) and culture for 24 h at 37°C, 5% CO_2, and 95% RH.

Note: The optimal number of cells for seeding may depend on different cell batches, passage numbers and mediums, and therefore, should be tested first.

(2) Thaw p-tau and other proteins, if needed, on ice before sample preparation.

(3) Mix tau protein solution at 10× concentration of final concentration with or without compounds of interest. Note: LD50 of p-tau should be addressed first before further experiments. Typically, LD50 of p-tau is around 0.5–0.8 μM depending on cell condition. Serial dose of p-tau can start from 0.2–1.0 μM.

(4) Add 10 μL of protein solution into each well and incubate for up to 20 h.

(5) Relative cell viability can be determined using either way below. See Fig. 2 for an example.

   a. Cell counting kit 8 (CCK8): Add 10 μL CCK8 solution into each well and incubate for 2–3 h measuring the absorbance at 450 nm using a plate reader. Calculate relative cell viability based on CCK8 signals.

   b. FDA/PI staining: Trypsinize the cells after the medium removal and transfer the cell solution to microcentrifuge tubes and spin down the cells at 1000 × *g* for 5 min at 4°C. Discard the supernatant and resuspend the cell pellet in 3 μL PBS (pH 7.4).

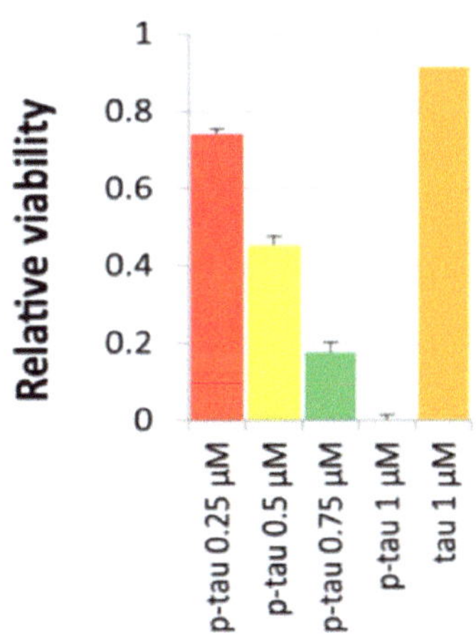

Fig. 2 Viability of 5Y cells after 20-h treatment of 0.25–1 μM of p-tau or 1 μM of unmodified tau pre-aggregated for 20 h. Cells were double-stained by fluorescein diacetate (FDA) and propidium iodide (PI) for the quantification of relative viability [16]

Incubate it with 1 µg/mL fluorescein diacetate (FDA) by adding 1 µL of 5 µg/mL FDA solution for 5 min followed by 2 µg/mL propidium iodide (PI) by adding 1 µL of 10 µg/mL PI solution at room temperature. Examine and count cells using a fluorescence microscope. Relative cell viability is calculated by FDA positive cells/(FDA positive cells + PI positive cells).

3. **PhotoInduced Cross-Linking of Unmodified Proteins (PICUP Assay)** [20].

   (1) Incubate 10 µM of tau protein with or without compounds of interest in 20 mM Tris buffer (pH 7.4) at 37°C for a desired amount of fibrillization time.

   (2) Mix tau solution, 20 mM APS, and 1 mM Tris(bipyridine) ruthenium(II) ($Ru(Bpy)_3$ in short) at 9:0.5:0.5 volume ratio to make the final concentrations of APS and $Ru(Bpy)_3$ to be 1 mM and 0.05 mM, respectively.

   (3) Place the reaction mixture approximately 10 cm from a 50 W, 5000 lumen LED work light exposed for 50 s. Note: The distance and the exposure time should be optimized for different proteins.

   (4) Quench the reaction by adding DTT to a final concentration 50 mM.

   (5) Analyze the cross-linked products by SDS-PAGE. See Fig. 3 for an example.

4. **Transmission Electron Microscopy (TEM)**

   (1) Incubate 10 µM of p-tau with or without compounds of interest in 20 mM Tris buffer (pH 7.4) at 37°C for a desired amount of fibrillization time.

   (2) Dilute the sample ten-fold with 20 mM Tris buffer and incubate with 2.5% glutaraldehyde for 5 min at RT.

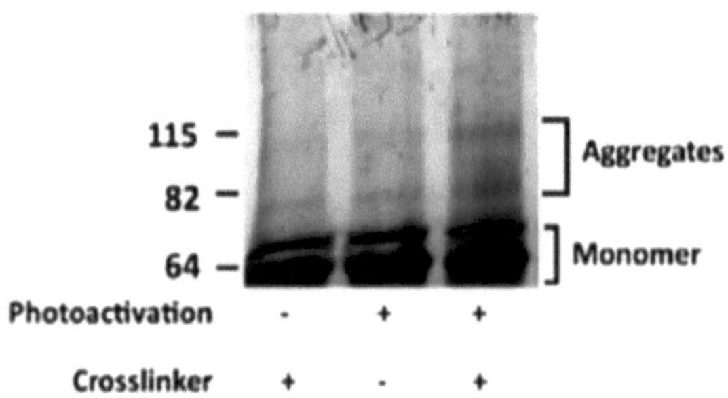

Fig. 3 PICUP (photoinduced cross-linking of unmodified protein) assays detecting high-molecular-weight p-tau species. Shown are overnight aggregation products [16]

(3) Put 20 μL of the sample on a 150-mesh square copper grid for 10 min for fixation.

(4) Wash the grid with deionized water three times and carefully dry the grid with filter paper.

(5) Negatively stain the sample by dropping 10 μL of 1% uranyl acetate on the grid for 10 s.

(6) Dry the grid with filter paper.

(7) Image the sample by JOEL 1400 Flash TEM. See Fig. 4 for an example.
Note: To verify the identity of the fibril, immunogold staining can be administrated [21].

5. **Examination of p-tau Uptake by Cells**

(1) Seed SH-SY5Y cells in chambered coverglass at the density of 4×10^4 cells/mL in DMEM/F12 media (10% FBS, 1% pen/strep) and culture for 24 h at 37 °C, 5% CO_2 and 95% RH.

(2) Label p-tau with fluorescence by Alexa Fluor protein labeling kit.

(3) Treat the cells with varying amounts of Alexa conjugated p-tau with or without compounds of interest for a desired amount time.

(4) Image the live cells by Nikon A1 confocal microscope with temperature and CO_2 controlled at 37°C, 5% CO_2. See Fig. 5 for an example.

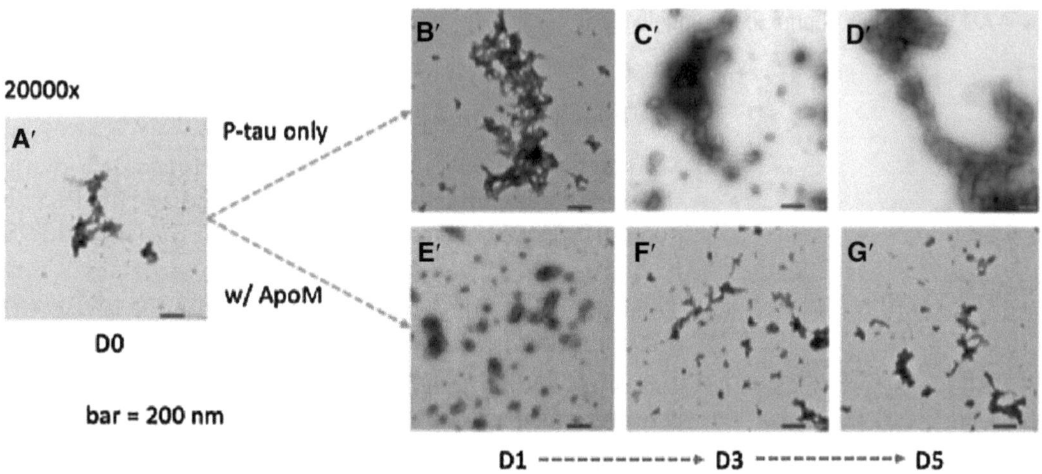

Fig. 4 p-tau sampled from 0, 1, 3, and 5 days of in vitro aggregation reactions with or without apomorphine. One micromolar of the sample was negatively stained and observed by TEM. (**A–D**) p-tau only; (**E–G**) p-tau aggregation in the presence of apomorphine, a p-tau aggregation inhibitor. Scale bar = 200 nm [15]

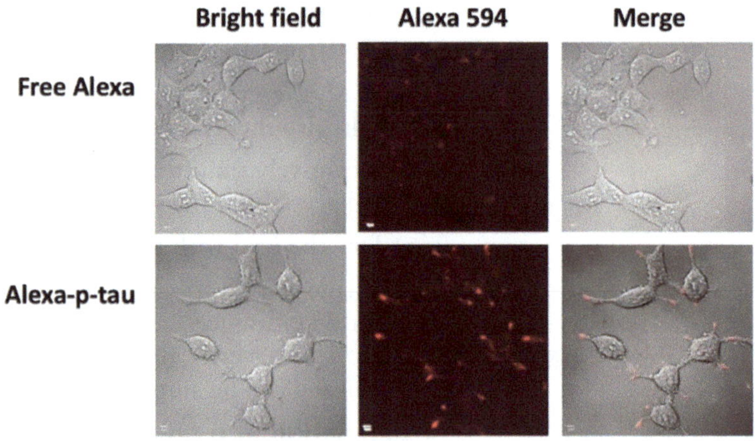

Fig. 5 SH-SY5Y cells were treated with or without 0.1 μM Alexa 594-conjugated p-tau for 20 h before imaging by the Olympus FV microscope, 40× objective [15]

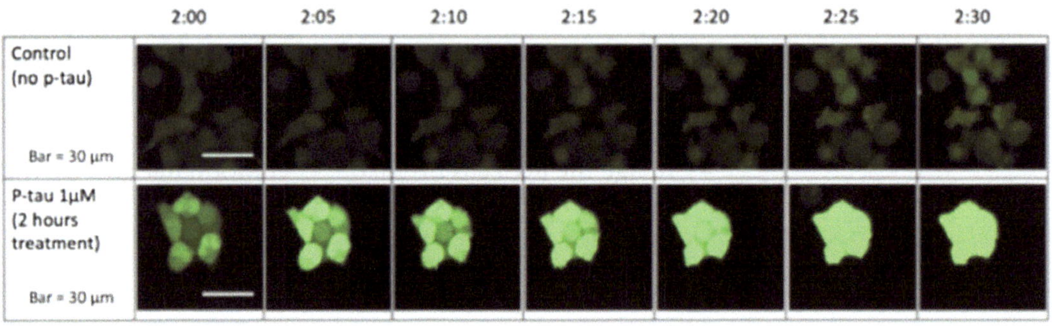

Fig. 6 SH-SY5Y cells were treated with 1 μM p-tau for 2 h and recorded by confocal microscopy for 30 min. Fluo-4 AM was added 1 hprior to the imaging [15]

6. **Intracellular Calcium Level Imaging**

(1) Seed SH-SY5Y cells in chambered coverglass at a density of 4×10^4 cells/mL in DMEM/F12 media (10% FBS, 1% pen/strep) and culture for 24 h at 37 °C, 5% CO_2 and 95% RH.

(2) Treat the cells with varying amounts of p-tau with or without compounds of interest for a desired amount time.

(3) 1 h prior to the imaging, add Fluo-4 AM to a final concentration 1 μM to the cells for intracellular free calcium staining.

(4) Image the live cells with Nikon A1 confocal microscope with temperature and CO_2 controlled at 37°C, 5% CO_2. See Fig. 6 for an example.

7. **Fourier-Transform Infrared Spectroscopy (FTIR)**

(1) Incubate 10 μM of p-tau with or without compounds of interest in 20 mM Tris buffer (pH 7.4) at 37°C for up to 5 days for fibrillization.
Note: Longer incubation may be needed for the detection of clear spectral shift.

(2) Add four protein sample volumes of cold acetone (−20°C) to the sample and mix to precipitate the protein, and then centrifuge the solution at 14,000 rpm in a microcentrifuge tube for 15 min.

(3) Wash the pellet with cold acetone once and centrifuge to remove residual liquid.

(4) Air-dry the protein pellet for 5 min.

(5) Dissolve the protein pellet in D_2O to bring the protein concentration to ~ 1 mg/mL and incubate at room temperature for 2 h.
Note: This step is critical for the H_2O surrounding the protein to exchange with D_2O. H_2O would affect protein secondary structure signal in FTIR.

(6) Load the sample to a demountable FTIR Liquid Cells with CaF_2 windows with a 100-μm spacer. Note: CaF_2 would not be dissolved by aqueous samples.

(7) Record the infrared spectrum with a Mattson Galaxy series GL5020 spectrometer between wave numbers 1900 and 1200 cm^{-1} at the resolution of 2 cm^{-1}, and average 128 scans for comparison. See Fig. 7 for an example.

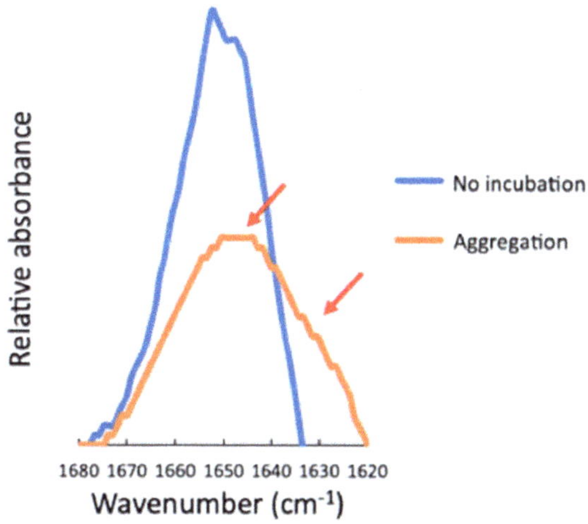

Fig. 7 FTIR (Fourier transformation infrared) spectroscopy analysis of p-tau after aggregation. Spectral shifts after aggregation are indicated by two arrows [16]

Acknowledgments

We thank Stacy Hovde, Mengyu Liu, and all the people in the Kuo group for their technical support and discussion throughout the progression of this work. This work was supported by NIH grants (R01NS135693, RF1AG077475, and R01AG062435) to MHK.

Conflict of Interest The authors have no conflicts of interest to declare that are relevant to the content of this chapter.

References

1. Goedert M et al (1988) Cloning and sequencing of the cDNA encoding a core protein of the paired helical filament of Alzheimer disease: identification as the microtubule-associated protein tau. Proc Natl Acad Sci U S A 85(11):4051–4055
2. Barbier P et al (2019) Role of tau as a microtubule-associated protein: structural and functional aspects. Front Aging Neurosci 11:204
3. Ke YD et al (2012) Lessons from tau-deficient mice. Int J Alzheimers Dis 2012:873270
4. Morris M et al (2015) Tau post-translational modifications in wild-type and human amyloid precursor protein transgenic mice. Nature Neurosci 18(8):1183–1189
5. Goedert M, Ghetti B, Spillantini MG (2012) Frontotemporal dementia: implications for understanding Alzheimer disease. Cold Spring Harb Perspect Med 2(2):a006254
6. Kovacs GG (2015) Invited review: neuropathology of tauopathies: principles and practice. Neuropathol Appl Neurobiol 41(1):3–23
7. Hanger DP, Anderton BH, Noble W (2009) Tau phosphorylation: the therapeutic challenge for neurodegenerative disease. Trends Mol Med 15(3):112–119
8. Jucker M, Walker LC (2018) Propagation and spread of pathogenic protein assemblies in neurodegenerative diseases. Nat Neurosci 21(10):1341–1349
9. Kuchibhotla KV et al (2014) Neurofibrillary tangle-bearing neurons are functionally integrated in cortical circuits *in vivo*. Proc Natl Acad Sci U S A 111(1):510–514
10. Clavaguera F et al (2009) Transmission and spreading of tauopathy in transgenic mouse brain. Nat Cell Biol 11(7):909–913
11. Kaufman SK et al (2016) Tau prion strains dictate patterns of cell pathology, progression rate, and regional vulnerability *in vivo*. Neuron 92(4):796–812
12. Clavaguera F et al (2013) Brain homogenates from human tauopathies induce tau inclusions in mouse brain. Proc Natl Acad Sci 110(23):9535–9540
13. Ward SM et al (2012) Tau oligomers and tau toxicity in neurodegenerative disease. Biochem Soc Trans 40(4):667–671
14. Sui D et al (2015) Protein interaction module-assisted function X (PIMAX) approach to producing challenging proteins including hyperphosphorylated tau and active CDK5/p25 kinase complex. Mol Cell Proteomics 14(1):251–262
15. Song Z et al (2024) Hyperphosphorylated tau inflicts intracellular stress responses that are mitigated by apomorphine. Mol Neurobiol 61(5):2653–2671
16. Liu M et al (2020) Hyperphosphorylation renders tau prone to aggregate and to cause cell death. Mol Neurobiol 57(11):4704–4719
17. Wolfe MS (2012) The role of tau in neurodegenerative diseases and its potential as a therapeutic target. Scientifica (Cairo) 2012:796024
18. Grundke-Iqbal I et al (1986) Abnormal phosphorylation of the microtubule-associated protein tau (tau) in Alzheimer cytoskeletal pathology. Proc Natl Acad Sci U S A 83(13):4913–4917
19. Wang Y, Mandelkow E (2016) Tau in physiology and pathology. Nat Rev Neurosci 17(1):5–21
20. Rahimi F, Maiti P, Bitan G (2009) Photo-induced cross-linking of unmodified proteins (PICUP) applied to amyloidogenic peptides. J Vis Exp (23):1071
21. Wang KW, Zhang G, Kuo MH (2023) Frontotemporal dementia P301L mutation potentiates but is not sufficient to cause the formation of cytotoxic fibrils of tau. Int J Mol Sci 24(19):14996

Chapter 5

Evaluation of the Kinetics of Tau Fibril Formation

Natalie G. Horgan, Taiwo A. Ademoye and Jessica S. Fortin

Abstract

Thioflavin S (ThS) fluorescence assays and transmission electron microscopy (TEM) are two common techniques used to detect tau fibril formation and monitor aggregation kinetics. Thioflavin S is a fluorescent dye that binds to β-sheet structures, displaying enhanced fluorescence upon binding to fibrils. Transmission electron microscopy uses an electron beam to generate high magnification images of fibrils, allowing further assessment by area and length measurements. This chapter outlines various protocols for assessing tau 2N4R fibril formation in the presence of test compounds, including performing thioflavin S fluorescence assays, conducting transmission electron microscopy, utilizing immunogold labeling for protein identification, and measuring fibril area and length.

Keywords Aggregation, β-sheet, Fibril, Immunogold, Tau, Thioflavin S (ThS), Transmission electron microscopy (TEM)

1 Introduction

Thioflavin S (ThS), a benzothiazole dye, is widely used to assess aggregation kinetics. Upon binding to β-sheet-rich structures, ThS exhibits enhanced fluorescence with peak emission at 455 nm [1]. Historically, thioflavin dyes have been used in both fluorescence spectroscopy assays and histological staining under microscopy to measure amyloid-like fibril formation *in vitro* and *in vivo* [2, 3]. The mechanism of its binding is debated, with studies suggesting that it binds to fibrils with their long axes aligned parallel to the fibril axis [4], docks onto β-sheet surfaces formed by a single tyrosine ladder [5], and forms micelles that bind to amyloid-like fibrils [6]. Although these dyes are specific for amyloid-like fibrils, they can also interact with other molecules, including DNA, hyalinous cartilage, and acetylcholinesterase [7].

Transmission electron microscopy (TEM) is a technique performed with an electron microscope that transmits a particle beam of electrons through a sample to form high magnification images. It can be used for length measurements to assess fibril formation,

Jessica S. Fortin (ed.), *Translational Research Methods for Tauopathies*, Neuromethods, vol. 233,
https://doi.org/10.1007/978-1-0716-5234-3_5,

disaggregation, and secondary aggregation pathways, as well as to determine aggregate composition via immunogold labeling [8]. Outside of these applications, TEM can also distinguish between different conformations of tau protein found in Alzheimer's disease, including paired helical filaments (PHFs) [9–11] and straight filaments (SFs) [11, 12]. These structures consist of two C-shaped protofilaments with a core of eight β-sheet [13], surrounded by a disordered "fuzzy coat," but differ in protofilament symmetry. In PHFs, the protofilaments have helical (2_1 screw) symmetry, while the protofilaments in SFs are packed asymmetrically. Both PHFs and SFs are composed of hyperphosphorylated, truncated tau, which make up neurofibrillary tangles that impair axonal transport and microtubule stability [14, 15]. Interestingly, PHFs contain all six tau isoforms [16].

This chapter outlines the protocol for evaluating the kinetics of tau fibril formation and visualizing fibrils using ThS fluorescence assays and TEM with test compounds. These compounds are assessed for their efficacy in reducing and potentially disaggregating fibrils formed by tau 2N4R. The TEM protocol also describes how to measure fibril area and length, as well as how to perform immunogold labeling to specifically identify tau protein.

2 Materials

2.1 Reagents and Equipment for Thioflavin S Fluorescence Assay

- Arachidonic acid (TCI America #A0781): This chemical should be stored at −20°C for up to 6 months.
- Black 384-well flat-bottom microplate (Greiner Bio-One #784076).
- Chelex beads (G-Biosciences #BTNM-0024).
- Dimethyl sulfoxide (Sigma-Aldrich #34869): This solvent should be stored at room temperature for up to 2 years.
- DL-dithiothreitol (DTT) (VWR Life Science #97061-340): This agent should be stored at 2–8°C for up to 3 years.
- Heparin sodium salt (Sigma-Aldrich #H3393): This compound should be stored at room temperature for up to 5 years from the date of manufacture.
- PBS (Gibco #10010-023, 1×, pH 7.4): 137 mM NaCl, 2.7 mM KCl, 10 mM Na_2HPO_4, and 1.8 mM KH_2PO_4. This buffer should be stored at room temperature for up to 24 months from the date of manufacture.
- Synergy HT multi-mode microplate reader (BioTek, Winooski, VT).
- Tau 2N4R (rPeptide #T-1001-2): This peptide should be stored between −20°C and −80°C for up to 2–3 years prior to

reconstitution. This peptide should be stored at −80°C for up to 1–2 weeks after reconstitution.

- Test compounds.
- Thioflavin S (Sigma-Aldrich #T1892): This powder should be stored at room temperature for up to 3–5 years.

2.2 Reagents and Equipment for Transmission Electron Microscopy

- 400-mesh Formvar/carbon-coated copper grid (Electron Microscopy Sciences #FCF400-CU).
- Arachidonic acid (TCI America #A0781): This chemical should be stored at −20°C for up to 6 months.
- Centrifuge tubes.
- DL-dithiothreitol (DTT) (VWR Life Science #97061-340): This agent should be stored at 2–8°C for up to 3 years.
- Dimethyl sulfoxide (Sigma-Aldrich #34869): This solvent should be stored at room temperature for up to 2 years.
- Filter paper.
- GraphPad Prism 9.
- Heparin sodium salt (Sigma-Aldrich #H3393): This compound should be stored at room temperature for up to 5 years from the date of manufacture.
- ImageJ software.
- JEOL 1400 flash electron microscope.
- PBS (Gibco #10010-023, 1×, pH 7.4): 137 mM NaCl, 2.7 mM KCl, 10 mM Na_2HPO_4, and 1.8 mM KH_2PO_4. This buffer should be stored at room temperature for up to 24 months from the date of manufacture.
- Tau 2N4R (rPeptide #T-1001): This peptide should be stored between −20°C and −80°C for up to 2–3 years prior to reconstitution. This peptide should be stored at −80°C for up to 1–2 weeks after reconstitution.
- Test compounds.
- Uranyl acetate (Electron Microscopy Sciences #22400-1): This solution should be stored at room temperature.

2.3 Reagents and Equipment for Immunogold Staining

- Bovine serum albumin (Aurion #25554): This compound should be stored at 4–8°C.
- Double-distilled water
- Lead citrate (Electron Microscopy Sciences #22410): This compound should be stored at room temperature.

- Normal goat serum (Aurion #25570): This solution should be stored at 4–8°C for up to 1 year.
- Primary antibody against tau, e.g. anti-tau-5 (rabbit or mouse).
- Goat anti-rabbit or mouse IgG antibody conjugated with 10 nm colloidal gold: This antibody should be stored at 2–8°C for up to 1 year.
- PBS (Gibco #10010-023, 1×, pH 7.4): 137 mM NaCl, 2.7 mM KCl, 10 mM Na_2HPO_4, and 1.8 mM KH_2PO_4. This buffer should be stored at room temperature for up to 24 months from the date of manufacture.
- Uranyl acetate (Electron Microscopy Sciences #22400-1): This solution should be stored at room temperature.

3 Methods

3.1 Thioflavin S Fluorescence Assay

This protocol describes the steps of performing a Thioflavin S (ThS) fluorescence assay. The purpose of this assay is to assess both the fluorescence intensity and aggregation kinetics of test compounds in the presence of varying concentrations of tau 2N4R (Fig. 1). In general, a higher fluorescence intensity is indicative of fibril formation and protein aggregation. Therefore, a test compound's ability to reduce ThS fluorescence suggests anti-fibrillar activity.

1. Prepare a 1 mM stock solution of ThS in PBS by combining 3.1886 mg ThS in 10 mL PBS.
2. Using a black 384-well flat-bottom microplate, pipette tau 2N4R into selected wells at a final concentration of 6 or 12 μM in a total volume of 10 μL per well.
   a. For 6 μM tau protein, add 1 μL of tau stock solution (~62 μM) and 3.1 μL Chelex-treated PBS.
   b. For 12 μM tau protein, add 2 μL of tau (stock solution at ~62 μM) and 2.1 μL Chelex-treated PBS.
3. For background wells (no protein), add 4.1 μL of Chelex-treated PBS.
4. To each well, except for background or control wells, add 2.5 μL of the 400 μM test compound (100 μM).
   a. For control wells, add 2.5 μL DMSO (0.25%) instead of the test compound.
   b. For background wells, add 2.5 μL PBS instead of the test compound.

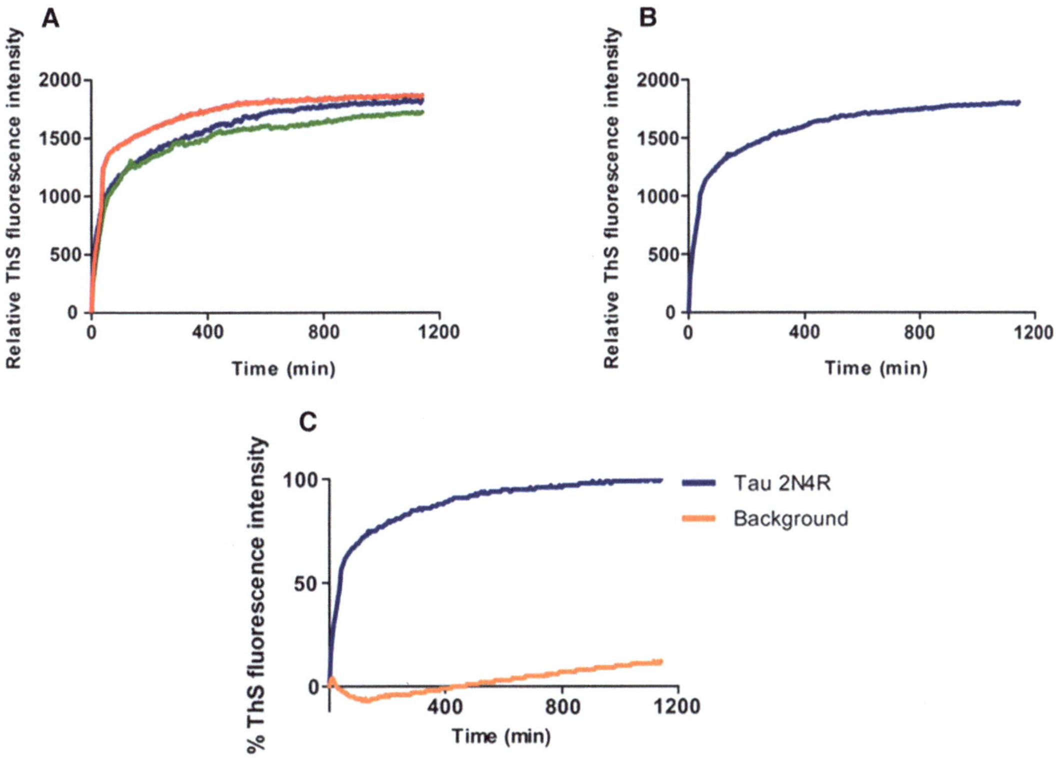

Fig. 1 Fibrillization kinetics of tau 2N4R assessed by the Thioflavin S (ThS) fluorescence assay. The experiment was performed using 12 μM tau 2N4R in Chelex-treated PBS, supplemented with 5 mM DTT, 150 μM heparin, and 0.092 μg/mL arachidonic acid. Samples were monitored at 37°C for 1200 min. Background fluorescence signal was subtracted from each measurement. (**A**) ThS fluorescence curves derived from triplicate measurements of tau 2N4R. (**B**) ThS fluorescence curve of the average of triplicate measurements of tau 2N4R. (**C**) ThS fluorescence curves of both tau 2N4R protein and background control, each expressed as a percentage of the highest fluorescence value observed in the tau sample

5. Into all wells, add 0.5 μL of 100 mM DTT (5 mM), 1.5 μL of 1 mM heparin (150 μM), 1 μL of 0.92 μg/mL arachidonic acid (0.092 μg/mL), and 0.4 μL of 1 mM ThS (40 μM).
6. Cover the plate with a plate seal.
7. Place the plate in a Synergy HT multi-mode microplate reader set to 37°C and shake the plate for 30 s before each read.
8. Measure tau aggregation by recording ThS fluorescence with an excitation wavelength of 440 nm and emission wavelength of 485 nm every 5 min over a period of 60 h.

3.2 Transmission Electron Microscopy

This protocol outlines the preparation of samples for subsequent TEM imaging and statistical analysis. The purpose of TEM is to directly visualize fibrils, which are characteristic of aggregation-prone proteins such as tau (Fig. 2).

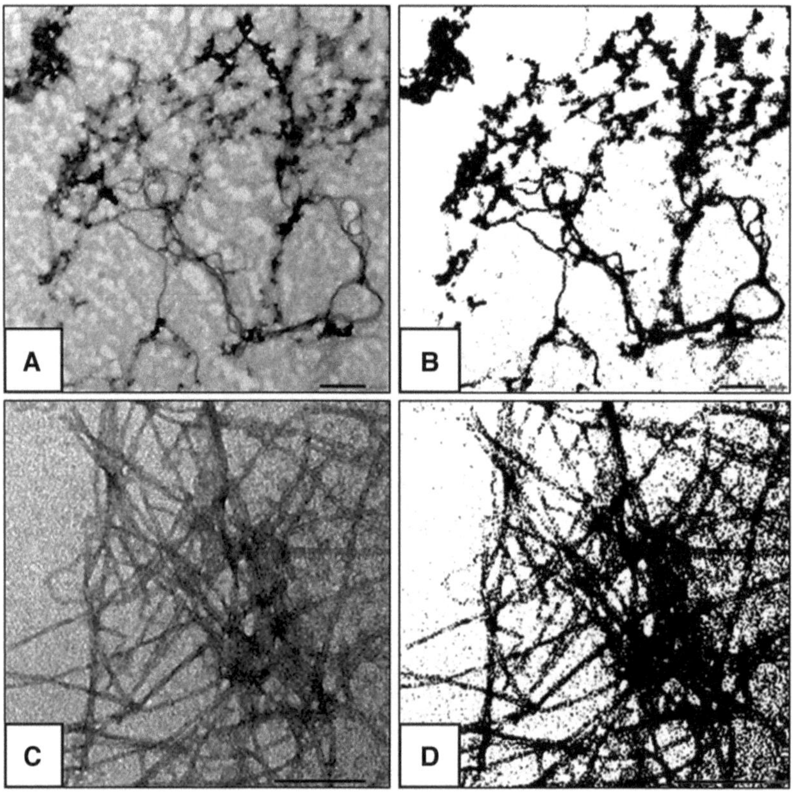

Fig. 2 Examples of fibrils from tau 2N4R surface area measured with ImageJ. (**A**) Photomicrograph of the tau 2N4R fibrils at 20K magnification. (**B**) Area captured by ImageJ from the tau 2N4R fibrils at 20K magnification. (**C**) Photomicrograph of the tau 2N4R fibrils at 40K magnification. (**D**) Area captured by ImageJ from the tau 2N4R fibrils at 40K magnification. Scale bar: 200 nm

The goal is to find test compounds that reduce fibril formation, length, and area.

1. Transfer 20 μL of each sample and 20 μL of the 0.25% DMSO control from the ThS assay into centrifuge tubes.
2. Or in the absence of samples resulting from a recent ThS experiment, prepare tau samples as explained in 3.1 section with the omission of ThS and incubate the samples in Eppendorf tubes at 37°C for 3 days to allow aggregation.
3. Deposit 10 μL of each sample on a 400-mesh Formvar/carbon coated copper grid and incubate for 1 min.
4. Wash the grid gently with distilled water.
5. Air-dry the grid, then stain for 1 min with 1% uranyl acetate.
6. Blot the grids with filter paper and allow it to dry.
7. Visualize the grids using a JEOL 1400 Flash electron microscope (100 kV).
8. Using ImageJ software, measure the surface areas (μm^2) of the fibrils and protofibrils.

9. Analyze eight different fibril/plaque regions per sample.
10. Plot the measurements and perform statistical analysis in GraphPad Prism 9, using an unpaired two-tailed t-test.

3.3 Immunogold Staining

Immunogold electron microscopy is a method that visualizes specific proteins by using antibodies conjugated to gold electron-dense gold nanoparticles (Fig. 3). This technique is particularly useful for characterizing fibrils, mapping protein-protein interactions, locating specific proteins, and identifying other materials present in a sample.

1. Repeat Steps 1–3 from the transmission electron microscopy protocol above.
2. Prepare the 5% BSA solution by weighing 2.5 mg of BSA and dissolving it in 50 μL of 0.01 M PBS.
3. Prepare the blocking solution by combining 45 μL of 0.01 M PBS, 2.5 μL of the prepared 5% BSA solution, and 2.5 μL normal goat serum (5%).
4. Prepare the incubation buffer system by combining 0.01 M PBS with 0.1% BSA.
5. Place the grids in the blocking system and incubate for 10–30 min at room temperature.
6. Wash the grids three times for 5 min each with the incubation buffer.

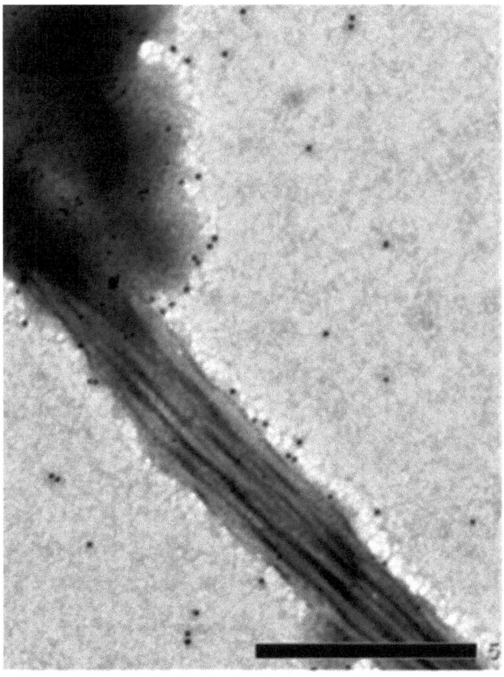

Fig. 3 Immunogold stained tau 1N4R fibrils (Image courtesy of Dr. Min-Hao Kuo, Michigan State University). Scale bar: 500 nm

7. Dilute the primary antibody 1:10 with the incubation buffer system. Incubate the grids with this diluted antibody for 1 h at room temperature.
8. Wash the grids three times for 5 min each with the incubation buffer.
9. Dilute the secondary antibody (anti-rabbit or mouse) conjugated with 10 nm colloid 1:20 with the incubation buffer system. Incubate the grids with this diluted antibody for 1 h at room temperature.
10. Wash the grids three times for 5 min each with the incubation buffer.
11. Wash the grids three times for 5 min each with double-distilled water.
12. Stain the grids with uranyl acetate and lead citrate.
13. Repeat Steps 6–10 from the transmission electron microscopy protocol above.

3.4 Measurement of Tau Fibril Length and Inter-Particle Distance

1. Open the image in ImageJ software (drag it in or use File → Open). Convert the image to 8-bit via Image → Type → 8-bit.
2. Select the magnifying glass icon and left-click to zoom in on the scale bar. Then, choose the straight-line tool, hold Shift to keep it straight, and draw the line along the known scale bar length.
3. Go to Analyze → Set Scale. In the dialog box, enter the known distance (in nanometers) from the scale bar. Click OK to calibrate the image.
4. Zoom in on the particles or fibrils. Use the straight-line tool to draw a line between the centers of two particles or along the length of a fibril. Then go to Analyze → Measure to record the length.
5. Document the measured distance. To overlay the line on the image, go to Edit → Fill. Repeat this measurement several times until there is enough data for averaging.
6. Once there is sufficient data, go to Results → Summarize to get the mean and standard deviation.
7. Click on File and save the results in an Excel format.

3.5 Threshold-Based Measurement of Tau Fibril Area

1. Repeat Steps 1–3 from the inter-particle distance protocol above.
2. Use the rectangular tool to select the desired area to analyze. This area is selected to exclude any imperfections from the portion being analyzed.

3. Go to Image → Duplicate → OK. The selected region opens up in a new window.
4. Go to Image → Adjust → Threshold and select the red overlay mode. Adjust the sliders until all fibrils are highlighted in red. Click Apply to convert the red overlay to a black-and-white image.
5. Close the window threshold. Click Analyze → Analyze Particles. Set the following options:
   a. Size: 0-infinity
      i. Adjust minimum size if you would like to exclude smaller particles.
   b. Circularity: 0.00–10.00
   c. Show: Outlines
   d. Tick display results, clear results, and add to manager, then click OK.
   e. If needed, to exclude particles or fibrils on the edges, tick Exclude on edges.
6. Go to Results → Summarize.
7. Click on File and save the results in an Excel format.

3.6 Manual Measurement of Tau Fibril Area

The following method is used to measure the fibril area when image contrast is poor and fibril boundaries are difficult depict.

1. Repeat Steps 1–3 from the inter-particle distance protocol above.
2. Use the rectangular tool to select the desired area to analyze. This area is selected to exclude any imperfections from the portion being analyzed.
3. Go to Image → Duplicate → OK. The selected region opens up in a new window.
4. Zoom in on the image, select the Freehand icon, and carefully trace around the area of a fibril.
5. Go to the Analyze → Measure to record the area. Then go to Edit → Fill to overlay the traced portion before moving onto the next fibril.
6. Repeat the tracing and measuring process until enough values are obtained for an average area measurement.
7. Go to Results → Summarize.
8. Click on File and save the results in an Excel format.

4 Recommendations

Unlike many proteins that self-aggregate under *in vitro* conditions, tau protein requires specific additives to induce its aggregation. These include DTT and heparin with or without arachidonic acid. Without these components, tau may not form detectable fibrils. Therefore, it is essential to include them during the ThS fluorescence assay to ensure proper aggregation. Additionally, it is recommended to use ThS concentrations in the range of 10–50 μM for maximum sensitivity in aggregation kinetic studies [17]. The experiment should be performed in triplicate to ensure the accuracy and reliability of the data.

5 Troubleshooting

Some compounds like polyphenols can interfere with the fluorescence emitted by thioflavin dyes [18]. Therefore, all results should be confirmed using alternative methods, such as transmission electron microscopy. In other words, thioflavin fluorescence assays alone should not be used to definitively confirm the presence of tau fibrils. Furthermore, pipetting errors can lead to inaccurate results. The presence of bubbles in the wells may interfere with fluorescence readings, resulting in false values. To ensure reliable data, avoid bubbles during pipetting and ensure the plate is clean and free from contaminants.

Acknowledgements

The authors would like to acknowledge the professional services of Alicia Withrow at the Center of Advanced Microscopy at Michigan State University. J.S.F. was supported by the National Institutes of Health (NIH) grant (AG071985).

Conflict of Interest The authors have no conflicts of interest to declare that are relevant to the content of this chapter.

References

1. MacKeigan TP, Morgan ML, Stys PK (2023) Quantitation of tissue amyloid via fluorescence spectroscopy using controlled concentrations of thioflavin-S. Molecules 28(11):4483. https://doi.org/10.3390/molecules28114483
2. Vassar PS, Culling CF (1959) Fluorescent stains, with special reference to amyloid and connective tissues. Arch Pathol 68:487–498
3. Naiki H, Higuchi K, Hosokawa M et al. (1989) Fluorometric determination of amyloid fibrils *in vitro* using the fluorescent dye, thioflavin T1. Anal Biochem 177(2):244–249. https://doi.org/10.1016/0003-2697(89)90046-8
4. Krebs MR, Bromley EH, Donald AM (2005) The binding of thioflavin-T to amyloid fibrils: localisation and implications. J Struct Biol

149(1):30–37. https://doi.org/10.1016/j.jsb.2004.08.002

5. Biancalana M, Makabe K, Koide A et al. (2009) Molecular mechanism of thioflavin-T binding to the surface of beta-rich peptide self-assemblies. J Mol Biol 385(4):1052–1063. https://doi.org/10.1016/j.jmb.2008.11.006
6. Khurana R, Coleman C, Ionescu-Zanetti C et al. (2005) Mechanism of thioflavin T binding to amyloid fibrils. J Struct Biol 151(3):229–238. https://doi.org/10.1016/j.jsb.2005.06.006
7. Gade Malmos K, Blancas-Mejia LM, Weber B et al. (2017) ThT 101: a primer on the use of thioflavin T to investigate amyloid formation. Amyloid 24(1):1–16. https://doi.org/10.1080/13506129.2017.1304905
8. Huseby CJ, Kuret J (2016) Analyzing tau aggregation with electron microscopy. Methods Mol Biol 1345:101–112. https://doi.org/10.1007/978-1-4939-2978-8_7
9. Barghorn S, Davies P, Mandelkow E (2004) Tau paired helical filaments from Alzheimer's disease brain and assembled *in vitro* are based on beta-structure in the core domain. Biochemistry 43(6):1694–1703. https://doi.org/10.1021/bi0357006
10. Kidd M (1963) Paired helical filaments in electron microscopy of Alzheimer's disease. Nature 197:192–193. https://doi.org/10.1038/197192b0
11. Crowther RA (1991) Straight and paired helical filaments in Alzheimer disease have a common structural unit. Proc Natl Acad Sci U S A 88(6):2288–2292. https://doi.org/10.1073/pnas.88.6.2288
12. Ksiezak-Reding H, Yang G, Simon M et al. (1998) Assembled tau filaments differ from native paired helical filaments as determined by scanning transmission electron microscopy (STEM). Brain Res 814(1-2):86–98. https://doi.org/10.1016/s0006-8993(98)01052-x
13. Fitzpatrick AWP, Falcon B, He S et al. (2017) Cryo-EM structures of tau filaments from Alzheimer's disease. Nature 547(7662):185–190. https://doi.org/10.1038/nature23002
14. Wiśniewski HM, Narang HK, Terry RD (1976) Neurofibrillary tangles of paired helical filaments. J Neurol Sci 27(2):173–181. https://doi.org/10.1016/0022-510x(76)90059-9
15. Xu S, Brunden KR, Trojanowski JQ et al. (2010) Characterization of tau fibrillization *in vitro*. Alzheimers Dement 6(2):110–117. https://doi.org/10.1016/j.jalz.2009.06.002
16. Goedert M, Spillantini MG, Cairns NJ et al. (1992) Tau proteins of Alzheimer paired helical filaments: abnormal phosphorylation of all six brain isoforms. Neuron 8(1):159–168. https://doi.org/10.1016/0896-6273(92)90117-v
17. Xue C, Lin TY, Chang D et al. (2017) Thioflavin T as an amyloid dye: fibril quantification, optimal concentration and effect on aggregation. R Soc Open Sci 4:160696. https://doi.org/10.1098/rsos.160696
18. Hudson SA, Ecroyd H, Kee TW et al. (2009) The thioflavin T fluorescence assay for amyloid fibril detection can be biased by the presence of exogenous compounds. FEBS J 276(20):5960–5972. https://doi.org/10.1111/j.1742-4658.2009.07307.x

Chapter 6

Tracking Tau Oligomer Formation In Vitro

Natalie G. Horgan, Taiwo A. Ademoye and Jessica S. Fortin

Abstract

Photoinduced cross-linking of unmodified proteins (PICUP) is a common technique used to produce and study stable tau oligomers by "freezing" proteins in their state just before cross-linking. Following oligomer formation, samples are separated by SDS-PAGE and stained by Coomassie blue dye to visualize the binds through hydrophobic, electrostatic, and heteropolar interactions. From there, the proteins are transferred onto a membrane and detected via Western blot using antibodies that generate a detectable enzymatic signal. In this chapter, we outline a protocol combining PICUP, Coomassie blue staining, and Western blotting with Tau-5 antibody to study the anti-oligomer activity of test compounds against tau 2N4R.

Keywords Coomassie blue, Photoinduced cross-linking of unmodified proteins (PICUP), Tau 2N4R, Western blot

1 Introduction

Photoinduced cross-linking of unmodified proteins (PICUP) is commonly used to assess protein–ligand interactions and study oligomer formation from aggregation-prone peptides. In this technique, tris(2,2'-bipyridyl)ruthenium(II), a photosensitizer with a λ_{max} of 452 nm ($\varepsilon = 14{,}700\ M^{-1}$), is accompanied by an electron acceptor, often ammonium persulfate, and exposed to visible light [1]. Ru(II) donates an excited electron to that acceptor, producing Ru(III), a strong one-electron oxidant [2, 3]. Ru(III) then oxidizes a neighboring protein, creating an unstable protein radical. This radical can react with a nearby protein to form higher-order species and eventually oligomers [4]. After this reaction, PICUP "freezes" the oligomeric states that exist just before cross-linking.

To visualize proteins on the PICUP gel, Coomassie brilliant blue staining is used [5]. There are two main variants of Coomassie dye, namely G-250 and R-250, which differ in their hue and structure. R-250 has a reddish tint and is more commonly used in sodium dodecyl sulfate polyacrylamide gel electrophoresis (SDS-PAGE)

Jessica S. Fortin (ed.), *Translational Research Methods for Tauopathies*, Neuromethods, vol. 233,
https://doi.org/10.1007/978-1-0716-5234-3_6,

staining protocols [6]. Alternatively, G-250 has a greenish tint, two additional methyl groups, and is typically used in the Bradford assay. Proteins interact with neutral Coomassie dye through hydrophobic interactions with tryptophan, tyrosine, and phenylalanine; electrostatic interactions between the dye's sulfonic group and basic amino acids; and heteropolar bonding between arginine and the dye's anionic species [7]. When bound to a protein, the dye displays an increased absorbance from 465 nm to 595 nm [8, 9].

To detect specific proteins, Western blotting is used [10]. It involves the transfer of proteins separated by SDS-PAGE onto a membrane. From there, a primary antibody is applied to recognize and bind to the target protein. Then, a secondary antibody that is conjugated to a reporter enzyme (e.g., horseradish peroxidase) binds to the primary antibody and allows for the visualization of the bound protein via a detectable enzymatic signal. For identifying tau protein, primary antibodies Tau-5 and TOMA-1 are widely used. The main difference lies in the tau species each antibody recognizes. Tau-5 is a pan-tau antibody that can detect both monomeric and oligomeric tau species, while TOMA-1 is selective for only tau oligomers.

This chapter outlines a protocol for assessing oligomer formation via PICUP, followed by Coomassie blue staining and Western blotting. From this method, test compounds can be evaluated for their anti-oligomer activity against tau 2N4R.

2 Materials

2.1 Reagents and Equipment for PhotoInduced Cross-Linking of Unmodified Proteins

- 0.2 mL PCR tubes.
- 16% SDS-PAGE gel.
- 2-mercaptoethanol (Sigma-Aldrich #M3148): This agent should be stored at 2–8°C for up to 3 years from the date of manufacture.
- 53 W (120 V) incandescent bulb.
- Acetic acid, glacial (Sigma-Aldrich #AX0073-9): This acid should be stored at room temperature for up to 1 year from the date of manufacture.
- Ammonium acetate (Sigma-Aldrich #238074): This compound should be stored at room temperature for up to 1 year from the date of manufacture.
- Ammonium persulfate (APS) (Sigma-Aldrich #A3678): This compound should be stored at room temperature for up to 1 year from the date of manufacture.
- Coomassie brilliant blue G-250 dye (Bio-Rad Laboratories #161-0400): This dye should be stored at room temperature for up to 12 months.

- Dimethyl sulfoxide (Sigma-Aldrich #34869): This solvent should be stored at room temperature for up to 2 years.
- Electrophoresis apparatus.
- G-BOX F3 imaging system (Syngene).
- Heating block.
- ImageJ software.
- Methanol (Sigma-Aldrich#179337).
- Molecular biology water.
- PBS (Gibco #10010-023, 1×, pH 7.4): 137 mM NaCl, 2.7 mM KCl, 10 mM Na_2HPO_4, and 1.8 mM KH_2PO_4. This buffer should be stored at room temperature for up to 24 months from the date of manufacture.
- Tau 2N4R (rPeptide #T-1001-2): Tau should be stored at −80°C for up to 2–3 years prior to reconstitution. This protein should be stored at −80°C for up to 1–2 weeks after reconstitution.
- Test compounds.
- Tris(2,2'-bipyridyl)dichlororuthenium(II) hexahydrate powder (Sigma-Aldrich #224758): This compound should be stored at room temperature in a cool and dark place (<15°C).

2.2 Reagents and Equipment for Western Blot

- Blocking buffer (5% nonfat milk in 1× tris-buffered saline with 0.1% Tween 20 detergent [TBST]): Prepare 100 mL of this buffer by combining 5 g nonfat dry milk powder (Sigma-Aldrich #M7409) in 100 mL of 1× TBST.
- ECL Western blotting substrate (Thermo Scientific #32209): This substrate should be stored at 2–8°C for up to 1 year.
- HRP-conjugated anti-mouse IgG antibody (Rockland Immunochemicals, Inc #610-1319-0100): This antibody should be stored at −20°C for up to 6–12 months.
- Incubation tray.
- Nitrocellulose membrane 0.2 μm (Cytiva Amersham™ Protran™ #10600019).
- Plastic sheet protector.
- Tau-5 monoclonal antibody (Invitrogen #AHB0042): This antibody should be stored at −20°C for up to 6–12 months.
- Test compounds.
- Wash buffer (1× Tris-buffered saline with 0.1% Tween 20 detergent): Prepare 1 L of this buffer by combining 100 mL tris-buffered saline (Thermo Scientific Chemicals #J62938.K7, 10×, pH 7.4), 900 mL Milli-Q water, and 1 mL Tween 20 detergent (Sigma-Aldrich #P9416).

3 Methods

3.1 Photoinduced Cross-Linking of Unmodified Proteins

PICUP is performed to produce stable tau oligomers (Fig. 1). After it is executed, it can be stained by Coomassie blue and followed by Western blotting to visualize the bands and determine the efficacy of compounds against tau oligomer formation.

Initiating the Cross-Linking Reaction

1. Prepare a 400 μM stock solution of test compounds by adding 1 μL of 40 mM test compound to 99 μL PBS.
2. Prepare a ~62 μM stock solution of tau 2N4R by adding 35 μL of molecular biology water to 100 μg tau 2N4R.
3. Prepare a 3 mM stock solution of tris(2,2'-bipyridyl)dichlororuthenium(II) hexahydrate ($[Ru(bpy)_3]^{3+}$) by adding 1 mL of PBS to 2.2 mg of $[Ru(bpy)_3]^{3+}$.
4. Prepare a 60 mM stock solution of ammonium persulfate (APS) by adding 1 mL of PBS to 13.7 mg APS.
5. In two 0.2 mL PCR tubes, add 2 μL of ~62 μM tau 2N4R (6 μM final concentration) and 9 μL PBS.

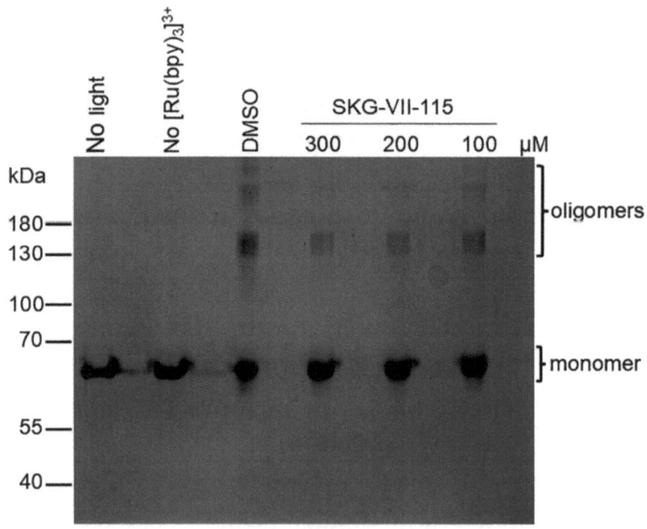

Fig. 1 SKG-VII-115, a coumarin-based compound, showed a reduction in the oligomeric formation of tau 2N4R as assessed by photo-induced cross-linking of unmodified proteins. The experiment was performed using 6 μM tau 2N4R, and oligomers were induced by tris(2,2'-bipyridyl)dichlororuthenium(II) hexahydrate ($[Ru(bpy)_3]^{3+}$) and ammonium persulfate (APS) under 2 min of light exposure. Tau 2N4R was cross-linked in the presence of **SKG-VII-115** (compound **13** in ACS Omega 2025) at concentrations of 100–300 μM in PBS. The samples were loaded onto an SDS-PAGE gel, separated by electrophoresis, and stained with Coomassie blue dye. Control samples with no light, no $[Ru(bpy)3]^{3+}$, and no compound (0.25% DMSO) were included. The 0.25% DMSO control shows oligomeric bands above 130 kDa, which are either absent or of lower intensity in the samples treated with **SKG-VII-115**. The monomeric band was formed around 65 kDa

6. To the treatment tubes, add 5 μL of test compound (100 μM final concentration), 2 μL of 3 mM tris(bipyridine)ruthenium(II) complex solution (300 μM final concentration), and 2 μL of 60 mM ammonium persulfate (6 mM final concentration).
7. To the control tube without $[Ru(bpy)_3]^{3+}$, add 5 μL of 1% DMSO, 2 μL of PBS, and 2 μL of 60 mM APS stock solution.
8. Place these tubes in a dark enclosure lined with aluminum foil to reflect the light.
9. Position a lamp with a 53 W (120 V) incandescent bulb approximately 10 cm from the samples and expose the tubes for 2 min (time may vary).
10. After light exposure, add 8.3 μL of 3X Laemmli buffer supplemeted with 15% 2-mercaptoethanol to stop the free radicals.
11. Heat the samples for 10 min at 95°C on a heating block before starting SDS-PAGE.

Performing SDS-PAGE

1. Load 25 μL of each sample onto a 16% SDS-PAGE gel.
2. Run the gel on an electrophoresis apparatus at a voltage of 65–70 V for approximately 5 h, or until the last three bands of the marker migrate off the gel.
3. After electrophoresis, fix the gel for 30–60 min with the fixing solution and then stain the gel with Coomassie blue.

Coomassie Blue Staining

1. Prepare the fixing solution by combining 100 mL deionized water, 250 mL methanol (50% final concentration), 50 mL acetic acid glacial (10% final concentration), and 0.96 g ammonium acetate (100 mM final concentration). Add deionized water to bring the total volume to 500 mL.
2. Prepare the 0.025% Coomassie solution by combining 450 mL deionized water, 50 mL acetic acid glacial, and 0.125 g Coomassie dye.
3. Incubate the gel in the fixing solution for 30–60 min. Of note, the gel can be maintained in the fixing solution for several days.
4. Stain the gel with 0.025% Coomassie solution (in 10% acetic acid) for 48 h. Of note, the gel can be maintained in the Coomassie staining solution for several days.
5. Destain the gel twice with 10% acetic acid. Each incubation should last 15–60 min.
6. Transfer the gel to deionized water.
7. Scan the gel using a G:BOX F3 imaging system (Syngene). If needed, quantify high molecular weight band with ImageJ software.

Measurement of the Optical Density of PICUP Gels

ImageJ software can be used to quantify protein bands and determine their optical density from the SDS-PAGE gel (Table 1). In doing so, ImageJ generates a graph with the peak areas for each band. It is important to run two separate ImageJ sessions, specifically one to measure the area under the monomer peaks and another to measure the area under the oligomer peaks. This is because ImageJ does not support drawing rectangles around two separate rows of bands in the same work session.

1. Open the image in ImageJ software (drag it in or use File → Open). Convert the image to 32-bit via Image → Type → 32-bit.
2. Adjust the image's brightness and contrast via Image → Adjust → Brightness/Contrast to optimize band visualization.
3. Use the rectangular tool to draw a tall rectangle around the monomer band in the first lane. Then, select Analyze → Gels → Select First Lane.
   a. After this step is completed, a number '1' will appear in the middle of the rectangle. If it does not appear, the rectangle height is likely too short and will need to adjusted to be longer than the width.
4. After the first lane is selected, a second identical rectangle will be placed on top of it. Drag this rectangle to the second lane's monomer band. Then go to Analyze → Gels → Select Next Lane.
5. Continue dragging the rectangle and selecting next lanes sequentially until all monomeric bands have been selected.

Table 1
SKG-VII-115, a coumarin-based compound, exhibits a concentration-dependent reduction of tau 2N4R oligomers

	Monomer area	Oligomer area	% Oligomer	% Oligomer reduction
No light	15,949	–	–	–
No $[Ru(bpy)_3]^{3+}$	19,956 (control 1)	–	–	–
0.25% DMSO	12,911	20,332 (control 2)	100%	0%
SKG-VII-115 100 μM	17,103	9340	46%	54%
SKG-VII-115 200 μM	17,877	5804	29%	71%
SKG-VII-115 300 μM	17,103	5238	26%	74%

ImageJ was used to quantify the protein bands and optical density of tau 2N4R oligomers from the photo-induced cross-linking of unmodified proteins assay

6. Once all lanes are selected, go to Analyze → Gels → Plot Lanes. This will generate a graph of the peaks of the selected bands, in the order picked.
7. Use the straight-line tool to draw a baseline across both sides of each peak. The line must touch both sides of the peak. Repeat this process for all graphs.
8. Use the wand tool and click inside each enclosed peak. After selected, the peak will be outlined in yellow. Repeat this process for all peaks.
9. After selecting the last peak, the Results window will pop up. This window shows the area values for each band.
10. Copy and paste these values into an Excel spreadsheet. Label the results as "Monomer Row Area."
    a. Of note, larger area values correspond to heavier, darker bands.
11. Repeat Steps 1–10 for the oligomer row of bands.
12. In Excel, calculate the ratio of each column by dividing the oligomer area by the monomer area control 1 (herein 19,956) or oligomer area control 2 (herein 20,332), then multiply by 100%. Round this resulting ratio to one decimal place to obtain the percentage of oligomer formation. The percentage of oligomer reduction is obtained by subtracting the percentage of oligomer formation from 100%.

3.2 Western Blot Analysis

Western blotting is used to identify specific proteins of interest and confirm the anti-oligomer effects of compounds using antibody binding (Fig. 2). If bands related to higher-molecular weight species are less intense or absent in compound-treated samples compared to the controls, the compound may be capable of preventing or reducing tau oligomer formation.

1. Prepare 100 mL of the blocking buffer by combining 5 g nonfat dry milk powder in 100 mL of 1× TBST.
2. Prepare 1 L of the wash buffer by combining 100 mL 10× tris-buffered saline, 900 mL Milli-Q water, and 1 mL Tween 20 detergent.
3. Incubate 6 μM tau 2N4R with desired test compounds at concentration of 100 μM in 20 μL of PBS for 3 days at 37°C. The PBS will contain the following to initiate the misfolding of tau protein: 5 mM DTT, 150 μM heparin, and 10 mg/mL arachidonic acid in Chelex-treated PBS.
4. Add 8.3 μL of 3X Laemmli buffer supplemented with 15% 2-mercaptoethanol to the samples. Heat the samples for 10 min at 95°C, load the samples (14 μL) on a gel and then transfer the tau protein onto a nitrocellulose membrane.

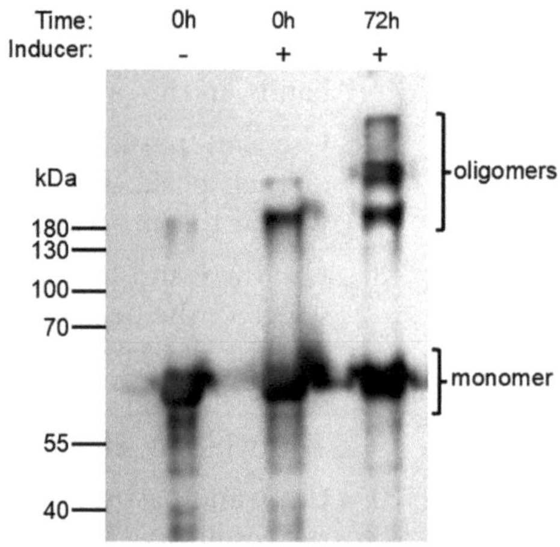

Fig. 2 Tau 2N4R forms several oligomeric species above 180 kDa in Western blot analysis. The experiment used 6 μM tau 2N4R in the presence of 0.25% DMSO, with or without incubation at 37°C for 72 h. The samples were separated on a 16% SDS-PAGE gel, transferred to a nitrocellulose membrane, and probed with a tau-5 anti-mouse antibody. The inducer contained 5 mM DTT, 150 μM heparin, and 10 mg/mL arachidonic acid in Chelex-treated PBS. The monomeric band was formed between 55 and 70 kDa

5. Incubate the membrane in blocking buffer (5% nonfat milk in 1× Tris-buffered saline with Tween 20 (TBST)) for 30–60 min at room temperature on a rocking platform.
6. Dilute the primary antibody (anti-tau-5) 1:1000 with the blocking buffer. Incubate the membrane protein side-up with this diluted antibody overnight at 4°C on a rocking platform.
7. Wash the membrane three times for 10 min each with the wash buffer (1× TBST) on a rocking platform.
8. Dilute the secondary antibody (HRP-conjugated anti-mouse antibody) 1:2500 with the blocking buffer. Incubate the membrane protein side-up with this diluted antibody for 1 h at room temperature on a rocking platform.
9. Wash the membrane three times for 10 min each with the wash buffer on a rocking platform to remove any unbound secondary antibodies.
10. Prepare the ECL substrate by combining equal parts peroxide solution and luminol enhancer solution.
11. Cover the membrane with the ECL substrate and incubate it for 1–5 min at room temperature.
12. Place the membrane in the G-BOX F3 imaging system (Syngene).

13. Take images of the membrane with an exposure time ranging from 2 to 5 min.

4 Recommendations

In the PICUP assay, both the monomeric and oligomeric bands must be visible on the gel. A major factor that can influence the appearance of these bands is the concentration of the cross-linking agent. A high concentration of $[Ru(bpy)_3]^{3+}$ may lead to over-crosslinking and disappearance of the monomeric band, while a low concentration may result in crosslinking failure. Therefore, a precision scale should be used to weigh the exact amount of $[Ru(bpy)_3]^{3+}$. Following the crosslinking reaction, appropriate staining is required to visualize the proteins on the gel. Although silver staining could be used due to its higher sensitivity, it is not necessary, since Coomassie blue staining is adequate to visualize tau proteins at a concentration of 6 μM. Coomassie blue offers several benefits, as it is easier and quicker to perform and lower in cost [11]. Additionally, the crosslinked peptide or protein can be stored at −20°C for up to 7 days before performing SDS-PAGE analysis. For membrane exposure, the ECL substrates must be applied to cover the entire surface of the membrane. If any part of the membrane is devoid of the ECL substrate, no visible signal will be observed in that area.

5 Troubleshooting

After performing PICUP and moving to Western blot, common errors, including unusual bands, absent/faint bands, and high background, can occur [12]. Unusual bands can result from protease degradation, excessive voltage, air bubbles, or buffer issues. To prevent unusual bands from forming, repeat the experiment with fresh samples, lower the voltage, ensure proper equipment usage, or prepare a new buffer, depending on the cause. In the case of absent or faint bands, low protein or antibody concentration, contaminated buffer, or overwashing can be of cause; therefore, verify that all proteins, antibodies, and buffers are fresh, correctly prepared, and used at the recommended concentrations. Furthermore, high background can arise from inadequate washing, excessive antibody concentration, or overly long exposure [13]. If a high background occurs, it is recommended to increase the number and volume of wash steps, reduce the antibody concentration, and shorten the exposure time.

Acknowledgments

J.S.F. was supported by the National Institutes of Health (NIH) grant (AG071985).

Conflict of Interest The authors have no conflicts of interest to declare that are relevant to the content of this chapter.

References

1. Fancy DA, Kodadek T (1999) Chemistry for the analysis of protein–protein interactions: rapid and efficient cross-linking triggered by long wavelength light. Proc Natl Acad Sci U S A 96(11):020–6024. https://doi.org/10.1073/pnas.96.11.6020
2. Bitan G (2006) Structural study of metastable amyloidogenic protein oligomers by photo-induced cross-linking of unmodified proteins. Methods Enzymol 413:217–236. https://doi.org/10.1016/S0076-6879(06)13012-8
3. Bitan G, Lomakin A, Teplow DB (2001) Amyloid beta-protein oligomerization: prenucleation interactions revealed by photo-induced cross-linking of unmodified proteins. J Biol Chem 276(37):35176–35184. https://doi.org/10.1074/jbc.M102223200
4. Rahimi F, Maiti P, Bitan G (2009) Photo-induced cross-linking of unmodified proteins (PICUP) applied to amyloidogenic peptides. J Vis Exp 23:1071. https://doi.org/10.3791/1071
5. Meyer TS, Lamberts BL (1965) Use of coomassie brilliant blue R250 for the electrophoresis of microgram quantities of parotid saliva proteins on acrylamide-gel strips. Biochim Biophys Acta 107(1):144–145. https://doi.org/10.1016/0304-4165(65)90403-4
6. Chevalier F (2010) Standard dyes for total protein staining in gel-based proteomic analysis. Materials (Basel) 3(10):4784–4792. https://doi.org/10.3390/ma3104784
7. Georgiou CD, Grintzalis K, Zervoudakis G (2008) Mechanism of Coomassie brilliant blue G-250 binding to proteins: a hydrophobic assay for nanogram quantities of proteins. Anal Bioanal Chem 391(1):391–403. https://doi.org/10.1007/s00216-008-1996-x
8. Kielkopf CL, Bauer W, Urbatsch IL (2020) Bradford assay for determining protein concentration. Cold Spring Harb Protoc 2020(4):102269. https://doi.org/10.1101/pdb.prot102269
9. Splittgerber AG, Sohl J (1989) Nonlinearity in protein assays by the Coomassie blue dye-binding method. Anal Biochem 179(1):198–201. https://doi.org/10.1016/0003-2697(89)90225-x
10. Towbin H, Staehelin T, Gordon J (1979) Electrophoretic transfer of proteins from polyacrylamide gels to nitrocellulose sheets: procedure and some applications. Proc Natl Acad Sci USA 76(9):4350–4354. https://doi.org/10.1073/pnas.76.9.4350
11. Sasse J, Gallagher SR (2009) Staining proteins in gels. Curr Protoc Mol Biol Chapter 10:Unit 10.6. https://doi.org/10.1002/0471142727.mb1006s85
12. Mahmood T, Yang PC (2012) Western blot: technique, theory, and trouble shooting. N Am J Med Sci 4(9):429–434. https://doi.org/10.4103/1947-2714.100998
13. Faoro V, Becker KF, Stanta G (2011) Western blotting. In: Stanta G (ed) Guidelines for molecular analysis in archive tissues. Springer, Berlin, Germany. https://doi.org/10.1007/978-3-642-17890-0_43

Part II

Isolation of Tau from Diseased Patients

Check for updates

Chapter 7

Isolation and Characterization of Amyloid-β Plaque Cores and Neurofibrillary Tangles from Alzheimer's Disease Brain

Germán Plascencia-Villa and George Perry

Abstract

Abnormal protein aggregation is the most distinctive pathological hallmark of neurodegenerative diseases. Amyloid-β plaque cores and tau neurofibrillary tangles are the main targets for the advancement of translational science in Alzheimer's disease (AD) and related dementia through the development of many therapeutic active compounds (small drugs and immunotherapies), diagnostic assays, and imaging-monitoring agents of disease progression. Here, we describe experimental protocols for high-yield isolation of amyloid-β plaque cores and tau neurofibrillary tangles used to elucidate molecular and cellular responses as cause or consequence of neurodegenerative processes. These complex protein aggregates are used for preclinical studies and advanced analytical characterization, including ultrastructural and morphological studies and advanced high-throughput methods, such as mass spectrometry. These advances aim to elucidate more details of the pathophysiology of AD and other neurodegenerative disorders.

Keywords Alzheimer's disease, Neurodegeneration, Neurons, Dementia, Amyloid-β plaque cores, Neurofibrillary tangles, Tau, Protein aggregates

1 Introduction

Neurodegenerative diseases show distinctive histopathological hallmarks, including abnormal protein aggregation, synaptic and neuronal dysfunction, altered proteostasis, cytoskeletal abnormalities, alterations in energy metabolism, DNA and RNA defects, neuroinflammation, and progressive neuronal cell death [1]. Abnormal protein aggregation is often the most distinctive pathological hallmark, such as in Alzheimer's disease (AD) and related dementia, where amyloid-β peptide (Aβ) and hyperphosphorylated tau (p-tau) form amyloid-β plaque cores (APC) and neurofibrillary tangles (NFT), respectively [2, 3]. Consequently, the deposition of APC and NFT in brain areas affected by AD is linked to clinical outcomes and progressive loss of neurons, such as deterioration of memory, cognition, and functional abilities over time, supporting their neuropathological involvement in disease progression.

Jessica S. Fortin (ed.), *Translational Research Methods for Tauopathies*, Neuromethods, vol. 233,
https://doi.org/10.1007/978-1-0716-5234-3_7,

The aggregation of Aβ into APC and p-tau as NFT indicates a loss of functions of these neuronal proteins and the gain of toxic functions [1]. In other words, normally functional neuronal proteins suffer a transformative mechanism into neurotoxic and reactive forms, often associated with inherited mutations in key proteins of their processing cellular pathways, which ultimately leads to abnormal protein aggregation. Production of APC occurs through the amyloidogenic pathway processing of amyloid-β precursor protein (APP), commonly known as the amyloid-β cascade hypothesis (Fig. 1) [4–9]. APP is sequentially cleaved by the enzymes β- and γ-secretases, releasing the transmembrane domain of APP, known as Aβ peptide. The Aβ peptide is predominantly 40–42 aa long, and the misfolded monomers follow aggregation into oligomers, protofibrils, and eventually fibrils, which aggregate into extracellular amyloid plaque cores over the course of AD. Furthermore, Aβ aggregation indirectly triggers

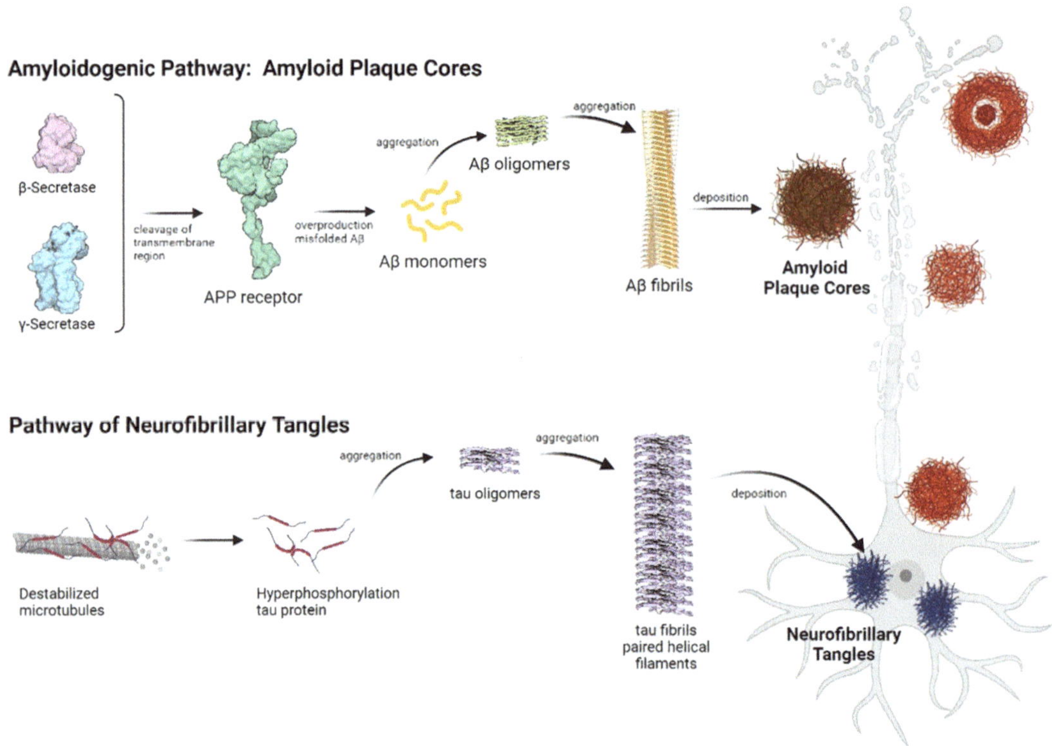

Fig. 1 Amyloid-β cascade pathways indicating the deposition of amyloid plaque cores and neurofibrillary tangles. Top: Amyloid-β precursor protein (APP) is a cellular receptor that is cleaved by the enzymes β-secretase and γ-secretase to release the Aβ peptide (40–42 aa) into the extracellular milieu in the affected areas of the brain. The Aβ peptide is overproduced and will aggregate into oligomers, fibrils and then be deposited into plaque cores. The amyloid-β plaque cores accumulate outside compromised neurons, triggering synapse disruption, inflammation, and neuronal loss. Bottom: Tau protein is involved in microtubule stabilization, during neurodegeneration tau undergoes hyperphosphorylation and forms oligomeric and fibrillar aggregates (paired helical filaments) in compromised neurons, then form intracellular deposits as neurofibrillary tangles

hyperphosphorylation of tau, a microtubule-associated protein. Hyperphosphorylated tau detaches from neuronal microtubules, destabilizing them and eventually leading to p-tau aggregation into paired helical filaments (PHF) and neurofibrillary tangles (NFT) inside affected neurons.

The overproduction and deposition of Aβ are related to neurotoxic effects through several mechanisms that disrupt neuronal functions, such as oxidative stress by the production of reactive oxygen species, mitochondrial dysfunction, calcium dysregulation, neuroinflammation, synaptic dysfunction, cell membrane disruption, activation of programmed cell death pathways, disruption of the cholinergic and glutamatergic systems [7, 9–12]. Moreover, tau aggregates seem to exert a synergistic effect with Aβ contributing to neurotoxic effects and neuronal degeneration through disruption of microtubule stability, impairment of axonal transport, activation of cellular stress responses, neuroinflammation, mitochondrial and synaptic dysfunction [13–15]. Overall, the accumulation of APC causes disruption of neuronal connectivity and induces pathophysiological changes, including synaptic dysfunction, inflammation, oxidative stress, and neuronal loss. Also, the formation of NFT contributes to neuronal dysfunction and neuronal death. The cumulative effects of Aβ toxicity and tau promote the activation of microglia and astrocytes, releasing proinflammatory cytokines and exacerbating neuronal and synaptic dysfunction. Finally, Aβ and p-tau deposition led to widespread neuronal loss and brain atrophy, particularly in the hippocampus and cortical regions, which are associated with memory and cognition.

Translational research and development of therapeutic interventions for AD have focused on treatments classified as disease-modifying small molecules, disease-modifying biologics, cognitive enhancers, and neuropsychiatric symptoms. Currently, 141 potential therapeutic compounds are under 187 clinical trials (phase 1, 2, or 3) [16, 17]. Among the most common targets is amyloid-β, with 16% (22 therapeutics agents) of the disease-modifying treatments under study and 13 compounds (9%) targeting tau, whereas the majority of AD treatments (28 agents, 29%) target neurotransmitters. The remaining agents have mechanisms of action in inflammation (24 agents, 17%), synaptic plasticity/neuroprotection (18 agents, 13%), metabolism and bioenergetics (ten agents, 7%), oxidative stress (seven agents, 5%), and proteostasis/proteinopathy (four agents, 3%). In fact, more than 30 anti-Aβ immunotherapies have been developed as potential treatments for AD, including active and passive immunotherapy and vaccines [17, 18]. Two therapeutic antibodies (aducanumab and lecanemab) that target Aβ aggregates received FDA approval. These immunotherapies are intended to treat mild cognitive impairment or mild AD by removing or reducing the amount of abnormal Aβ aggregates in the brain.

Alzheimer's disease can be diagnosed or monitored with brain imaging techniques to identify structural and functional changes in the brain, including the location of deposits of amyloid plaques and tau tangles. Positron emission tomography (PET) imaging uses radiotracers to bind APC and NFT. Several radiotracers have been developed to bind specifically APC, including (^{11}C)PiB (Pittsburgh Compound B), (^{18}F)Florbetapir (Amyvid), (^{18}F)Flutemetamol (Vizamyl), and (^{18}F)Florbetapen (Neursceq), that allow detection of APC in patients for early diagnosis and in differential diagnosis of AD and related dementia [19, 20]. Tau PET imaging is also a significant tool for AD diagnostics and monitoring of disease progression, including the use of (^{18}F)Flortaucipir (AV-1451, Tauvid), (^{18}F)MK-6240, (^{18}F)RO-948, (^{18}F)PI-2620, and (^{11}C)PBB (Pyridil-butadienyl-benzothiazole 3) [20, 21]. Furthermore, Aβ and tau are used as plasma or CSF biomarkers of AD through the development of ultrasensitive assays to determine the Aβ40/42 ratio, p-tau181/Aβ42, and %p-tau217 (ratio of phosphorylated-tau217 top non-phosphorylated) [22, 23]. Particularly, the FDA-approved blood test from Roche, commercialized as Elecsys, and others in preclinical and clinical stages: Lumipulse G System, PrecivityAD2, and other electrochemiluminescence-based and mass spectrometry assays.

Given the central role of APC and NFT in AD and related dementia, the isolation and advanced characterization of these protein aggregates is essential for the rational development of therapeutic compounds, diagnostic assays, and elucidating complex cellular and molecular responses in the brain in neurodegenerative diseases. Herein, we present the procedures for isolating amyloid-β plaque cores and tau neurofibrillary tangles from postmortem brains and their characterization with advanced microscopy and mass spectrometry.

2 Materials

2.1 Brain Tissue

Fresh or frozen brain tissue collected postmortem. Human brains are collected from donors or patients diagnosed with Alzheimer's disease and stored in tissue banks. Coronal, whole hemisphere slices of 1 cm are cut at standardized landmarks and frozen with a cooling device at −90°C (or dry ice) to preserve anatomical structures. Also, a small block of the brain is processed for histopathological analysis, confirmation of disease biomarkers, and other anatomical abnormalities [24, 25].

Alternatively, this protocol can be used with the brains of rodents used as transgenic models for neurodegenerative diseases.

2.2 Materials and Equipment for Processing of Tissue Brain

Buffers and solutions

- Phosphate-buffered saline (PBS). pH 7.4, sterile-filtered.
- Buffer A. 2 % Sodium dodecyl sulfate (SDS), 50 mM Tris-HCl, pH 7.6.
- Buffer B. 0.1% SDS, 150 mM NaCl, 0.02 % NaN_3.
- Sodium dodecyl sulfate (SDS), solution 2% w/v SDS in PBS.
- Protease inhibitors. Halt Protease Inhibitor Cocktail (100×) (Thermo Scientific, #78429).
- Sucrose, ACS grade.
- Congo red solution (1% w/v in 80% ethanol, with 0.01 M NaOH).
- Thioflavin-S solution (10 mg/mL in 50% ethanol).
- Xylene, histology grade.
- Ethanol, histology grade.
- Paraformaldehyde 40%, histology grade.
- Glutaraldehyde 50%, histology grade.
- Hematoxylin staining solution (optional).
- Propylene oxide.
- LX112 resin.
- Uranyl acetate.
- Trypsin.
- Acetic acid.
- Ammonium bicarbonate.
- Matrix α-cyano-4-hydroxycinnamic acid.

Tools and materials

- Dissecting instruments (tweezers).
- Disposable stainless-steel razor blades.
- Disposable scalpel.
- Petri dishes (plastic sterile).
- Dounce glass homogenizers (A and B pestles).
- Mechanical homogenizer (optional).
- Conical sterile tubes, 50 and 15 mL.
- Ultracentrifuge tubes, 50 mL.
- Sterile tubes, 1.5 or 2 mL.
- 300 and 100 μm mesh filters.
- Micropipettes.
- Pipet aid.

- Disposable pipettes.
- Glass slides and coverslips.
- Filter paper.
- Silicon wafer chips (TedPella).
- ITO-coated glass slides.

Equipment

- Dry bath or hot plate.
- Vortex.
- Oven.
- Centrifuge. Eppendorf or equivalent, with rotors for 15- and 50-mL conical tubes.
- Ultracentrifuge. Beckman L8-80m or equivalent, with SW28 Ti rotor for 50 mL tubes.

2.3 Optical Microscopy

Fluorescent optical microscope with GFP (green) filter and polarizer (optional).

2.4 Electron Microscopy

Transmission electron microscope, for imaging at 80–200 keV.

Scanning electron microscope, for imaging at 5–30 keV.

2.5 Optional or Additional Equipment

Critical point dryer.

Gold sputter coater.

Ultramicrotome/diamond knife.

FACS cell sorter.

MALDI-TOF mass spectrometer.

3 Methods

3.1 Processing of Brain Tissue and Purification of APC and NFT

APC and NFT were isolated using a modified method based on [26] and [27].

Brain Tissue Preparation

Obtain postmortem human brain tissue, preferably from the temporal, entorhinal cortex, or hippocampus, where APC and NFT are most abundant in subjects diagnosed with AD. Keep the tissue sections in the ultrafreezer (−80°C) in individual plastic bags if stored, and thaw on ice prior to processing.

Transfer tissue to a disposable Petri dish and remove leptomeninges and blood vessels with dissecting tweezers, then separate the white matter section with a scalpel or razor blade. Mince the cleaned gray matter tissue with a razor blade to obtain small pieces (1–2 mm), weigh and transfer to a 50-mL conical tube, add 5× buffer A, and incubate for 2 h at room temperature.

Tissue Homogenization

Place the brain tissue in a Dounce homogenizer, and add ten volumes of Buffer A relative to the tissue weight (this part can be performed in batches depending on the total amount of tissue and capacity of the homogenizer). Homogenize with 10–15 strokes of the loose (A) pestle and then with 5–10 strokes with the tight (B) pestle. Alternatively, use a mechanical homogenizer or a sonicator, performing pulses of 2–5 s on low power.

The homogenized tissue is heated at 100°C for 10 min, once the solution cools down to room temperature, it is filtered through a 300 μm mesh filter to remove debris and then with 100 μm mesh filters. The retained fraction can be passed again through the Dounce homogenizer to increase yield. All filtrated tissue homogenates are transferred to 50 mL conical tubes.

Concentration and Washing

Perform low-speed centrifugation (300–500 g for 30 min at 4°C) of the filtrate to concentrate the amyloid-rich fraction. Separate the supernatant collected containing soluble protein fractions and neurofibrillary tangles, and store at 4°C for next steps (see Step 3.1.6).

Resuspend the pellet in Buffer B, vortexed, and centrifuged 300–500 g for 30 min. This process was repeated three times. Alternatively, resuspend the pellet in Buffer B and sieve through 50 μm mesh filters.

High-Speed Ultracentrifugation

Resuspend the pellet in Buffer B. Prepare a discontinuous sucrose gradient (1.8, 1.6, 1.4, and 1.2 M) in buffer B in 50-mL ultracentrifugation tubes. Layering of the sucrose gradient should be performed to maintain each layer by slowly pipetting 10 mL layers and adding 1 mL at a time, starting from the highest concentration to the lowest concentration; the sample containing the amyloid plaques is loaded on top. The total volume should be around 45–48 mL, including the gradient and protein sample.

Centrifuge at 72,000 g for 60 min at 10°C, using a Beckman SW28 Ti rotor or similar. Amyloid plaque cores will collect at the interface between each layer. Recover each fraction of the proteins between layers using a pipette; the amyloid cores will have a characteristic dark brown to light brown coloration. Dilute in 5× buffer B to remove excess sucrose, and centrifuge at 1200 rpm for 30–45 min. Remove the supernatant and resuspend the protein pellet in a small volume (~500 μL) of buffer B for further analysis or store at 4°C.

Cell Sorting

Optional step. The recovered amyloid plaque fractions can be further purified through fluorescence-activating cell sorting (FACS). The most abundant fractions were processed in FACS Aria-II Cell Sorter (BD Biosciences) under forward angle light scattering and autofluorescence mode, separating the protein aggregates by size using the nozzles of 130 and 100 μm. Protein aggregates of the subpopulation of 5–30 μm were recovered and concentrated by centrifugation (5000 rpm for 30 min).

Recovering of NFT

Supernatant collected in Step 3.1.3 contains soluble protein fractions and neurofibrillary tangles [26, 28]. Protein aggregates were concentrated by ultracentrifugation at 100,000 g for 30 min at 10°C. The pellet was resuspended in buffer B and then loaded to a discontinuous sucrose gradient (1.8, 1.6, 1.4, and 1.2 M) in buffer B in 50-mL ultracentrifugation tubes.

Centrifuge at 100,000 g for 2 h at 25°C, using a Beckman SW28 rotor or similar. NFT were collected from the 1.4 and 1.8 M interface. Each fraction was recovered, diluted with 5× buffer B, and centrifuged at 10,000 rpm for 30 min. Remove the supernatant and resuspend the pellet in a small volume (~500 μL) of buffer B for further analysis or store at 4°C.

This protocol allowed to isolate APC efficiently, removing soluble proteins, small fibrillar aggregates, lipids, lipofuscin granules, and the majority of NFT (Fig. 2). The sucrose gradient produced enriched fractions with APC of variable sizes, from 5 to 30 μm, whereas the FACS cell sorting helped to separate small (<5 μm) from medium to large (5–30 μm) amyloid cores [27]. Besides discontinuous sucrose gradients ultracentrifugation, APC could be isolated from leptomeningeal vessels and microvessels, as well as directly from brain tissue sections by laser capture microdissection [29].

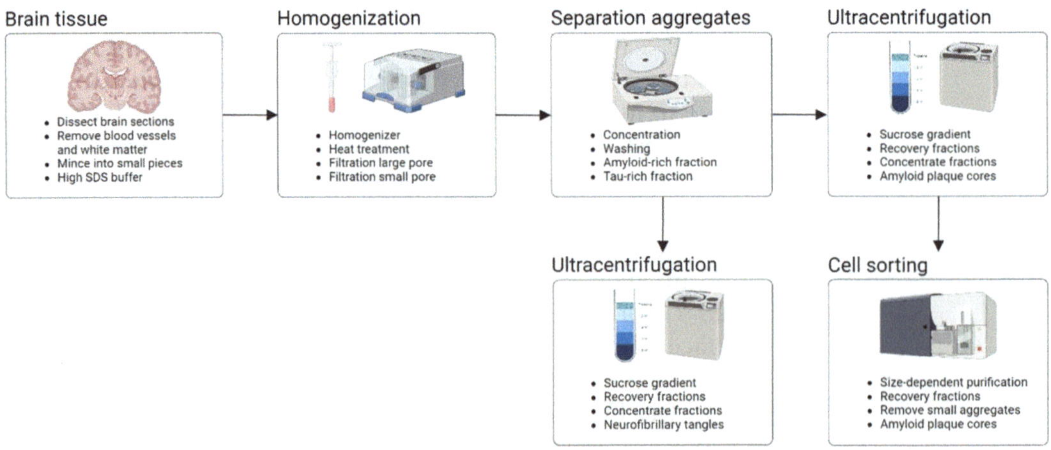

Fig. 2 Workflow for isolation of amyloid plaque cores (APC) and neurofibrillary tangles (NFT) from the human brain. APC and NFT were isolated using a modified method based on [26] and [27]

3.2 Optical Microscopy of APC and NFT

Proteins isolated from the ultracentrifugation are analyzed with optical fluorescent microscopy. Congo red and thioflavin-S are the most commonly used histological methods used to detect amyloid deposits. These dyes bind to amyloid fibrils, producing characteristic red-orange coloration under polarized light or with a green fluorescent filter.

Congo Red Staining

For tissue sections. Human or animal brain is fixed with formaldehyde and embedded in paraffin or fresh frozen. Cut tissue sections (5–10 μm) using a microtome or cryostat. Mount the sections onto pretreated positive-charged glass slides. To remove paraffin, submerge the slides in xylene for 5 min, and repeat xylene wash for two times until all paraffin is removed. Rehydrate sections using serial ethanol series (100%, 95%, 70%, and ddH_2O) for 5 min each.

For isolated proteins. Place 5–10 μL of protein solution onto a pretreated positive-charged glass slide and dry at 100°C (over a hot plate or dry bath) for 10–15 min. Remove the excess buffer with filter paper. Add 20 μL of 4% paraformaldehyde and incubate for 10 min. Remove excess with filter paper.

Staining. Immerse the glass slides in the Congo red solution for 30 min at room temperature. Rinse in ddH_2O for 2 min. Optionally, immerse glass slides in 0.1 M NaOH for 10–15 s to enhance staining and remove background excess staining. Counterstaining with hematoxylin is optional by incubation samples for 1–2 min and washing with ddH_2O for 5 min.

Mounting. Dehydrate with serial ethanol series (70%, 95%, and 100%) for 5 min each. Submerge in xylene for 5 min each and repeat. Remove excess solvent by touching the side of the glass slides with absorbent paper. Apply a water-based mounting medium (Permount, Fisher Scientific) over the tissue sections or proteins and cover them with a coverslip.

Imaging. Analyze samples with an optical microscope; the amyloid plaques will show a bright red-orange birefringent property under polarized light. Alternatively, a fluorescence microscope (green filter) can be used to observe a strong fluorescence signal from amyloid deposits.

Thioflavin-S Staining

Similar to Congo red staining, thioflavin-S staining is commonly used histological technique for detecting amyloid fibrils, but also neurofibrillary tangles [30].

Follow the same steps for mounting tissue sections or proteins (see Step 3.2.1).

Staining. Prepare a fresh solution of 0.5% thioflavin-S in ddH_2O or 50% ethanol. Submerge samples in thioflavin-S solution for 15–30 min at room temperature. Wash with ddH_2O for 2–5 min to remove excess dye. Counterstaining with hematoxylin is optional by incubation samples for 1–2 min and washing with ddH_2O for 5 min.

Mounting. Dehydrate with serial ethanol series (70 %, 95%, and 100%) for 5 min each. Submerge in xylene for 5 min each and repeat. Remove excess solvent by touching the side of the glass slides with absorbent paper. Apply a water-based mounting medium over the tissue sections or proteins and cover them with a coverslip.

Imaging. Analyze samples under a fluorescence microscope using a green (GFP) filter. Thioflavin-S emits a bright yellow-green fluorescence when bound to amyloid fibrils, allowing the visualization of amyloid deposits.

Congo red or thioflavin-s staining was used to verify the presence of APC in the fractions recovered from ultracentrifugation gradients. Optical microscopy revealed the characteristic morphology of amyloid plaque cores, both in tissue sections and isolated from brain tissue. APC are often spherical or with a slightly irregular shape with a diameter of 10–100 μm, with a highly compact structure of densely aggregated fibrils, and with a diffuse or less compact surrounding (Fig. 3) [31, 32]. In tissue sections, the APC are associated with

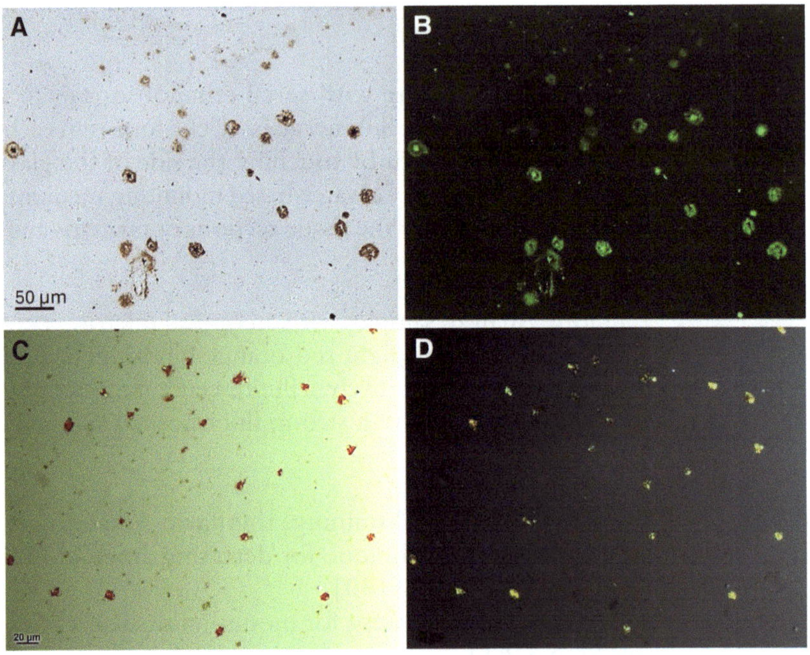

Fig. 3 Optical microscopy. (**A**) Bright field imaging of brain tissue section. (**B**) Fluorescence imaging of brain tissue section. (**C**) Bright field imaging of isolated amyloid plaque cores (gradient ultracentrifugation). (**D**) Polarized light imaging of isolated amyloid plaque cores

dystrophic neurites, corresponding to damaged segments of neurons. In comparison, NFT are located inside of pyramidal neurons of the hippocampus and cortex, with a filamentous structure resembling a flame or spindle-shaped appearance, or in some cells with a rounded shape.

3.3 Advanced Electron Microscopy Characterization

The morphology and ultrastructure of APC and NFT are revealed through high-resolution electron microscopy imaging.

Scanning Electron Microscopy

Sample processing. Take 10–20 μL of protein aggregates isolated from the human brain (Sect. 3.1). Centrifuge at 3000 rpm for 15 min and then remove the buffer B, which contains SDS and NaCl. Add 100 μL of PBS and centrifuge at 3000 rpm for 15 min to pellet the proteins. Repeat the PBS washing. Dehydrate through a graded ethanol series (50%, 75%, 95%, and 100%) for 15 min each step.

Mount 20 μL of protein sample over an ultraflat silicon wafer (5 × 5 mm chips, TedPella) and incubate for 30 min at room temperature covered with a petri dish. Transfer the mounted samples to a porous pot for a critical point dryer. Run the critical point dryer (Leica EM CPD300) for 16 cycles at low speed. Optional step: to enhance the contrast, perform coating with a gold sputter coater.

Imaging. Perform imaging with a high-resolution scanning electron microscope. Optimize voltage to observe the samples with clarity and no charging effects (these factors vary depending on the instrument used). With UHR-FE-SEM (Hitachi S-5500), proteins were effectively visualized with an accelerated voltage of 5–10 keV (Fig. 4).

Scanning electron microscopy imaging revealed with high detail the morphology of APC, consisting of a compact dense core and with protruding fibrillar arrangement, this is usually described as pompons with protruding spikes and tufts (Fig. 4). Also, small vessels or blood capillaries intertwined with amyloid fibrils were sometimes

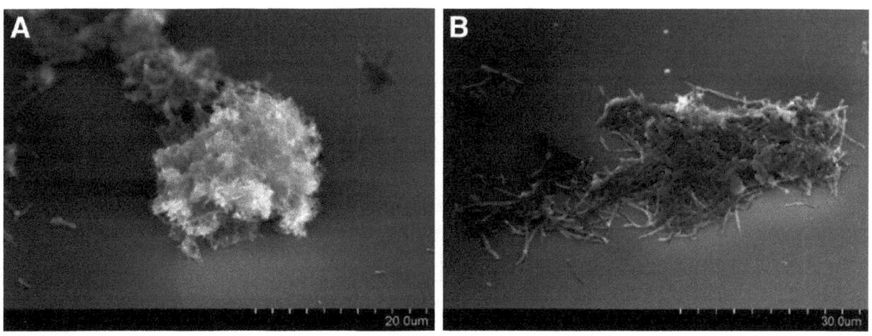

Fig. 4 Scanning electron microscopy imaging. (**A**) Amyloid-β plaque core. (**B**) Neurofibrillary tangle

observed, and in some cases with elongated filaments. NFT showed a more planar structure with compact paired helical filaments forming an elongated or arrow-like assembly, sometimes described as flame-shaped or as globular inclusions when inside the neurons.

Transmission Electron Microscopy

Sample processing. Take 10–20 μL of protein aggregates isolated from the human brain (Sect. 3.1). Centrifuge at 10,000 rpm for 15 min and then remove the buffer B, which contains SDS and NaCl, without disrupting the pellet formed. Fix with 4% formaldehyde/1% glutaraldehyde in PBS for 1 h at room temperature. Optional step: postfixation with osmium tetroxide 0.5% for 1 h. Rinse two times with ddH_2O for 15 min. Dehydrate with graded ethanol series (50%, 75%, 95%, and 100%) for 15 min each step. Wash once in propylene oxide to remove any remaining ethanol. Infiltrate protein pellets with 50% LX112 resin (Ladd Research) in propylene oxide for 1 h. Transfer pellets to beem capsules, infiltrate with 100% LX112 resin, remove any air bubbles, and cure resin for 48 h at 65°C.

Cut ultrathin sections (70–95 nm) with ultramicrotome (Leica Ultracut) using a 45° diamond knife (Diatome). Ultrathin sections are mounted on 300 mesh regular nickel or copper grids (Electron Microscopy Sciences) and placed onto a hot plate at 37°C to dry.

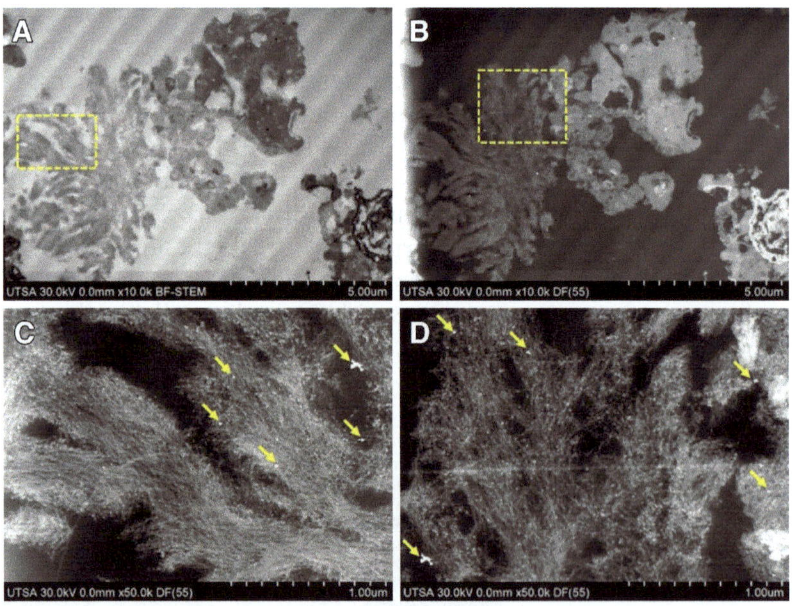

Fig. 5 Transmission electron microscopy imaging. Imaging of ultrathin sections of amyloid plaque cores from AD human brain. (**A**) Low voltage BF-STEM revealed the ultrastructure and arrangements of amyloid fibers. (**B**) Low voltage ADF-STEM showed the presence of high-contrast areas and spots within amyloid cores. (**C** and **D**) High magnification ADF-STEM imaging revealing presence of electrodense aggregates linked to amyloid fibers, indicated with yellow arrows (Reproduced with permission from Plascencia-Villa et al. [32])

Optionally, stain the samples with 1% uranyl acetate solution, adding 10 μL of staining solution for 5 min, remove excess with filter paper and wash with ddH_2O for 2–5 min, then let dry for 1–2 h at room temperature, covered with a Petri dish.

Imaging. Perform imaging with a high-resolution transmission or scanning-transmission electron microscope. With UHR-FE-SEM (Hitachi S-5500), proteins were effectively visualized with Duo-BF/DF-STEM detector an accelerated voltage of 30 keV (Fig. 5). Furthermore, the thin sections are suitable for conventional TEM imaging (60–100 keV), high-resolution TEM (200 keV) or atomic resolution STEM (80–200 keV), also for elemental chemical analysis with energy dispersive X-ray spectrometry (EDX), electron energy loss spectroscopy (EELS), and X-ray spectro-microscopy.

Imaging of ultrathin sections with TEM/STEM showed the APC ultrastructure, indicating a compact arrangement of fibrils of 7–12 nm in diameter protruding in all directions and some areas of electro-dense materials (Fig. 5). The compact Aβ fibrils showed a stellated morphology. In our analysis, we discovered the presence of electrodense materials using a HAADF detector (arrows in Fig. 5C, D). Moreover, these aggregates resulted in iron-enriched materials within APC and were identified as magnetite (Fe_3O_4) and maghemite (Fe_2O_3) through aberration-corrected atomic resolution STEM, energy dispersive X-ray spectroscopy and electron energy loss spectroscopy. Figure reproduced from [32]. Additionally, using synchrotron-based X-ray spectromicroscopy and X-ray circular dichroism allowed to determine the speciation and location of calcium compounds, as well as iron and copper aggregates of variable oxidative states within APC from AD [33–35].

3.4 Mass Spectrometry Proteomics Profiling of APC

Analysis of AD-Derived Protein Aggregates Through Mass Spectrometry

Sample preparation for MS MALDI-TOF is described in [36]. Prepare trypsin solution (0.01 mg/mL, pH 8.2) by dissolving 20 μg of MS-grade trypsin in 20 μL of 50 mM acetic acid and 100 μL of 100 mM ammonium bicarbonate. A small volume of 10–20 μL of isolated amyloid plaque cores is incubated with the same volume of trypsin at 37°C overnight. Mount processed samples onto an ITO-coated glass slide, two spots of 5 μL each per sample, and dry in a desiccator under vacuum. Wash the slides with ddH_2O and then ethanol for 2 min each. Carefully add 5 μL of 10 mg/mL solution of α-cyano-4-hydroxycinnamic acid (MALDI matrix) over each of the protein spots, and let it dry under vacuum.

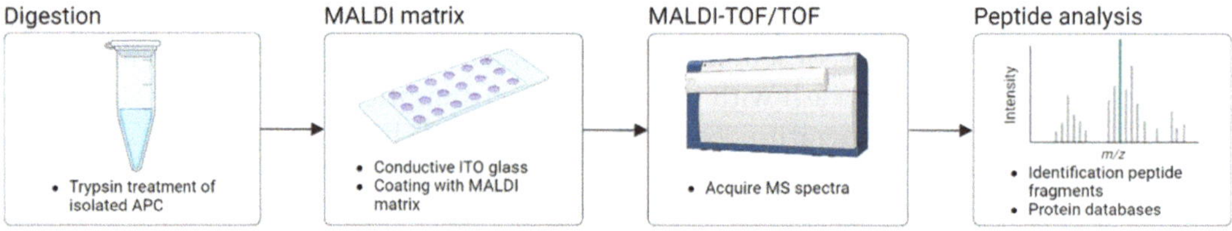

Fig. 6 Mass spectrometry. Simplified workflow for mass spectrometry analysis of amyloid plaque cores through MALDI-TOF/TOF

Acquisition and analysis of mass spectrometry. Mass spectrometry was performed with an UltrafleXtreme MALDI-TOF/TOF (Bruker). The system is calibrated with a standard peptide mixture and matrix of α-cyano-4-hydroxycinnamic acid. Spectra are collected over a m/z range of 900–5000 in reflectron positive mode, using a 1 kHz laser and over 500 laser spots per spectra.

MS/MS analysis is performed for each peptide species resolved in the full reflectron TOF spectra; a lower signal-to-noise of 4 and resolution of 4000 were selected for peptide peak spectrum analysis. Peptide fragments obtained were analyzed with MASCOT and NCBI Blast databases, with peptide tolerance of ±150 ppm, fragment tolerance of ±2 Da, and up to two missed cleavages.

Mass spectrometry (MALDI-TOF/TOF) characterization of isolated APC from the human brain was able to identify 33 different components, including some proteins related to immune and inflammatory responses, cytoplasmic and axonal components, extracellular matrix, and neuronal receptor (Fig. 6) [36]. MS confirmed the idea that APC are complex structures; even though Aβ is the main component, a plethora of biological components are codeposited within or around APC, including proteins, inorganic materials, and lipids. In comparison, the characterization of APC isolated by laser capture microdissection and analyzed by LC-MS coupled with tandem MS identified up to 256 proteins co-isolated with the APC from multiple AD cases, but only 26 of them constituted the main enriched proteins in the amyloid aggregates [37]. Confirming that APC are complex heterogeneous neuropathological structures and that the composition varies among AD patients. Remarkably, the main components besides Aβ include cell adhesion, membrane trafficking and cytoskeleton proteins, receptors, kinase and regulators, chaperones, inflammatory, immune response, and proteolysis. Using tandem mass tag mass spectrometry (TMT-MS)-based proteomic analysis of AD brains identified novel plaque proteins, including M42 matrisome module, cerebrovascular amyloid (CAA), dystrophic neuronal responses, midkine, and pleiotrophin, which could regulate amyloid deposition [38].

4 Conclusions

This protocol outlines a comprehensive approach for isolating amyloid plaque cores and neurofibrillary tangles from human brain tissue. The successful execution of this procedure allows for obtaining highly enriched protein aggregates and preserves their structural integrity for further biochemical, histochemical, advanced imaging, and high-throughput analyses, contributing to many integrative aspects of basic and translational research on neurodegenerative diseases.

Acknowledgments

We thank the Kleberg Foundation, Lowe Foundation, Semmes Foundation, San Antonio Medical Foundation, Alzheimer's Association, and NIH National Institute on Aging for their support of our research programs. Administrative support and coordination from UTSA Brain Health Consortium.

Conflict of Interest The authors have no conflicts of interest to declare that are relevant to the content of this chapter.

References

1. Wilson 3rd DM, Cookson MR, Van Den Bosch L, Zetterberg H, Holtzman DM, Dewachter I (2023) Hallmarks of neurodegenerative diseases. Cell 186(4):693–714. https://doi.org/10.1016/j.cell.2022.12.032
2. 2024 Alzheimer's disease facts and figures (2024). Alzheimers Dement 20(5):3708–3821. https://doi.org/10.1002/alz.13809
3. Knopman DS, Amieva H, Petersen RC, Chételat G, Holtzman DM, Hyman BT, Nixon RA, Jones DT (2021) Alzheimer disease. Nat Rev Dis Primers 7(1):33. https://doi.org/10.1038/s41572-021-00269-y
4. Hardy JA, Higgins GA (1992) Alzheimer's disease: the amyloid cascade hypothesis. Science 256(5054):184–185. https://doi.org/10.1126/science.1566067
5. Selkoe DJ (2002) Alzheimer's disease is a synaptic failure. Science 298(5594):789–791. https://doi.org/10.1126/science.1074069
6. Masters CL, Selkoe DJ (2012) Biochemistry of amyloid β-protein and amyloid deposits in Alzheimer disease. Cold Spring Harb Perspect Med 2(6):a006262. https://doi.org/10.1101/cshperspect.a006262
7. Hampel H, Hardy J, Blennow K, Chen C, Perry G, Kim SH, Villemagne VL, Aisen P, Vendruscolo M, Iwatsubo T, Masters CL, Cho M, Lannfelt L, Cummings JL, Vergallo A (2021) The amyloid-β pathway in Alzheimer's disease. Mol Psychiatry 26(10):5481–5503. https://doi.org/10.1038/s41380-021-01249-0
8. Selkoe DJ, Hardy J (2016) The amyloid hypothesis of Alzheimer's disease at 25 years. EMBO Mol Med 8(6):595–608. https://doi.org/10.15252/emmm.201606210
9. Castellani RJ, Plascencia-Villa G, Perry G (2019) The amyloid cascade and Alzheimer's disease therapeutics: theory versus observation. Lab Invest 99(7):958–970. https://doi.org/10.1038/s41374-019-0231-z
10. Lambert MP, Barlow AK, Chromy BA, Edwards C, Freed R, Liosatos M, Morgan TE, Rozovsky I, Trommer B, Viola KL, Wals P, Zhang C, Finch CE, Krafft GA, Klein WL (1998) Diffusible, nonfibrillar ligands derived from Abeta1-42 are potent central nervous system neurotoxins. Proc Natl Acad Sci U S A 95(11):6448–6453. https://doi.org/10.1073/pnas.95.11.6448
11. Butterfield DA, Halliwell B (2019) Oxidative stress, dysfunctional glucose metabolism and Alzheimer disease. Nat Rev Neurosci 20(3):148–160. https://doi.org/10.1038/s41583-019-0132-6

12. Castellani RJ, Plascencia-Villa G, Perry G (2022) Pathogenesis of Alzheimer's disease. In: Kostrzewa RM (ed) Handbook of neurotoxicity. Springer International Publishing, Cham, pp 1709–1728. 10.1007/978-3-031-15080-7_162
13. Rapoport M, Dawson HN, Binder LI, Vitek MP, Ferreira A (2002) Tau is essential to beta -amyloid-induced neurotoxicity. Proc Natl Acad Sci U S A 99(9):6364–6369. https://doi.org/10.1073/pnas.092136199
14. Mandelkow EM, Mandelkow E (2012) Biochemistry and cell biology of tau protein in neurofibrillary degeneration. Cold Spring Harb Perspect Med 2(7):a006247. https://doi.org/10.1101/cshperspect.a006247
15. Ballatore C, Lee VM, Trojanowski JQ (2007) Tau-mediated neurodegeneration in Alzheimer's disease and related disorders. Nat Rev Neurosci 8(9):663–672. https://doi.org/10.1038/nrn2194
16. Cummings J, Zhou Y, Lee G, Zhong K, Fonseca J, Cheng F (2024) Alzheimer's disease drug development pipeline: 2024. Alzheimer's Dementia 10(2):e12465. https://doi.org/10.1002/trc2.12385
17. Plascencia-Villa G, Perry G (2022) 3.04 - Alzheimer's Disease Pharmacology. In: Kenakin T (ed) Comprehensive Pharmacology. Elsevier, Oxford, pp 34–63 https://doi.org/10.1016/B978-0-12-820472-6.00018-9
18. Plascencia-Villa G, Perry G (2023) Lessons from antiamyloid-β immunotherapies in Alzheimer's disease. In: Espay AJ (ed) Handbook of clinical neurology, vol 193. Elsevier, Amsterdam, The Netherlands, pp 267–291. https://doi.org/10.1016/B978-0-323-85555-6.00019-9
19. Chapleau M, Iaccarino L, Soleimani-Meigooni D, Rabinovici GD (2022) The role of amyloid PET in imaging neurodegenerative disorders: a review. J Nucl Med 63(Suppl 1):13s–19s. https://doi.org/10.2967/jnumed.121.263195
20. Bao W, Xie F, Zuo C, Guan Y, Huang YH (2021) PET neuroimaging of Alzheimer's disease: radiotracers and their utility in clinical research. Front Aging Neurosci 13:624330. https://doi.org/10.3389/fnagi.2021.624330
21. Smith R, Hägerström D, Pawlik D, Klein G, Jögi J, Ohlsson T, Stomrud E, Hansson O (2023) Clinical utility of tau positron emission tomography in the diagnostic workup of patients with cognitive symptoms. JAMA Neurol 80(7):749–756. https://doi.org/10.1001/jamaneurol.2023.1323
22. Pais MV, Forlenza OV, Diniz BS (2023) Plasma biomarkers of Alzheimer's disease: a review of available assays, recent developments, and implications for clinical practice. J Alzheimers Dis Rep 7(1):355–380. https://doi.org/10.3233/adr-230029
23. Barthélemy NR, Salvadó G, Schindler SE, He Y, Janelidze S, Collij LE, Saef B, Henson RL, Chen CD, Gordon BA, Li Y, La Joie R, Benzinger TLS, Morris JC, Mattsson-Carlgren N, Palmqvist S, Ossenkoppele R, Rabinovici GD, Stomrud E, Bateman RJ, Hansson O (2024) Highly accurate blood test for Alzheimer's disease is similar or superior to clinical cerebrospinal fluid tests. Nature Med 30(4):1085–1095. https://doi.org/10.1038/s41591-024-02869-z
24. Vonsattel JPG, Aizawa H, Ge P, DiFiglia M, McKee AC, MacDonald M, Gusella JF, Landwehrmeyer GB, Bird ED, Richardson Jr.EP, Hedley-Whyte ET (1995) An improved approach to prepare human brains for research. J Neuropathol Exper Neurol 54(1):42–56. https://doi.org/10.1097/00005072-199501000-00006
25. Plascencia-Villa G, Perry G (2022) Neuropathologic changes provide insights into key mechanisms of alzheimer disease and related dementia. Am J Pathol 192(10):1340–1346. https://doi.org/10.1016/j.ajpath.2022.07.002
26. Ihara Y, Abraham C, Selkoe DJ (1983) Antibodies to paired helical filaments in Alzheimer's disease do not recognize normal brain proteins. Nature 304(5928):727–730. https://doi.org/10.1038/304727a0
27. Selkoe DJ, Abraham CR, Podlisny MB, Duffy LK (1986) Isolation of low-molecular-weight proteins from amyloid plaque fibers in Alzheimer's disease. J Neurochem 46(6):1820–1834. https://doi.org/10.1111/j.1471-4159.1986.tb08501.x
28. Iqbal K, Zaidi T, Thompson CH, Merz PA, Wisniewski HM (1984) Alzheimer paired helical filaments: bulk isolation, solubility, and protein composition. Acta Neuropathol 62(3):167–177. https://doi.org/10.1007/bf00691849
29. Rostagno A, Ghiso J (2009) Isolation and biochemical characterization of amyloid plaques and paired helical filaments. Curr Protoc Cell Biol Chapter 3:Unit 3.33 33.33.31-33. https://doi.org/10.1002/0471143030.cb0333s44
30. Sun A, Nguyen XV, Bing G (2002) Comparative analysis of an improved thioflavin-s stain, Gallyas silver stain, and immunohistochemistry for neurofibrillary tangle demonstration on the same sections. J Histochem Cytochem 50(4):463–472. https://doi.org/10.1177/002215540205000403
31. Plascencia-Villa G, Perry G (2022) Neuropathologic changes provide insights into key mechanisms of Alzheimer disease and related dementia. Am J Pathol 192(10):1340–1346. https://doi.org/10.1016/j.ajpath.2022.07.002

32. Plascencia-Villa G, Ponce A, Collingwood JF, Arellano-Jiménez MJ, Zhu X, Rogers JT, Betancourt I, José-Yacamán M, Perry G (2016) High-resolution analytical imaging and electron holography of magnetite particles in amyloid cores of Alzheimer's disease. Sci Rep 6(1):24873. https://doi.org/10.1038/srep24873
33. Everett J, Collingwood JF, Tjendana-Tjhin V, Brooks J, Lermyte F, Plascencia-Villa G, Hands-Portman I, Dobson J, Perry G, Telling ND (2018) Nanoscale synchrotron X-ray speciation of iron and calcium compounds in amyloid plaque cores from Alzheimer's disease subjects. Nanoscale 10(25):11782–11796. https://doi.org/10.1039/c7nr06794a
34. Everett J, Lermyte F, Brooks J, Tjendana-Tjhin V, Plascencia-Villa G, Hands-Portman I, Donnelly JM, Billimoria K, Perry G, Zhu X, Sadler PJ, O'Connor PB, Collingwood JF, Telling ND (2021) Biogenic metallic elements in the human brain? Sci Adv 7(24):eabf6707. https://doi.org/10.1126/sciadv.abf6707
35. Everett J, Brooks J, Tjendana Tjhin V, Lermyte F, Hands-Portman I, Plascencia-Villa G, Perry G, Sadler PJ, O'Connor PB, Collingwood JF, Telling ND (2024) Label-free in situ chemical characterization of amyloid plaques in human brain tissues. ACS Chem Neurosci 15(7):1469–1483. https://doi.org/10.1021/acschemneuro.3c00756
36. Kelley AR, Perry G, Bach SBH (2018) Characterization of proteins present in isolated senile plaques from Alzheimer's diseased brains by MALDI-TOF MS with MS/MS. ACS Chem Neurosci 9(4):708–714. https://doi.org/10.1021/acschemneuro.7b00445
37. Liao L, Cheng D, Wang J, Duong DM, Losik TG, Gearing M, Rees HD, Lah JJ, Levey AI, Peng J (2004) Proteomic characterization of postmortem amyloid plaques isolated by laser capture microdissection*. J Biol Chem 279(35):37061–37068. https://doi.org/10.1074/jbc.M403672200
38. Levites Y, Dammer EB, Ran Y, Tsering W, Duong D, Abreha M, Gadhavi J, Lolo K, Trejo-Lopez J, Phillips J, Iturbe A, Erquizi A, Moore BD, Ryu D, Natu A, Dillon K, Torrellas J, Moran C, Ladd T, Afroz F, Islam T, Jagirdar J, Funk CC, Robinson M, Rangaraju S, Borchelt DR, Ertekin-Taner N, Kelly JW, Heppner FL, Johnson ECB, McFarland K, Levey AI, Prokop S, Seyfried NT, Golde TE (2024) Integrative proteomics identifies a conserved Aβ; amyloid responsome, novel plaque proteins, and pathology modifiers in Alzheimer's disease. Cell Rep Med 5(8):101556. https://doi.org/10.1016/j.xcrm.2024.101669

Check for updates

Chapter 8

Isolation and Proteomic Analysis of Pathological Tau-Seeds

Henika Patel, Pablo Martinez, Armando Amaro and Cristian A. Lasagna-Reeves

Abstract

A unifying pathological signature of tauopathies is the aggregation of tau, which contributes to the disease-specific patterns of neurodegeneration. The "tau-seeds" are intermediate soluble tau aggregates capable of seeding and propagating in a prion-like manner. We developed a protocol using size-exclusion chromatography, FRET-based tau seeding assays, and affinity purification to isolate and analyze tau seeds from various biological sources, including tissues from humans, mice, and Drosophila. Furthermore, we utilized mass spectrometry to identify the protein interactome of tau-seed, thereby facilitating the discovery of novel tau-interactors for therapeutic interventions.

Keywords Tau-seed, Proteomics, Tau-seeding, Interactome

1 Introduction

The abnormal aggregation of tau constitutes a hallmark histopathological feature across a spectrum of neurodegenerative diseases, collectively termed tauopathies, which include Alzheimer's disease (AD), progressive supranuclear palsy (PSP), and other related conditions. Emerging evidence indicates that the soluble intermediate tau aggregates, "tau-seed," are the primary toxic entities with the ability to recruit and convert "normal" tau into pathological forms and propagate in a prion-like manner [1]. A major challenge has been to characterize and understand tau pathogenesis due to the complexity of tau conformations, aggregation, and pathological manifestations. It remains unclear which mechanisms cause isoform-specific tau aggregation and result in diverse neuropathological features. Furthermore, even when derived from the same isoform, tau aggregates exhibit diverse morphological and biochemical features across individuals with the same tauopathy. For instance, in Alzheimer's Disease, the soluble hyperphosphorylated tau species display significant heterogeneity in seeding capability, which correlates with the disease severity

Jessica S. Fortin (ed.), *Translational Research Methods for Tauopathies*, Neuromethods, vol. 233,
https://doi.org/10.1007/978-1-0716-5234-3_8,

and clinical outcomes [2]. In PSP, the phospho-tau profiles differ between brain regions and patients, indicating the presence of an ensemble of variably phosphorylated tau-species [3]. Such variations in conformations and seeding activity may result from diverse posttranslational modifications (PTMs) and tau's interaction with different molecular partners inside the cell, such as mitochondria, synapses, nucleus, and modulators of protein homeostasis [4, 5]. Interactions with different proteins may contribute to the diversity of tau conformations with distinct properties.

Tau pathology has been proposed to spread to synaptically connected regions through trans-synaptic propagation. In both patients with Alzheimer's disease, as well as in mouse models of tauopathy, soluble tau aggregates with seeding properties have been observed to accumulate at synapses and interact with synaptic proteins, supporting the idea that tau may exploit the synaptic machinery for interneuronal transfer [4–8]. Despite this knowledge, the nature of this "tau-seed" involved in spreading and seeding remains unknown. While a soluble high-molecular-weight (HMW) tau species has been implicated in the trans-synaptic propagation [9], the precise composition of this species remains unknown; it is unclear if it is made of solely tau or includes other interacting proteins that contribute to tau toxicity and ability to spread. We developed a protocol using size-exclusion chromatography and FRET-based tau seeding assay to isolate and analyze tau seeds and determine the protein interactome to identify proteins that can contribute to tau pathogenesis and toxicity (Fig. 1). This protocol can be used to compare bioactive tau-seed with other tau species (e.g., monomeric tau) from the same tissue or across diseases [8, 10]. With this method, we identified Bassoon, a presynaptic protein, as a crucial interactor of tau-seed, which exacerbates tau toxicity and propagation [8]. Identifying the molecular players that can stabilize tau-seed and contribute to distinct disease pathogenesis is crucial to developing targeted interventions.

2 Materials and Equipment

2.1 Preparation of Tissue Lysates

- Protease and Phosphatase inhibitor (e.g., Thermo Scientific).
- Bradford Assay kit.
- Dounce homogenizer and sonicator.

2.2 Size-Exclusion Chromatography (SEC)

- 500 μL and 1 mL Hamilton gas-tight Syringe (Hamilton).
- Liquid chromatography system, e.g., ÄKTA pre 25 L (GE/Cytiva).
- Chromatography Fraction collector (e.g., F9-C, GE/Cytiva).
- Superose 6 increase 10/300 GL column (GE/Cytiva).

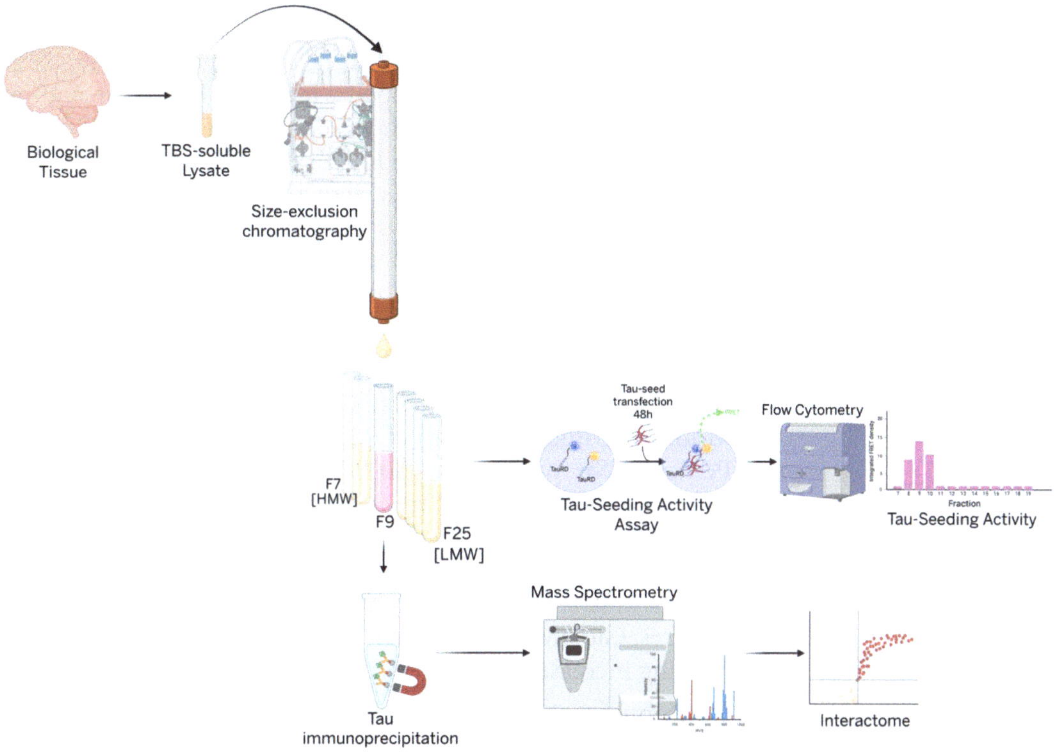

Fig. 1 Schematic illustration of the pipeline to isolate tau-seed. TBS-soluble lysates from biological tissues, such as patient brains, are resolved using a size exclusion chromatography column to separate high and low molecular weight tau-complexes (HMW and LMW, respectively). The seeding activity of fractions is analyzed using the FRET-based tau seeding assay. Tau-seeds are isolated from fractions using an immunoprecipitation assay and analyzed through mass spectrometry to determine the protein interactome of the tau-seed

- Amicon 2 mL 3 K ultrafiltration filter (Millipore Sigma).
- 2 mL sample loop.
- Size exclusion column (SEC) buffer: To prepare 500 mL, combine 5 mL of 5 M NaCl (final concentration: 50 mM) and 25 mL of 1 M Tris-HCl, pH 8.0 (final concentration: 50 mM) with ultrapure water. Adjust the final volume to 500 mL. Always prepare a fresh buffer and filter it through a 0.22 μm pore filter before use.

2.3 FRET-Based Tau Biosensor Cell Culture

- Lipofectamine 2000 (Thermo Scientific).
- TauRD P301S FRET biosensor (ATCC).
- HEK293T (ATCC).
- TauRD P301S CFP (UT Southwestern/Marc Diamond).
- TauRD P301S YFP (UT Southwestern/Marc Diamond).
- Opti-MEM (Gibco).

- CO_2 incubator.
- FlowJo v10 software (BD Biosciences).
- BD LSRFortessa X-20 equipped with a BD High Throughput Sampler (HTS) (BD Biosciences). To detect individual fluorophores and FRET signal, the FITC, BV421, and BV510 channels were selected. YFP was detected in the FITC channel [488 nm excitation, 505 long-pass (LP) mirror, and 530/30 band pass (BP) filter]. CFP was measured using the BV421 channel (405 nm excitation, 450/50 BP filter). FRET signal was measured using the BV510 channel (405 nm excitation and 505 LP mirror, and 525/50 BP filter).
- DMEM Cell Culture Media: Combine 220 mL Dulbecco's Modified Eagle's Medium (DMEM, Invitrogen) with 2.5 mL Penicillin-Streptomycin (10,000 units/mL of penicillin and 10,000 μg/mL of Streptomycin), 2.5 mL of GlutaMax Supplement (Gibco, final concentration: 2 mM), and 25 mL of heat-inactivated fetal bovine serum (FBS). The antibiotics penicillin and streptomycin prevent bacterial contamination of cell cultures. GlutaMax supplement is an alternative to L-glutamine, offering increased stability and improved cell health.
- FACS buffer: Add 2 mL of heat-inactivated fetal bovine serum (FBS) and 186 mg of EDTA sodium salt dihydrate (final concentration: 5 mM) to 1× PBS. Adjust the final volume to 100 mL with 1× PBS. Filter the buffer with a 0.22 μm pore filter and store at 4°C or lower temperature.

2.4 Affinity-Based Isolation of Tau-Seeds

- Biotin mIgG (2 μg per 100 ng of human tau) (BioLegend).
- Biotin HT7 (2 μg per 100 ng of human tau) (Invitrogen).
- Streptavidin magnetic beads (Invitrogen).
- Rotating mixer (e.g., ELMI Intelli-Mixer).
- Magnetic separation rack (Invitrogen).

2.5 Tandem Mass Tag (TMT)-Based Proteomics

- TMT pro labels (Thermo Scientific): Add 5 mg of TMT pro reagent to 250 μL of acetonitrile. Divide into ten tubes and speed vac immediately (final concentration: 0.5 mg dry). Store aliquots at −80°C for up to 1 year. Avoid freeze-thaw cycles and keep dry.
- 1 M Triethylammonium bicarbonate (TEAB) (Thermo Scientific).
- 50% hydroxylamine (Thermo Scientific). Dilute ten times in water to make 5% solution.
- Acetonitrile (ACN) (Thermo Scientific).
- 0.1% formic acid in water (Fisher Scientific).
- 0.1% formic acid, 80% acetonitrile (Fisher Scientific).

- Proteome Discoverer 2.5 (Thermo Scientific).
- 8 M Urea buffer: Add 4.8 g of urea to 10 mL of 100 mM Tris-HCl, pH 8.5. The buffer can be stored at 25°C for up to 3 months.
- TECP solution: For a 1 M solution, add 2 g of tris(2-carboxyethyl) phosphine (TCEP) in 6.9 mL of ultra-pure water. Aliquot this solution in 1 mL and 100 µL aliquots and store at −80°C for up to 2 years. Alternatively, premade solutions can be sourced from a commercial vendor. For a 100 mM solution, add 100 µL of 1 M TCEP solution to 900 µL of ultrapure water. Make aliquots of the solution and store at −20°C for up to 2 months. Avoid repeated freeze-thaw cycles of the aliquots.
- 2-chloroacetamide (CAA) solution: Add 48.3 mg of 2-chloroacetamide to 1 mL of ultrapure water. Make aliquots and store at −20°C for up to 2 months. Avoid repeated freeze-thaw cycles of the aliquots.
- Trypsin-LysC Stock solution: Add 100 µg of Promega Trypsin/LysC (final concentration: 0.1 µg/µL) to 1 mL of resuspension buffer (provided with trypsin). Make aliquots and store at −80°C for up to 1 year. Avoid repeated freeze-thaw of the aliquots.

3 Methods

3.1 TBS-Soluble Tissue Lysate Preparation

The preparation of detergent-free, Tris-buffered saline (TBS)-soluble brain lysates provides a protein fraction suitable for downstream biochemical and proteomic analyses. This protocol can be applied to fresh or frozen brain tissue from mouse or human origin, as well as to cultured cells, and can be adapted for use with other tauopathy models. All steps should be carried out on ice or at 4°C to preserve protein integrity [8, 11, 12].

1. If using fresh brain tissue, the region of interest should be carefully dissected on an ice-cold surface. For frozen tissue, samples should be allowed to thaw on ice until just pliable. Once prepared, weigh the tissue to determine the appropriate volume of buffer for homogenization. Homogenization is performed in a prechilled Dounce homogenizer by adding TBS buffer at a ratio of 10:1 (v/w; 10 µL of TBS per mg of tissue), supplemented with protease and phosphatase inhibitor cocktails according to the manufacturer's recommendations (*see* **Note 1**). Homogenize the sample using 10–15 strokes with a loose-fitting pestle. For enhanced disruption, subject the homogenate to sonication in a cold-water bath at 50% power,

using 5-s pulses followed by 3-s pauses for a total of 60 s (*see* **Note 2**).

2. Following homogenization, transfer the lysate to a centrifuge tube and centrifuge at 20,000 × *g* for 15 min at 4°C to separate the soluble fraction.
3. To collect the TBS-soluble protein fraction, transfer the supernatant to a clean tube. The pellet containing insoluble material can be retained for further analyses, if desired.
4. Quantify the protein concentration using a bicinchoninic acid (BCA) or Bradford assay.
5. Prepare aliquots of the TBS-soluble lysate and store at −80°C for long-term use, avoiding repeated freeze-thaw cycles.

This protocol is readily adaptable to other species and experimental systems, including brain tissue from *Drosophila melanogaster*, *Caenorhabditis elegans*, or other model organisms, with minor modifications to homogenization parameters as needed. Minor adjustments to the homogenization or centrifugation conditions may also be necessary, depending on tissue type or specific experimental requirements.

3.2 SEC-Based Fractionation of Tissue for Tau Species Separation

The use of size exclusion chromatography (SEC) enables the separation of native protein complexes based on their hydrodynamic radius, preserving their conformation and interactions. This protocol is compatible with TBS-soluble lysates derived from fresh or frozen mouse or human brain tissue and can be adapted for tauopathy models or other proteinopathies.

Preparation Before Fractionation

1. Thaw the TBS-soluble brain lysate on ice and determine protein concentration using a standard quantification assay (e.g., Bradford or BCA). Calculate the volume of sample required to load onto the column, ensuring an equal amount of protein loading for all samples to maintain comparability. We recommend using a total of 4 mg for mouse and 1 mg for human tissue in a maximum volume of 500 μL (*see* **Note 3**).
2. Prepare size exclusion chromatography (SEC) buffer. If the buffer was prepared previously, filter it again using a 0.22 μm sterile filtration unit. For optimal reproducibility, a freshly prepared and filtered buffer is recommended (All reagents should be of molecular biology grade).

AKTA Protocol Setup

1. Create a chromatography method using the AKTA system software. A representative protocol includes:
   a. Using a 2 mL injection loop and injecting the sample at a rate of 1 mL/min. The loop volume should ideally be at least 2× the sample injection volume to ensure complete filling of the loop and consistent sample delivery.

b. Discarding the first 6 mL of the elution.
c. Collecting 16 fractions (1 mL per fraction) at a flow rate of 0.3 mL/min.
d. After fraction collection, flushing the column with 1.5 column volumes (CV) (36 mL) of SEC buffer.

Pre-Fractionation Column Preparation

1. Before use, remove the storage buffer (20% ethanol) from the column by flushing with 2 CV (48 mL) of molecular-grade water.
2. Equilibrate the column with 1.5–2 CV (36–48 mL) of SEC buffer.

Sample Fractionation and Storage

1. Place a 1 mL tube rack or box in the fraction collector, ensuring proper alignment with the collection positions. Label all tubes or boxes to avoid sample loss or misidentification.
2. Add 20 μL of 50× protease and phosphatase inhibitor cocktail directly to each collection tube (final concentration will be 1× after 1 mL fraction collection) before starting the run to prevent proteolysis during fractionation.
3. Load the prepared sample into the injection loop using a Hamilton syringe, ensuring no air is injected into the system.
4. Run the protocol as described, collecting 16 fractions (1 mL each) at the specified flow rate.
5. Immediately store collected fractions at −80°C for further analysis.

3.3 FRET-Based Tau Aggregation Assay

The TauRD P301S FRET biosensor cells are used to detect biochemically active tau species based on their "seeding activity. These cells stably express the tau repeat domain (RD) fused to cyan fluorescent protein (CFP) and yellow fluorescent protein (YFP). When material containing tau-seed is added to the cell culture media of the biosensor cells, the exogenous material is internalized, where it nucleates the aggregation of the tau reporter proteins. This brings CFP and YFP in close proximity, resulting in a fluorescence resonance energy transfer (FRET) signal that can be quantitatively measured using flow cytometry, microscopy, or plate readers. The FRET-based tau aggregation assay will enable the analysis of the seeding activity of different SEC fractions from the tissue lysate (*see* **Note 4**).

Plating of TauRD Biosensor Cells (Day 1)

1. Prepare fresh cell cultures of Tau RD FRET biosensor, HEK293T, TauRD CFP, and TauRD YFP cells according to the appropriate protocol [11].
2. Dilute the FRET biosensor cells to ~269,000 cells/mL and plate 130 μL (~13,000 cells) per well in a 96-well cell culture plate (e.g., Thermo Scientific, Cat. No. 167425). Such plates

are surface-treated to promote optimal cell attachment and growth.

3. Incubate the cells for 24 h in a CO_2 chamber maintained at 37°C and 5% CO_2.

Transfecting Biosensor Cells with Tau-Seed (Day 2)

1. Calculate the total number of samples to be tested (*see* **Note 5**), including the following essential experimental controls:
   i. No treatment control—FRET biosensor cells without any sample or transfection.
   ii. Mock transfection control—Cells to be treated with transfection reagent alone.
   iii. Positive control—A sample known to contain tau-seed, such as brain lysate from a PS19 tauopathy mouse.
   iv. Negative control—A sample confirmed to lack tau-seeding activity, such as brain lysates from a wild-type or tau-knockout mouse.
2. Prepare the transfection reagent by mixing 1.25 μL Lipofectamine 2000 with 8.75 μL Opti-MEM (Gibco) per well. Incubate at 25°C for 5 min.
3. Prepare the final transfection complex by mixing 10 μL of transfection reagent with 10 μL of each SEC fraction (*see* **Note 6**). Incubate at 25°C for 20 min.
4. Add 20 μL of the transfection complex into each well, ensuring the pipette tip is submerged in the cell culture medium without touching the adherent cells.
5. Incubate the cells for 48 h in a CO_2 chamber maintained at 37°C and 5% CO_2.

Measuring FRET-Based Seeding Using Flow Cytometry (Day 4)

The protocol for cell collection, preparation for flow cytometry, and analysis is as follows.

1. To collect the biosensor cells from the wells, aspirate the media from the wells using a multichannel pipette, ensuring that the cells at the bottom are not detached. The use of a multichannel pipette allows for efficient, consistent handling across wells while minimizing cell disturbance and processing times.
2. To detach the adherent cell layer, add 50 μL of warm trypsin-EDTA 0.05% per well and incubate for 3–5 min at room temperature. Add 100 μL of cell culture medium to each well for quenching the trypsin activity.
3. Resuspend the cells by pipetting the medium up and down, ensuring complete detachment. Transfer the cell suspension

to a 96-well V-bottom plate (*see* **Note 7**) and seal the plate using sealing film with removable tabs.

4. Centrifuge the plate with cells at 500 × *g* for 5 min.
5. Remove the sealing film and discard the supernatant. Add 25 μL of FACS buffer to each well. It is not necessary to resuspend the cell pellet in FACS buffer at this stage. If using an automated sampler, such as the BD HTS, cell resuspension can be performed before the sample injection stage (*see* **Note 8**). If manual resuspension is necessary, pipette gently to avoid bubble formation, which may interfere with the flow cytometer analysis.
6. Prepare the compensation controls–HEK293T, TauRD CFP, and TauRD-YFP cells—by trypsinizing and collecting each cell type into separate 15 mL conical tubes. Centrifuge the cells at 500 × *g* for 5 min to pellet the cells. Discard the supernatant and resuspend the cell pellet in FACS buffer.
7. Add compensation control cells to empty wells in the 96-well V-bottom plate to analyze them with the rest of the samples.
8. Until ready for the run, keep cells in a dark place at 4°C.
9. Configure the flow cytometer by adjusting the voltage and acquisition parameters as previously detailed [11]. Use compensation controls to adjust the voltage and create a gating strategy to select Cells, Single cells, and FRET populations. To configure the HTS sampler, adjust the loader settings as previously outlined [11]. Enter the sample information and plate acquisition settings, including sample annotations, in the HTS acquisition. Configure the experiment to collect at least 10,000 events in the Single Cell gate. This can be adjusted to collect a higher number of events for samples with very low seeding activity.
10. Analyze the plate(s) with the experimental setup and export the data into .fcs files for downstream analysis.
11. Import the .fcs files in FlowJo software and label the files with Sample IDs if not performed in Step 10. Create gates to select the following populations in order:
    a. Select the "Cells" population using SSC-A and FSC-A parameters.
    b. Within the "Cells" population, plot FSC-H vs FSC-W to identify and exclude doublets (events with increased width) and define this gate as "Single Cells FSC."
    c. Similarly, use SSC-H vs SSC-W to remove remaining doublets from the "Cells" population, and label the resulting gate as "Single Cells SSC."

d. Combine the two gates from 11b and 11c using a Boolean "AND" gate of "Single Cells FSC" and "Single Cells SSC" and label the resulting gate as "Single Cells."

12. To correct for spillover compensation, use the appropriate single-color compensation samples. Use the TauRD-YFP expressing cells to adjust for "% FITC in BV421" and "%FITC in BV510." Use the TauRD-CFP expressing cells to adjust compensation for "% BV421 in FITC." Use the mock transfection control cells to adjust for "% BV421 in BV510." For a detailed compensation procedure, refer to [11].
13. From the "Single Cell" population, plot Comp-FITC-A vs Comp-BV421-A to gate the main fluorescent population while excluding the low fluorescent debris, and label it as "Fluorescent Cells." Create a derived parameter "BV510_BV421" using the formula Comp-BV510-A/ Comp-BV421-A.
14. To gate the "FRET" population within the "Fluorescent Cells" population, plot BV510_BV421 vs Comp-BV421-A and create a rectangular gate around the FRET-positive population. To ensure correct gating boundaries, use the positive control with known tau-seeding activity to ensure all positive FRET signal is included, and use a negative control to exclude any background or spillover from the FRET-negative population.
15. Create a data table with the following parameters for exporting:
    a. Keyword: PLATE NAME
    b. Keyword: WELL ID
    c. Keyword: SAMPLE ID
    d. Statistic—"% FRET": Select Frequency of Parent and Population: FRET.
    e. Statistic—"Median Fluorescent Intensity (MFI)": Select Median, Population: FRET, and Parameter: Comp-BV510-A.
    f. Formula—"Integrated FRET": Under the formula tab, perform the calculation %FRET x MFI.
16. Export the data table and use "Integrated FRET" as a measure of seeding activity. To isolate tau-seed, select the SEC fraction(s) with the highest seeding activity.

3.4 Affinity-Based Isolation of Bioactive Tau-Seeds

Tau species with seeding activity can be selectively enriched from biological samples through immunoprecipitation using biotinylated antibodies targeting human tau, as previously established [8]. This method is compatible with a variety of protein sources, including HEK293T lysates, size-exclusion chromatography (SEC) fractions from PS19 mouse brains, and human postmortem tissue from Alzheimer's disease (AD) or progressive supranuclear palsy (PSP) cases. The protocol can be completed in approximately 2 days and is optimized to preserve the biochemical activity of tau-seed for downstream analyses, such as seeding assays, mass spectrometry, or Western blot.

1. To begin, thaw the SEC fractions (or other tissue lysates) on ice and reserve a small aliquot as input material for downstream comparison. Divide each sample into two microcentrifuge tubes, one for the specific tau antibody (HT7) and one for an isotype control (IgG).
2. Add biotinylated HT7 monoclonal antibody at a concentration of 2 μg per 100 ng of total human tau, as quantified by ELISA. An equivalent amount of mouse IgG1 isotype control antibody is added to the control tube.
3. Bring the total volume to a minimum of 350 μL with TBS buffer to ensure proper mixing inside the tube and incubate the samples overnight at 4°C on a rotator.
4. The following day, prepare streptavidin-conjugated magnetic beads by washing 20 μL of bead suspension (corresponding to 200 μg of beads) three times with 1× TBS+0.05% Tween-20 (TBST) buffer in a 1.5 mL microcentrifuge tube using a magnetic rack to retain the beads.
5. After the final wash, resuspend the beads in TBST to the original volume (20 μL).
6. Add 20 μL of bead suspension per 2 μg of HT7 antibody to each tube containing the antibody-bound sample. Repeat this step for IgG control.
7. Incubate the immunocomplexes with the beads for 4 h at 4°C on a rotator to allow binding of the complexes to the beads.
8. After incubation, spin the samples briefly and place them on a magnetic rack. Carefully remove the supernatant (flow-through) and wash the beads three times with 1 mL of TBST, followed by a final wash with 1× TBS.
9. To minimize plastic-derived contamination, resuspend the beads in fresh 1× TBS and transfer to a new tube for a final wash step.
10. At this stage, beads can either be stored at −80°C for subsequent proteomic analysis or resuspended in Western blot sample buffer and denatured at 95°C for 5 min before SDS-PAGE and immunoblotting (*see* **Note 9**).

Throughout the procedure, it is critical to prevent the magnetic beads from drying out, as this can impair protein recovery and activity. Samples including input, flow-through, eluates, and bead-bound fractions can all be used for subsequent analyses such as Western blotting, mass spectrometry, or functional tau seeding assays.

3.5 Quantitative Proteomic Profiling of Tau-Seeds

This protocol, which utilizes liquid chromatography with tandem mass spectrometry (LC-MS/MS), will enable the identification of the proteins coisolated with tau during immunoprecipitation, thereby facilitating the discovery of tau interactors. The immunoprecipitated proteins are digested, barcoded with TMT (to enable multiplexing), and then analyzed using liquid chromatography-tandem mass spectrometry. This experiment can be adapted to other LC-MS/MS configurations, provided they offer adequate mass resolution to resolve the isobaric reporter ions accurately. Lower-resolution instruments, such as time-of-flight mass spectrometers, can typically distinguish up to nine different tandem mass tags (TMT), thereby facilitating multiplexing up to nine samples in a single run. High-resolution instruments can resolve the full set of 18 TMT labels, thereby enabling higher sample throughput.

Protein Digestion and TMT Labeling

This protocol describes on-bead protein digestion, followed by peptide cleanup and labeling with isobaric tags (TMT) to enable multiplexed analysis of samples, allowing for increased throughput and reduced variability across experimental conditions.

1. To denature the proteins and break disulfide bonds, add 30 μL of 8 M Urea, 100 mM Tris pH 8.5, to the beads, and then add 1.5 μL of 100 mM TCEP solution. Incubate at 32°C for 15 min with constant shaking (*see* **Notes 10 and 11**).
2. Add 0.6 μL of 500 mM chloroacetamide (CAA) solution to alkylate cysteine residues and prevent disulfide bond reformation. Vortex briefly, spin down the samples, and incubate at 25°C in the dark for 30 min (*see* **Note 11**).
3. To digest the proteins, perform a two-step LysC-trypsin digestion (*see* **Note 12**):
   a. Dilute the urea in samples to 4 M by adding 28 μL of 50 mM Tris, pH 8.5.
   b. Add 4 μL of 0.1 μg/μL LysC/trypsin stock solution to the samples and incubate at 35°C for 2 h with constant shaking at 600 RPM. LysC cleaves the proteins under harsh denaturing conditions, reducing proteins in size and complexity.
   c. Dilute the urea in samples to 2 M by adding 60 μL of 50 mM Tris pH 8.5. Incubate at 35°C for 12–16 h with constant shaking at 600 RPM. This will allow trypsin to

complete the digestion of proteins in milder reducing conditions.

d. To quench digestion activity, add 6 μL of 10% trifluoroacetic acid, vortex, and spin down.

4. To clean up the peptides, use reverse-phase spin columns [11]:

   a. Place the spin column in a 2 mL tube. Condition the column by washing twice with 200 μL of 0.1% formic acid in acetonitrile, centrifuging at 1500×*g* for 1 min each time, discarding the flowthrough.

   b. Equilibrate the column with two washes of 200 μL 0.5% trifluoroacetic acid (TFA) in water, centrifuging as in Step 4a. Load the acidified peptide digest onto the column and centrifuge again. Repeat loading the flowthrough to maximize peptide binding.

   c. Wash twice with 200 μL 0.5% TFA in water.

   d. Elute peptides twice with 20 μL of 70% acetonitrile containing 0.1% formic acid into a low-bind tube. Dry the eluate using a vacuum centrifuge and store peptides at −80°C for up to 2 years.

5. Label peptides with isobaric tags.

   a. Resuspend the peptides in 25 μL of 100 mM TEAB.

   b. Dissolve TMT aliquots (0.2–0.5 mg) in 25 μL of acetonitrile and add them to the corresponding peptide sample. Mix well and incubate for 1 h at 25°C.

   c. Quench TMT labeling by adding 3 μL of 5% hydroxylamine and incubating for 15 min at 25°C.

The samples should be labeled and mixed strategically to enable comparison across samples or groups. For optimal identification and quantification, it is advised to run the HT7 and IgG samples in the same injection.

Mass Spectrometry

This protocol uses the Orbitrap Tribrid instrument for MS2-based quantification of the TMT-labeled proteins. Reversed-phase nano-chromatography separates the TMT-labeled peptides, which are then analyzed by high-resolution mass spectrometry, enabling the identification and quantification of peptides from the sample.

1. Resuspend the mixed dried peptides in Buffer A (25–100 μL) depending on the LC-MS being used.
2. LC-MS/MS methods

   a. LC Buffer A: 0.1% formic acid in water.

   b. LC Buffer B: 0.1% formic acid, 80% acetonitrile.

The sample loading is calculated using the formula: 2n + 2 μL, where n is the desired injection volume. For example, for a 4 μL injection, the sample loading volume will be 10 μL. The overfilling strategy helps ensure that the entire loop is saturated with the injected material, thereby preventing sample loss.

c. Perform LC-MS/MS using a linear gradient from 8% to 95% Buffer B over ~170 min at a 400 nL/min flow rate. The full gradient is shown in Table 1.

d. Mass spectrometer Global settings:
   - Instrument type = LC
   - Expected peak width = 30 s
   - Advanced peak determination: On
   - Default charge state = 2
   - Easy-IC (Internal Calibration) = On
   - Spray Voltage = Static, 1800 V
   - Ion Transfer Tube Temperature = 350°C
   - FAIMS Mode = Standard resolution, static

e. FAIMSpro Settings: Each LC-MS experiment was conducted using three FAIMS compensation voltages (CVs) in a three-cycle loop, with 1.5 s per CV:
   - CV = −45 V
   - CV = −55 V
   - CV = −65 V

f. All LC-MS experiments used identical MS1 and MS2 settings. The MS1 and MS2 settings used are as follows:

MS1 Settings:
   - Obitrap Resolution: 120,000
   - Scan Range: 400–1600
   - RF lens (%): 30

Table 1
LC gradient conditions

Time (min)	Flow rate (nL/min)	% Buffer B
0	400	8
160	400	35
170	400	95
172	400	95
180	400	5

- AGC Target: Standard
- Maximum inject time: 50 ms
- Data type: Profile
- Polarity: Positive

Precursor ion filtering included Monoisotopic peak determination restricted to peptide, an intensity threshold of 2.5×10^4, and charge states between +2 and +6. Dynamic exclusion was enabled with precursors excluded after one time for 60 s with a mass tolerance of ±5 ppm, excluding isotopes.
MS2 Settings:

- Isolation Mode: Quadrupole
- Isolation width (m/z): 0.7
- Activation type (fixed): HCD
- HCD collision energy: 34
- Orbitrap resolution: 50,000
- AGC target: Custom, 200%
- Maximum inject time: Dynamic
- Scan Range: Define first mass
- First mass (m/z): 100

Data Analysis

This protocol describes the analysis of mass spectrometry data using Proteome Discoverer 2.5. Alternative free software such as MSFragpipe and MaxQuant can also be utilized.

1. Obtain the uncompressed ".fasta" file of the reference protein database from UniProt.org (or NCBI) for Homo sapiens (or other appropriate organism used for the samples). Also, download a database of common laboratory contaminants such as keratin and trypsin. Only reviewed and canonical sequences are required. Isoforms can be omitted except for Tau. Load the protein database in Proteome Discoverer.
2. Develop a quantification method based on the total number of samples, as described in detail in [11]. Additionally, include isotopic impurity information related to the TMT labels used in the experiment.
3. Create a new project and add the quantification method, categorical study factors to the "Study definition" window of the project. Refer to the Proteome Discoverer Manual for details on categorical and biological replicates for such analyses.
4. Add the.raw files from the mass spectrometer to the input files window and select the quantification method from the pull-down menu.

5. Under the "Samples" window, add the categorical and/ or biological factors to describe each TMT label. Refer to Proteome Discoverer User Manual and [11] for more details.
6. Start a new analysis by selecting "New Analysis" in the software. Load the template located at *common template > consensusWF > CWF_Comprehensive_EnhancedAnnotation_ Rprotein_Quan*, designed for comprehensive data annotation and reporter-based quantification. In the "Processing Step" window, load the processing template from *ProcessingWF_ Tribrid > PWF_Reproter_Based_SequestHT_Percolator*, which uses the Sequest HT search engine and Percolator algorithm for reporter-based quantification.
7. Modify the processing workflow as follows:
   a. *Spectrum files RC Node*
   Select the imported FASTA sequence, set the enzyme to *Trypsin (Full)*, and apply static modifications: *TMTpro (304.207 Da)* on peptide N-termini and lysines.
   b. *Reporter Ions Quantifier Node*
   Set peak integration to *20 ppm*, choose the *most confident centroid*, use *FTMS, MS2*, and *HCD* for analyzer, MS order, and activation type.
   c. *Sequest HT Node*
   For the Protein database, use the organism and contaminant FASTA, Enzyme: *Trypsin (full)*, Max missed cleavages: *3*, Peptide length: *6–144 aa*, Precursor tolerance: *10 ppm*, Fragment tolerance: *0.02 Da*, Max modifications per peptide: *3*, Dynamic mods: *Oxidation (M, + 15.995), Phospho (S/T/Y, + 79.996), protein N-terminal acetylation, Met-loss, Met-loss+acetyl* and Static mods: *TMTpro on N-terminus and Lys, Carbamidomethyl on Cys.*
8. For FDR cutoffs, use the Percolator FDR node.
9. For phosphorylation and other PTMs, include the IMP-ptmRS node after Percolator.
10. Modify the consensus workflow:

Reporter Ion Quantifier Node: Set Reporter Ion Abundance to *Automatic*, *enable* Quant Value Corrections *(True)*, set Coisolation Threshold to *30*, and Average Reporter S/N threshold to *6*. Set both Normalization Mode and Scaling Mode to *None* (*see* **Note 13**).

4 Notes

1. If using cultured cells, homogenization should be performed using approximately 15 mL of TBS per cm² of surface area.

2. For cell cultures, sonication may be performed at 30% power under the same pulsing conditions.
3. Avoid loading excessive protein in the column to prevent band-broadening, aggregation/precipitation, tailing, etc. Loading <5 mg/mL is ideal for columns of 10 mm diameter. Alternate columns can be utilized to obtain an increased yield of output materials. For samples with volumes >500 μL, concentrate before loading to obtain a loading volume of 250–500 μL with protein concentration below ~10 mg/mL.
4. Tau-seed containing material can be obtained from other sources such as mouse brains [8, 12], recombinant proteins, cell lysates [8], *C. Elegans*, *D. Melanogaster*, etc. The FRET biosensor cells are sensitive to both detergents and pH. To preserve viability and ensure assay reliability, prepare biological samples in gentle buffers such as Tris-buffered saline (TBS) at physiological pH (~7.4). Avoid excessive detergent concentrations or extreme pH, as these can compromise cell integrity and assay outcome.
5. Due to the inherent variability of the assay, it is recommended to include at least two technical replicates per condition to ensure reliability and reproducibility.
6. To adapt the protocol for other types of biological material beyond SEC fractions, normalize transfections based on total protein concentration. Use approximately 5–20 μg total protein per well, diluted to a final volume of 10 μL in Opti-MEM, to ensure consistent input across samples.
7. To analyze cells at a later time point, the cells can be fixed using 100 μL of 2% paraformaldehyde (PFA) in 1× PBS for 10 min at room temperature. Fixed biosensor cells can be stored up to 24 h at 4°C. Similarly, fixed compensation cells (HEK293T, TauRD CFP, and TauRD-YFP) can be stored at 4°C up to 1 month.
8. When using the BD HTS in high throughput mode, take extra care to avoid air bubbles and sample clumping, as these can obstruct the sampler. Ensure sufficient FACS buffer is used to fully resuspend the cell pellet and verify that the HTS mixing speed and loader settings are optimized to prevent bubble formation, evenly disperse the pellet, and maintain consistent sample flow [11].
9. If native tau-seed are desired for further biochemical assays, an optional elution step can be performed by resuspending the beads in 50 μL of 0.1 M glycine (pH 2.8) and incubating for 10 min at room temperature with occasional vortexing. After brief centrifugation and magnetic separation, the supernatant containing the eluted tau-seed can be transferred to a fresh tube and neutralized by adding 5 μL of 1 M Tris-HCl

(pH 8.0) to reach a final pH of 7.0–7.5. Eluates can be used for downstream applications such as atomic force microscopy (AFM) or in vitro seeding assays.

10. Avoid overheating (>35°C) of samples containing proteins and beads in the presence of urea. At elevated temperatures, the amino acids can get carbamylated in the presence of urea, which modifies the side chain and N-termini of amino acids, potentially interfering with the enzymatic digestion and downstream analysis.
11. Reduction and alkylation can be performed with alternate reagents such as DTT or iodoacetamide. Ensure that the fixed modification specified during data analysis matches the reagents used to ensure proper peptide identification.
12. LysC and trypsin proteases from alternate sources can be used. However, it is important to maintain consistency of proteases within and across experiments to minimize variations in digestion efficiency and peptide identification. If < 1 mg protein is used for immunoprecipitation, the protease quantity can be reduced to 0.1–0.2 μg.
13. For PTM investigation, add the "*Modification Sites*" and "*Peptide Isoform Grouper*" nodes to the consensus workflow.

Conflict of Interest

The authors have no conflicts of interest to declare that are relevant to the content of this chapter.

References

1. Holmes BB, Diamond MI (2014) Prion-like properties of tau protein: the importance of extracellular Tau as a therapeutic target. J Biol Chem 289(29):19855–19861. https://doi.org/10.1074/jbc.R114.549295
2. Dujardin S, Commins C, Lathuiliere A, Beerepoot P, Fernandes AR, Kamath TV, De Los Santos MB, Klickstein N, Corjuc DL, Corjuc BT, Dooley PM, Viode A, Oakley DH, Moore BD, Mullin K, Jean-Gilles D, Clark R, Atchison K, Moore R, Chibnik LB, Tanzi RE, Frosch MP, Serrano-Pozo A, Elwood F, Steen JA, Kennedy ME, Hyman BT (2020) Tau molecular diversity contributes to clinical heterogeneity in Alzheimer's disease. Nature Med 26(8):1256–1263. https://doi.org/10.1038/s41591-020-0938-9
3. Puig B, Rey MJ, Ferrer I (2005) Individual and regional variations of phospho-tau species in progressive supranuclear palsy. Acta Neuropathol 110(3):261–268. https://doi.org/10.1007/s00401-005-1046-0
4. Tracy TE, Madero-Perez J, Swaney DL, Chang TS, Moritz M, Konrad C, Ward ME, Stevenson E, Huttenhain R, Kauwe G, Mercedes M, Sweetland-Martin L, Chen X, Mok SA, Wong MY, Telpoukhovskaia M, Min SW, Wang C, Sohn PD, Martin J, Zhou Y, Luo W, Trojanowski JQ, Lee VMY, Gong S, Manfredi G, Coppola G, Krogan NJ, Geschwind DH, Gan L (2022) Tau interactome maps synaptic and mitochondrial processes associated with neurodegeneration. Cell 185(4):712–728 e714. https://doi.org/10.1016/j.cell.2021.12.041
5. McInnes J, Wierda K, Snellinx A, Bounti L, Wang YC, Stancu IC, Apostolo N, Gevaert K, Dewachter I, Spires-Jones TL, De Strooper B, De Wit J, Zhou L, Verstreken P (2018)

Synaptogyrin-3 mediates presynaptic dysfunction induced by tau. Neuron 97(4):823–835 e828. https://doi.org/10.1016/j.neuron.2018.01.022

6. DeVos SL, Corjuc BT, Oakley DH, Nobuhara CK, Bannon RN, Chase A, Commins C, Gonzalez JA, Dooley PM, Frosch MP, Hyman BT (2018) Synaptic tau seeding precedes tau pathology in human Alzheimer's disease brain. Front Neurosci 12:267. https://doi.org/10.3389/fnins.2018.00267
7. Yoshiyama Y, Higuchi M, Zhang B, Huang SM, Iwata N, Saido TC, Maeda J, Suhara T, Trojanowski JQ, Lee VM (2007) Synapse loss and microglial activation precede tangles in a P301S tauopathy mouse model. Neuron 53(3):337–351. https://doi.org/10.1016/j.neuron.2007.01.010
8. Martinez P, Patel H, You Y, Jury N, Perkins A, Lee-Gosselin A, Taylor X, You Y, Viana Di Prisco G, Huang X, Dutta S, Wijeratne AB, Redding-Ochoa J, Shahid SS, Codocedo JF, Min S, Landreth GE, Mosley AL, Wu YC, McKinzie DL, Rochet JC, Zhang J, Atwood BK, Troncoso J, Lasagna-Reeves CA (2022) Bassoon contributes to tau-seed propagation and neurotoxicity. Nat Neurosci 25(12):1597–1607. https://doi.org/10.1038/s41593-022-01191-6
9. Takeda S, Wegmann S, Cho H, DeVos SL, Commins C, Roe AD, Nicholls SB, Carlson GA, Pitstick R, Nobuhara CK, Costantino I, Frosch MP, Muller DJ, Irimia D, Hyman BT (2015) Neuronal uptake and propagation of a rare phosphorylated high-molecular-weight tau derived from Alzheimer's disease brain. Nat Commun 6: 8490. https://doi.org/10.1038/ncomms9490
10. Jury-Garfe N, Redding-Ochoa J, You Y, Martinez P, Karahan H, Chimal-Juarez E, Johnson TS, Zhang J, Resnick S, Kim J, Troncoso JC, Lasagna-Reeves CA (2024) Enhanced microglial dynamics and a paucity of tau seeding in the amyloid plaque microenvironment contribute to cognitive resilience in Alzheimer's disease. Acta Neuropathol 148(1):15. https://doi.org/10.1007/s00401-024-02775-1
11. Martinez P, Patel H, You Y, Doud EH, Mosley AL, Lasagna-Reeves CA (2024) Protocol for the isolation and proteomic analysis of pathological tau-seeds. STAR Protoc 5(3):103185. https://doi.org/10.1016/j.xpro.2024.103185
12. Patel H, Martinez P, Perkins A, Taylor X, Jury N, McKinzie D, Lasagna-Reeves CA (2022) Pathological tau and reactive astrogliosis are associated with distinct functional deficits in a mouse model of tauopathy. Neurobiol Aging 109:52–63. https://doi.org/10.1016/j.neurobiolaging.2021.09.006

Part III

Mimicking Neurodegenerative Processes In Vitro

Chapter 9

Neurotoxic Tau Oligomers: Isolation, Preparation, and Characterization

Sheeza Khan, Md. Anzarul Haque and Rakez Kayed

Abstract

A large group of neurodegenerative diseases (ND), grouped as tauopathies, are related to the misfolded tau protein. The neurofibrillary tangles (NFTs) are considered as the hallmark for the tauopathies, but these are not the toxic species that cause the disease. Numerous studies have implicated tau oligomers (TauO), an intermediate entity in the formation of NFTs, for the commencement of the disease. Hence, it becomes essential to understand the mechanisms underlying the production of toxic TauO to help design strategies to prevent their formation and check the progression of ND. Therefore, we have developed streamlined methods and protocols for reliable detection, isolation, amplification and characterization of TauO. We have also designed methods for the seeding and propagation of TauO both under in vitro and in vivo conditions. This chapter taken together provides very simple and reliable steps and procedures to produce TauO, starting from their detection and isolation from the diseased brain to their amplification followed by their characterization and assessment of their biological activity.

Keywords Amplified brain-derived tau oligomers, Recombinant tau oligomer, Immunofluorescence, Fluorescent amyloid multi-emission spectra microscopy, Fourier transform infrared spectroscopy, Tau oligomer specific antibodies, Tau oligomer bioactivity, High-performance liquid chromatography

1 Introduction

Misfolding and aggregation of tau have been implicated in tauopathies [1–5]. Hyper-phosphorylation of tau causes it to detach from the microtubules and increases its tendency to bind to the monomeric tau, thus leading to the formation of TauO [6–8]. TauO is a detergent-soluble nonfibrillar multimeric form of protein. This TauO keeps on increasing in length by further addition of monomeric tau [5, 9–13]. During this time, it increases its β-sheet content which imparts granular appearance to the tau aggregates under the atomic force microscopy (AFM) [14–16]. This granular TauO ultimately forms the fibrils which further fuse together to give rise to the NFTs [17, 18]. Presence of NFTs serves as the distinctive histopathological

Jessica S. Fortin (ed.), *Translational Research Methods for Tauopathies*, Neuromethods, vol. 233,
https://doi.org/10.1007/978-1-0716-5234-3_9,

feature of tauopathies [19]. However, numerous studies have indicated that mature tangles (NFTs) are not the toxic entities, but it is the TauO which is toxic [20–33]. Therefore, understanding the mechanisms causing toxic TauO formation can give insights into finding strategies to control its formation or at least reduce its toxicity [16, 34]. The process of initiation of TauO formation is quite complex under in vitro conditions. A variety of molecules and approaches are being used to induce in vitro aggregation of tau protein [35–44]. To facilitate the study of tau aggregation mechanisms and toxicity, we have outlined and validated methods and procedures for effective production of TauO under in vitro conditions [10, 23, 34, 45–48]. Recombinant monomeric tau (rTauM) has become an essential tool to study tau aggregation mechanisms. We express and purify the rTauM and subject it to biochemical and biophysical characterization. This rTauM is further used to amplify the recombinant TauO (rTauO) and the brain-derived TauO (BDTO). rTauO is produced by subjecting the TauM to various molecules, for example, heparin and arachidonic acid [49–53], or simply keeping it at 37°C for 72 h. These methods yield tau filaments which are not appropriate to carry out TauO toxicity studies. Therefore, we isolate TauO from the diseased brain using different methods. TauO-specific novel antibodies created in our lab are used in conjunction with biochemical and immunohistochemistry assays to detect and characterize the BDTO (Alzheimer's disease, progressive supranuclear palsy, Parkinson's disease, dementia with Lewy bodies, Huntington's disease, and transgenic mouse models) [34, 45, 47, 48, 54–62]. Additionally, we perform biophysical studies and proteolytic digestion analysis to further confirm the BDTO. The BDTO, thus obtained, is then used as seeds to amplify the amount of BDTO, using the 3-repeat (3R) TauM and 4-repeat (4R) TauM [5, 63–65]. The amplified BDTO is further characterized through biochemical and biophysical techniques for further downstream processing [5, 34, 57]. The amplified BDTO are also characterized for their biological activity (i.e., seeding potency in cell lines and primary cortical neurons, toxicity to the cells, effect on synaptic function and synaptic plasticity (long-term potentiation) [5, 10, 34, 66]. Further, in vivo seeding potency of these BDTO and TauO is determined by stereotactic injections of these in transgenic mice for tauopathy. The injected transgenic mice are then evaluated for the tau pathology [10, 45, 66, 67].

In this chapter, we will discuss the methods used in our laboratory to generate samples enriched for rTauO and BDTO as well as highlight some methods, protocols and assays to amplify, characterize and check their biological activity.

2 Materials

2.1 Expression of TauM

1. LB broth.
2. Antibiotic kanamycin.
3. pET29b/Tau2N4R plasmid for Tau2N4R (full-length TauM) or pET29b/tau2N3R plasmid for Tau2N3R, the two isoforms for the TauM (*see* **Note 1**).
4. *Escherichia coli* (*E. coli*) BL21 (DE3) cells.
5. 1.00 M IPTG (isopropyl-1-thio-D-galactopyranoside).
6. Lysis buffer: 20 mM Tris, 500 mM NaCl, 1 mM EDTA, 100 mM PMSF (phenylmethylsulphonyl fluoride).
7. Protease inhibitor tablets.

2.2 Purification of TauM

1. Kontes® flex-column.
2. HisPur™ Ni-NTA resin (Thermo Scientific).
3. Wash buffer 1: 20 mM Tris, 500 mM NaCl, and pH 8.0.
4. Wash buffer 2: 20 mM Tris, 500 mM NaCl, 10 mM imidazole and pH 8.0.
5. Elution buffer: 20 mM Tris–HCl, 100 mM NaCl, 100 mM Imidazole, 8 M urea pH 8.0.
6. SEC buffer: 20 mM Tris–HCl, 100 mM NaCl (pH 7.4).
7. Superdex 75 10/300 GL, HiLoad 16/600 Superdex 75 pg and HisTrap FF from Cytiva.
8. Superdex 200 3.2/300 from Cytiva.
9. AKTA Explorer/ Pure fast flow liquid chromatography (FPLC).

2.3 Assays and Techniques for Characterization of TauM

SDS-PAGE

1. BIO-RAD, 12% Mini-PROTEAN TGX™ Precast Protein Gels.
2. 10× electrophoresis buffer, containing 25 mM Tris, 192 mM glycine, 0.1% SDS, pH 8.3. This is diluted with DI water to make 1×.
3. Loading dye containing 250 mM Tris–HCl (pH 6.8), 10% SDS, 30% glycerol, 0.05% bromophenol blue and reducing agent (dithiothreitol or β-mercaptoethanol).
4. Molecular weight marker containing a mixture of ten different recombinant proteins with molecular masses ranging from 10 to 250 kDa.

5. Electrophoresis chamber (Mini-PROTEAN Tetra Vertical Electrophoresis Cell, Bio-Rad).
6. Staining solution containing 0.1% Coomassie Brilliant Blue R-250 (CBB R-250), 40% methanol and 10% glacial acetic acid. Dissolve 0.25 g of CBB R-250 in 30 mL of methanol. Add 10 mL of glacial acetic acid. Make volume to 100 mL by adding DI water.
7. De-staining solution containing 50% methanol, 10% glacial acetic acid and the remaining 40% water.
8. Protein (TauM).

Western Blot (WB)

1. Protein (TauM).
2. Invitrogen NuPAGE Bis-Tris Gels, 4%–12%, 1.0–1.5 mm.
3. Nitrocellulose membrane with a pore size of 0.45 μm.
4. Transfer pads. We require two transfer pads for each nitrocellulose membrane.
5. Reagents:
   (i) Invitrogen NuPAGE 4× loading buffer.
   (ii) Invitrogen NuPAGE MES SDS 20× running buffer stock (50 mM MES, 50 mM Tris base, 0.1% SDS, 1 mM EDTA at pH 7.3). This is made to 1× for the experiment.
   (iii) Transfer buffer stock (1.92 M Glycine, 250 mM Tris at pH 9). This is made to 1× (800 mL DI water, 100 mL transfer buffer stock and 100 mL methanol).
   (iv) Methanol.
   (v) Deionized (DI) water.
   (vi) 1× TBST.
   (vii) 5% and 10% milk solution.
   (viii) Pierce ECL Western Blotting Substrate.
   (ix) Western Bright ECL HRP substrate from Advansta Inc.
6. Antibodies:
   (i) Primary antibodies: Total tau antibodies (Tau 5, Tau 13), Conformational antibodies (T18, T22).
   (ii) Secondary antibodies: Goat anti-mouse IgG HRP (Bio-Rad), Goat anti-rabbit IgG HRP (Bio-Rad).
7. Miscellaneous:
   (i) Filter paper.
   (ii) Clean container.
   (iii) Li-Cor Odyssey Pen.
   (iv) Gel Knife.
   (v) Gel Roller.

(vi) Ice Pack.

(vii) Mini Trans-Blot Cell.

High-Performance Liquid Chromatography (HPLC)

1. Protein (TauM).
2. Superdex 200 10/300 GL from Cytiva.
3. Akta Explorer/Pure fast flow liquid chromatography (FPLC).
4. Gel filtration standard (Catalog # 1511901) from Bio-rad.

Dot-Blot Assay

1. Nitrocellulose membrane 0.45 μm pore size (Thermo Fischer Scientific).
2. TauM.
3. Reagents:
   (i) 1× TBS with 1% Tween (TBST).
   (ii) 5% and 10% non-fat dry milk in 1× TBST.
   (iii) Pierce ECL Western Blotting Substrate.
4. Antibodies:
   (i) Primary antibodies: Total tau antibodies (Tau 5, Tau 13), Conformational antibodies (T18, T22).
   (ii) Secondary antibodies: Goat anti-mouse IgG HRP (Bio-Rad), Goat anti-rabbit IgG HRP (Bio-Rad).
5. Miscellaneous:
   (i) Flat ended tweezers.
   (ii) Film.

Dynamic Light Scattering (DLS)

1. TauM.
2. DLS (Zetasizer Malvern).
3. ZEN2112 Ultra-Micro Cell.

Mass Spectrometry

1. TauM.
2. Nanoflow Easy nLC100 UHPLC system (Thermo Scientific).
3. Q Exactive Orbitrap mass spectrometer (Thermo Scientific, San Jose, CA).
4. Capillary peptide trap column (Acclaim® Pepmap 100, 75 μm ×2 cm, C18, 3 μm, 100 Å, ThermoScientific).
5. UHPLC reversed-phase.column (Acclaim® Pepmap 100, 75 μm ×25 cm, C18, 2 μm, 100 Å, Thermo Scientific).

2.4 Preparation of Tau Oligomers (TauO)

Preparation of Recombinant TauO (rTauO)

1. Purified TauM (0.6 mg/ml).
2. 1× PBS.
3. Eppendorf Thermomixer Shaker with 1.5 mL Incubator Block.

Extraction and Purification of Brain Derived TauO (BDTO)

Preparation of Diseased Brain Homogenate

1. Frozen diseased brain.
2. Reagents:
   (i) 1× PBS at pH 7.4.
   (ii) Ethanol (70%).
   (iii) Complete mini EDTA-free Protease Inhibitor Cocktail Tablets.
   (iv) PhosStop, Phosphatase Inhibitor Cocktail Tablets.
3. Miscellaneous:
   (i) Stainless Steel Beads (5 mm).
   (ii) Micro Spoon (Cole-Parmer).
   (iii) Tissue lyser (Qiagen).

Co-Immunoprecipitation Method

1. Diseased brain homogenate.
2. DI water.
3. Pierce Co-Immunoprecipitation Kit from Thermo Fischer Scientific. The kit contains the following:
   (i) Resins:
      (a) AminoLink Plus Coupling Resin.
      (b) Pierce Control Agarose Resin.
   (ii) Reagents:
      (a) 20× coupling buffer, we prepare 1× from it using DI. The 1× coupling buffer is composed of 0.01 M sodium phosphate and 0.15 M sodium chloride at pH 7.2.
      (b) Sodium Cyanoborohydride solution (5 M).
      (c) Quenching buffer composed of 1 M Tris–HCl.
      (d) 100× conditioning buffer (*see* **Note 2**).
      (e) 1 M NaCl used as wash solution.
      (f) 0.1 M glycine at pH 2.8 is used as elution buffer.
      (g) IP lysis buffer.
      (h) 5× Lane Marker Sample buffer.
      (i) 20× Modified Dulbecco's Phosphate Buffered Saline (PBS).
4. Miscellaneous:
   (i) 2.0- and 1.5 mL Microcentrifuge collection tubes.
   (ii) Pierce Spin Columns.
   (iii) 1.5 mL Microcentrifuge Sample tubes.

Sucrose Step Gradient Method

1. PBS soluble fraction of the diseased brain homogenate.
2. Water (Molecular Biology Grade).
3. Reagents:
   (i) Sucrose (different percentage concentrations to create step gradient).
   (ii) 10% Sarkosyl (*N*-lauroylsarcosine) solution. It was made 1× by adding PBS.
   (iii) 1× PBS.
4. Miscellaneous:
   (i) Beckman XL-80 ultracentrifuge (or Beckman Optima XE-90) with a Beckman SW 28 Ti rotor.
   (ii) Hypodermic Disposable Needles 30 G.
   (iii) Hypodermic Disposable Needles 18 G 1 ½.
   (iv) 5.00 mL Ultra-Clear Tube.

Preparation of Amplified BDTO (aBDTO)

Preparation of Cross-Seeded aBDTO

1. Proteins:
   (i) 0.6 mg/mL (12 μM) Purified TauM (either of the two isoforms-2N4R or 2N3R).
   (ii) 0.006 mg/mL (0.12 μM) oligomers of Aβ42, α-synuclein or TDP-43.
2. Reagents:
   (i) 8.00 M urea.
   (ii) 1× PBS, pH 7.4.
3. Pierce BCA Protein Assay Kit for total protein estimation (Thermo Scientific).
4. Miscellaneous:
   (i) Microcon centrifugal filter device with a cut-off of 25 kDa (Millipore).
   (ii) TSK-GEL G3000 SWXL (30 cm × 7.8 mm) column (Supelco).
   (iii) LC-6AD Shimadzu high-performance liquid chromatography (HPLC) system.

Preparation of aBDTO

1. Proteins:
   (i) 0.6 mg/mL (12 μM) Purified TauM (either of the two isoforms-2N4R or 2N3R).
   (ii) 0.006 mg/mL (0.12 μM) BDTO.
2. Reagents
   (i) 8.00 M urea.
   (ii) 1× PBS, pH 7.4.

3. Pierce BCA Protein Assay Kit for total protein estimation (Thermo Fischer Scientific).
4. Miscellaneous:
   (i) Microcon centrifugal filter device with a cut-off of 25 kDa (Millipore).
   (ii) TSK-GEL G3000 SWXL (30 cm×7.8 mm) column (Supelco).
   (iii) LC-6AD Shimadzu high-performance liquid chromatography (HPLC) system.

2.5 Characterization of TauO (rTauO, BDTO, and aBDTO)

Assays and Techniques for Biochemical Characterization of TauO

Indirect ELISA

1. 96-well ELISA plates (Nunc polysorp, MaxiSorp).
2. TauO.
3. Reagents:
   (i) 0.05 M sodium bicarbonate at pH 9.6.
   (ii) 1× TBS low tween (TBST) buffer (19.98 mM Tris, 136 mM NaCl and 0.01% Tween-20) at pH 7.4.
   (iii) 5% and 10% milk solution. For this, 5% and 10% non-fat dry milk is mixed in 1× TBST.
   (iv) 2 M Hydrochloric acid (HCl).
   (v) TMB+ One-Step Substrate System, procured from Agilent DAKO. TMB is 3,3′-5,5′ tetramethylbenzidine.
4. Antibodies:
   (i) Primary antibodies: These are diluted in 5% fat-free milk and used at different dilution ratios as stated: T22 (1:250), TTC99 (1:250), TTCM1 (1:100), TTCM2 (1:1000), TOMA1-4 (1:1000), Tau 5 (1:3000), Tau13 (1:3000), Tau 12 (1:3000), HT7 (1:1000), Tau2 (1:1000), Tau 46 (1:3000), Tau 1 (1:3000) (*see* **Note 3**).
   (ii) Secondary antibodies: Mouse Ig, HRP-linked Whole Ab, Sheep (Cytiva, Catalog: NA931-1ML); Rabbit Ig, HRP-linked Whole Ab, Donkey (Cytiva, Catalog: NA9314-1ML); Goat anti-mouse IgG Horseradish-peroxidase (HRP) (Bio-Rad), Goat anti-rabbit IgG HRP (Bio-Rad).
5. Miscellaneous: Reservoirs.

Sandwich ELISA

1. TauO (2 μg).
2. 96 wells ELISA plate (NUNC MaxiSorp).
3. Reagents:
   - (i) 0.05 M sodium bicarbonate buffer at pH 9.6.
   - (ii) 1× TBST.
   - (iii) 5 % and 10 % nonfat dry milk solutions in 1× TBST.
   - (iv) TMB (3,3-5,5′ tetramethylbenzidine) One-Step Substrate System (DAKO).
   - (v) 2 M hydrochloric acid (HCl).
4. Antibodies:
   - (i) Primary antibodies: These are diluted in 5% milk and used at different dilution ratios as stated: T22 (1:250), TTC99 (1:250), TTCM1 (1:100), TTCM2 (1:1000), TOMA1-4 (1:1000), Tau 5 (1:3000), Tau13 (1:3000), Tau 12 (1:3000), HT7 (1:1000), Tau2 (1:1000), Tau 46 (1:3000), Tau 1 (1:3000).
   - (ii) Secondary antibodies: Mouse Ig, HRP-linked Whole Ab, Sheep (Cytiva, Catalog: NA931-1ML); Rabbit Ig, HRP-linked Whole Ab, Donkey (Cytiva, Catalog: NA9314-1ML); Goat anti-mouse IgG Horseradish-peroxidase (HRP) (Bio-Rad), Goat anti-rabbit IgG HRP (Bio-Rad).
5. Miscellaneous:
   - (i) Incubator at 37°C.

Dot-Blot Assay

1. Nitrocellulose membrane 0.45 μm pore size (Thermo Fischer Scientific).
2. TauO.
3. Reagents:
   - (i) 1× TBS with 1% Tween (TBST).
   - (ii) Pierce ECL Western Blotting Substrate.
   - (iii) 5% and 10% non-fat dry milk in 1× TBST.
4. Antibodies:
   - (i) Primary antibodies: These are diluted in 5% milk and used at different dilution ratios as stated: T22 (1:250), TTC99 (1:250), TTCM1 (1:100), TTCM2 (1:1000), TOMA1-4 (1:1000), Tau 5 (1:3000), Tau13 (1:3000), Tau 12 (1:3000), HT7 (1:1000), Tau2 (1:1000), Tau 46 (1:3000), Tau 1 (1:3000).
   - (ii) Secondary antibodies: Mouse Ig, HRP-linked Whole Ab, Sheep (Cytiva, Catalog: NA931-1ML); Rabbit Ig, HRP-linked Whole Ab, Donkey (Cytiva, Catalog: NA9314-1ML); Goat anti-mouse IgG Horseradish-peroxidase (HRP) (Bio-Rad), Goat anti-rabbit IgG HRP (Bio-Rad).

5. Miscellaneous:
   - (i) Flat-ended tweezers.
   - (ii) Film.

WB

1. TauO.
2. Invitrogen NuPAGE Bis-Tris Gels, 4%–12%, 1.0–1.5 mm.
3. Nitrocellulose membrane with a pore size of 0.45 μm.
4. Transfer pads. We require two transfer pads for each nitrocellulose membrane.
5. Reagents:
   - (i) Invitrogen NuPage 4× loading buffer
   - (ii) Invitrogen NuPAGE MES SDS 20× running buffer stock (50 mM MES, 50 mM Tris base, 0.1% SDS, 1 mM EDTA at pH 7.3). This is made to 1× for the experiment.
   - (iii) Transfer buffer stock (1.92 M Glycine, 250 mM Tris at pH 9). This is made to 1× (800 mL DI water, 100 mL transfer buffer stock and 100 mL methanol).
   - (iv) Methanol.
   - (v) DI water.
   - (vi) 1× TBST.
   - (vii) 5% and 10% milk solution.
   - (viii) Pierce ECL Western Blotting Substrate.
   - (ix) WesternBright ECL HRP substrate from Advansta Inc.
6. Antibodies:
   - (i) Primary antibodies: These are diluted in 5% milk and used at different dilution ratios as stated: T22 (1:250), TTC99 (1:250), TTCM1 (1:100), TTCM2 (1:1000), TOMA1-4 (1:1000), Tau 5 (1:3000), Tau 13 (1:3000), Tau 12 (1:3000), HT7 (1:1000), Tau 2 (1:1000), Tau 46 (1:3000), Tau 1 (1:3000)
   - (ii) Secondary antibodies: Mouse Ig, HRP-linked Whole Ab, Sheep (Cytiva, Catalog: NA931-1ML); Rabbit Ig, HRP-linked Whole Ab, Donkey (Cytiva, Catalog: NA9314-1ML); Goat anti-mouse IgG Horseradish-peroxidase (HRP) (Bio-Rad), Goat anti-rabbit IgG HRP (Bio-Rad).
7. Miscellaneous:
   - (i) Filter paper.
   - (ii) Clean container.
   - (iii) Li-Cor Odyssey Pen.
   - (iv) Gel Knife.
   - (v) Gel Roller.

(vi) Ice Pack.
(vii) Mini Trans-Blot Cell.

Filter Trap Assay (FTA)

1. Bio-Rad Bio-Dot SF microfiltration apparatus.
2. TauO.
3. Bio-Rad Nitrocellulose membrane.
4. Reagents:
   (i) DI water.
   (ii) 1× TBST.
   (iii) 5% and 10% nonfat dry milk solutions in 1× TBST.
   (iv) Pierce ECL Western Blotting Substrate.
5. Antibodies:
   (i) Primary antibodies: These are diluted in 5% milk and used at different dilution ratios as stated: T22 (1:250), TTC99 (1:250), TTCM1 (1:100), TTCM2 (1:1000), TOMA1-4 (1:1000), Tau 5 (1:3000), Tau 13 (1:3000), Tau 12 (1:3000), HT7 (1:1000), Tau 2 (1:1000), Tau 46 (1:3000), Tau 1 (1:3000).
   (ii) Secondary antibodies: Mouse Ig, HRP-linked Whole Ab, Sheep (Cytiva, Catalog: NA931-1ML); Rabbit Ig, HRP-linked Whole Ab, Donkey (Cytiva, Catalog: NA9314-1ML); Goat anti-mouse IgG Horseradish-peroxidase (HRP) (Bio-Rad), Goat anti-rabbit IgG HRP (Bio-Rad).
6. Miscellaneous:
   (i) Filter paper (Whatman gel blotting paper, Grade GB003, Millipore-Sigma).
   (ii) Vacuum source.

Proteinase K (PK) Digestion

1. TauO.
2. Reagents:
   (i) Molecular grade water.
   (ii) 50 mM sodium chloride solution stock solution. It will be diluted to 5 mM working solution.
   (iii) 1 M Tris HCl stock solution. It will be diluted to 100 mM working solution.
   (iv) Invitrogen 4× NuPage LDS sample buffer.
   (v) 10 μg/mL Proteinase K enzyme from Sigma Aldrich.
3. Miscellaneous:
   (i) Incubator at 37°C.
   (ii) Heat block at 95°C.
4. Materials for SDS-PAGE and WB (see Section: SDS-PAGE and Section: Western Blot (WB), respectively).

Immunohisto-Chemistry in Frozen Sections

1. Humidity chamber.
2. Mice brain sections.
3. Goat serum from Millipore Sigma.
4. BSA (Bovine Serum Albumin) from Millipore Sigma.
5. VECTASTAIN ABC-HRP Kit (Mouse IgG) (PK-4002) from Vector Laboratories
6. DAB Substrate kit, Peroxidase (HRP), with Nickel, (3,3'-diaminobenzidine) (SK-4100) from Vector Laboratories.
7. Antibodies:
   (i) Primary antibodies diluted in antibody diluent: T22, TOMA 1 or2, T18, Total Tau (Tau 13 or Tau 5), Phospho-Tau (AT8, AT100 or pThr231; for fibrils).
   (ii) Secondary antibodies: AR, AM, AR, AM, AM or AR.
8. Reagents:
   (i) Methanol.
   (ii) Ethanol (80%, 95%, and 100%).
   (iii) Xylene.
   (iv) 3% hydrogen peroxide (H_2O_2).
   (v) 1× PBS.
   (vi) 0.25% Triton X-100 from Millipore Sigma.
   (vii) 1× PBST (1× PBS and 0.25% Triton X-100).
   (viii) Antibody diluent (5% BSA in 1× PBST).
   (ix) Blocking buffer (5% BSA and 5% goat serum).
   (x) 1× PBST for MOM (Mouse-on-Mouse) (1× PBS and 0.01% Tween-20).
   (xi) MOM blocking buffer (one drop of MOM IgG in 4 mL 1× PBST).
   (xii) Molecular grade water.
9. Miscellaneous:
   (i) Cytoseal mounting media from Epredia.
   (ii) Hematoxylin counterstain from Vector Labs.

Immuno-Fluorescence in Frozen Sections

1. Humidity chamber.
2. Mice brain sections.
3. Reagents:
   (i) 1× PBS.
   (ii) 0.25% Triton X-100 from Millipore Sigma.
   (iii) 1× PBST (1× PBS and 0.25% Triton X-100).
   (iv) Antibody diluent (5% BSA in 1× PBST).

(v) Blocking buffer (5% BSA and 5% goat serum).
(vi) 1× PBST for MOM (Mouse-on-Mouse) (1× PBS and 0.01% Tween-20).
(vii) MOM blocking buffer (one drop of MOM IgG in 4 mL 1× PBST).
(viii) Molecular grade water.
(ix) 4% paraformaldehyde or chilled methanol.

4. 5% BSA in 1× PBST.
5. 5% BSA and 5% goat serum (Sigma-Aldrich).
6. Invitrogen Goat anti-mouse/rabbit IgG Alexa-568/488.
7. TrueBlack Lipofuscin Autofluorescence Quencher from Biotium.
8. Invitrogen ProLong Gold Antifade Mountant with DAPI from Thermo Fischer Scientific.
9. Miscellaneous:
   (i) Clear nail polish.
   (ii) PAP pen.

Immunohisto-Chemistry in Paraffin-Embedded Sections

1. Mice brain sections.
2. Goat serum from Millipore Sigma.
3. BSA (Bovine Serum Albumin) from Millipore Sigma.
4. VECTASTAIN ABC-HRP Kit (Mouse IgG) (PK-4002) from Vector Laboratories
5. DAB Substrate kit, Peroxidase (HRP), with Nickel, (3,3'-diaminobenzidine) (SK-4100) from Vector Laboratories.
6. Antibodies:
   (i) Primary antibodies diluted in antibody diluent: T22, TOMA 1 or2, T18, Total Tau (Tau 13 or Tau 5), Phospho-Tau (AT8, AT100 or pThr231; for fibrils).
   (ii) Secondary antibodies: AR, AM, AR, AM, AM or AR.
7. Reagents:
   (i) Xylene.
   (ii) Different concentrations of ethanol (80%, 95%, and 100%).
   (iii) Target retrieval solution from Dako.
      (a) Citrate-based Antigen Unmasking Solution, or
      (b) Tris-based Antigen Unmasking Solution.
   (iv) 1× PBS.
   (v) Different concentrations of H_2O_2 (3%).

(vi) 3% BSA in 1× PBS/5% goat serum procured from Sigma-Aldrich.
(vii) 1× PBST (1× PBS and 0.25% Triton X-100).
(viii) Antibody diluent (5% BSA in 1× PBST).
(ix) Blocking buffer (5% BSA and 5% goat serum).
(x) 1× PBST for MOM (Mouse-on-Mouse) (1× PBS and 0.01% Tween-20).
(xi) MOM blocking buffer (one drop of MOM IgG in 4 mL 1× PBST).
(xii) Molecular grade water.
(xiii) Triton-X100 was procured from Sigma-Aldrich.

8. Miscellaneous:
   (i) Cytoseal mounting media from Epredia.
   (ii) Hematoxylin counterstain from Vector Labs.

Immuno-Fluorescence in Paraffin-Embedded Sections

1. Humidity chamber.
2. Mice brain sections.
3. Antibodies:
   (i) Primary antibodies diluted in antibody diluent: T22, TOMA 1 or 2, T18, Total Tau (Tau 13 or Tau 5), Phospho-Tau (AT8, AT100 or pThr231; for fibrils).
   (ii) Secondary antibodies: AR, AM, AR, AM, AM or AR.
4. Reagents:
   (i) Xylene.
   (ii) Different concentrations of ethanol (70%, 80%, 95%, and 100%).
   (iii) Target retrieval solution from Dako.
      (a) Citrate-based Antigen Unmasking Solution, or
      (b) Tris-based Antigen Unmasking Solution.
   (iv) 1× PBS.
   (v) 3% BSA in 1× PBS/ 5% goat serum procured from Sigma-Aldrich.
   (vi) 3% H_2O_2.
   (vii) 1× PBST (1× PBS and 0.25% Triton X-100).
   (viii) Antibody diluent (5% BSA in 1× PBST).
   (ix) Blocking buffer (5% BSA and 5% goat serum).
   (x) 1× PBST for MOM (Mouse-on-Mouse) (1× PBS and 0.01% Tween-20).
   (xi) MOM blocking buffer (one drop of MOM IgG in 4 mL 1× PBST).

(xii) Molecular grade water.

5. Invitrogen ProLong Gold Antifade Mountant with DAPI from Thermo Fischer Scientific.
6. Invitrogen Goat anti-mouse/rabbit IgG Alexa-568/488.
7. Fluoromount G produced from Southern Biotech.
8. TrueBlack Lipofuscin Autofluroscence Quencher.
9. Miscellaneous:
   (i) Clear nail polish.
   (ii) PAP pen.

Assays and Techniques for Biophysical Characterization of TauO

Bis-ANS Fluorescence Assay

1. TauO.
2. Clear bottom black 96-well plate.
3. Bis-ANS procured from Millipore-Sigma.
4. POLARstar Omega Plate reader.

Thioflavin T (ThT) Fluorescence assay

1. TauO.
2. Clear bottom 96-well black plates.
3. Thioflavin T (ThT) procured from Millipore-Sigma.
4. POLARstar Omega Plate reader.

Circular Dichroism Spectroscopy

1. TauO.
2. Circular Dichroism Spectropolarimeter (Jasco J-815) equipped with a temperature controller (Jasco PTC-3515).
3. Quartz cuvette with 0.1 mm path.
4. Reagents:
   (i) 1× PBS.
   (ii) DI Water.
   (iii) Ethanol (70%).

DLS

1. TauO.
2. DLS (Zetasizer Malvern).
3. ZEN2112 Ultra-Micro Cell.

Mass Spectrometry

1. TauO.
2. Nanoflow Easy nLC100 UHPLC system (Thermo Scientific).
3. Q Exactive Orbitrap mass spectrometer (Thermo Scientific, San Jose, CA).

4. Capillary peptide trap column (Acclaim® Pepmap 100, 75 μm ×2 cm, C18, 3 μm, 100 Å, ThermoScientific).
5. UHPLC reversed-phase.column (Acclaim® Pepmap 100, 75 μm ×25 cm, C18, 2 μm, 100 Å, Thermo Scientific).

HPLC

1. TauO.
2. Superdex 200 3.2/300 from Cytiva.
3. Akta Explorer/Pure fast flow liquid chromatography (FPLC).
4. Gel filtration standard (Catalog # 1511901) from Bio-rad.

Atomic Force Microscopy (AFM)

1. TauO.
2. Metallic disc, 15 mm.
3. Mica plate, 12 or 9 mm.
4. DI water.
5. Multimode 8 AFM machine (Veeco, San Jose, CA, USA).
6. Nanoscope Analysis version 1.20rl AFM data processing software.

Transmission Electron Microscopy (TEM)

1. TauO.
2. Transmission electron microscope (FEI Tecnai G2) operated at 80 kV with an ORIUS SC200 CCD camera and Gatan Digital Micrograph software (Gatan Inc.).
3. Grid box.
4. Carbon-coated grids.
5. Whatman Qaulitative 90 mm cellulose filter paper.
6. Reagents:
   (i) Aqueous solution of 2% (w/v) Uranyl acetate (UA).
   (ii) Aqueous solution of 25% glutaraldehyde.
7. Miscellaneous:
   (i) Parafilm.
   (ii) Forceps.

Fourier Transform Infrared Spectroscopy (FTIR)

1. TauO.
2. NICOLET 6700 FT-IR (Thermo Scientific).
3. Reagents:
   (i) DI water.
4. Ethanol (70%).

Fluorescent Amyloid Multi-Emission Spectra (FLAMES) Microscopy

1. TauO.
2. Zeiss LSM 880 confocal microscope.
3. Nikon 63× oil immersion objective.

4. Grenier Bio-One UV-Star UV-Transparent Microplates (96-well).
5. Reagents:
   (i) Low-melting-point agarose (2-Hydroxyethyl agarose) procured from Sigma-Aldrich.
   (ii) Distilled water.
   (iii) Fluorescent dyes (50 μM ThT, 20 μM curcumin, 10 μ M FSB (3,3'-[(2-fluoro-1,4-phenylene) di-(1E)-2,1-ethenediyl] *bis*-6-hydroxy-benzoic acid). The dyes are dissolved in 1× PBS ThT and curcumin was procured from Sigma-Aldrich. FSB was procured from Santa Cruz Biotechnology.
6. Miscellaneous:
   (i) Heating plate with stir bar.
   (ii) Syringe filter.

Assays and Techniques for Estimation of aBDTO Bioactivity

Primary Cortical Neurons

1. aBDTO
2. Mouse primary neuronal cultures.
3. Gibco Neurobasal medium procured from Thermo Fischer Scientific.
4. Gibco B-27 Plus supplement (50×) from Thermo Fischer Scientific.
5. Gibco GlutaMAX supplement (100×) from Thermo Fischer Scientific.
6. Gibco Antibiotic-Antimycotic (100×) from Thermo Fischer r Scientific. It contains penicillin, streptomycin, and Gibco Amphotericin B to prevent cell culture contamination.
7. Miscellaneous:
   (i) 24-well coverslips.

Cell Lines

1. aBDTO.
2. SH-SY5Y human neuroblastoma cells.
3. Gibco DMEM (Dulbecco's modified Eagle's medium), high glucose from Thermo Fischer Scientific.
4. Gibco Fetal bovine serum from Thermo Fischer Scientific.
5. Gibco Penicillin-Streptomycin (10,000 U/mL) from Thermo Fischer Scientific. This solution contains 10,000 units/mL of penicillin and 10,000 μg/mL of streptomycin.

Lactate Dehydrogenase (LDH) Assay

1. aBDTO.
2. SH-SY5Y human neuroblastoma cells.
3. Roche Cytotoxicity Detection kit (LDS) from Millipore Sigma.
4. 96-well plates for tissue culture.
5. 96-well plate reader.
6. Miscellaneous:
   (i) Multichannel pipettes.
   (ii) Incubator (37°C, 5% CO_2, 90% humidity).

MTT (3-(4, 5-Dimethylthiazolyl-2)-2, 5-Diphenyltetrazolium Bromide) Assay

1. aBDTO.
2. SH-SY5Y human neuroblastoma cells.
3. Promega CellTiter 96AQ$_{ueous}$ Non-Radioactive Cell Proliferation Assay.
4. 96 well plates for tissue culture.
5. 96 well plate reader.
6. Miscellaneous:
   (i) Multichannel pipettes.
   (ii) Incubator (37°C, 5% CO_2, 90% humidity).

Immunocyto-Chemistry of Dendritic Spine

1. aBDTO.
2. Mouse primary neuronal cultures.
3. Antibodies:
   (i) Abcam anti-PSD95-synaptic marker.
   (ii) Abcam anti-beta III tubulin antibody-neuronal marker.
4. Miscellaneous:
   (i) 24-well coverslips.

Long-Term Potentiation (LTP) Measurement

1. Fresh mouse brain slices (375 μm thick).
2. Precisionary Instruments Compresstome VF-300.
3. Reagents:
   (i) aCSF (artificial cerebrospinal fluid) with the composition, 24 mM NaCl, 2.5 mM KCl, 1.2 mM NaH_2PO_4, 24 mM $NaHCO_3$, 5 mM HEPES, 12.5 mM glucose, 2 mM $MgSO_4{\cdot}7H_20$ and 2 mM $CaCl_2{\cdot}2H_20$.
   (ii) NMDG-aCSF (*N*-methyl-D-gluconate-aCSF) with the composition 92 mM NMDG, 2.5 mM KCl, 1.25 mM NaH_2PO_4, 30 mM $NaHCO_3$, 20 mM HEPES, 25 mM glucose, 2 mM thiourea, 5 mM Na-ascorbate, 3 mM Na-pyruvate, 0.5 mM $CaCl_2.4H_2O$ and 10 mM $MgSO4.7H_2O$.
   (iii) HEPES holding aCSF with the composition, 92 mM NaCl, 2.5 mM KCl, 1.2 NaH_2PO_4, 30 mM $NaHCO_3$, 20 mM HEPES, 25 mM glucose, 5 mM sodium

ascorbate, 2 mM thiourea, 3 mM sodium pyruvate, 2 mM $MgSO_4.7H_20$, $2CaCl_2.2H_20$, 12 *N*-Acetyl L-Cysteine.

Propagation in Transgenic Mice

1. aBDTO
2. hTau mice
3. C57BL/6 J mice (as non-transgenic control)
4. Stereotaxic apparatus.
5. Saline

3 Methods

3.1 TauM Preparation

1. To start the primary culture, take 20 mL LB broth and add kanamycin (50 µg/mL per liter). Inoculate with a glycerol stock of *E. coli* BL21 DE3 harboring pET28a (+)/MAPT by taking a small scoop (1%) using a micropipette tip. Incubate at 37°C overnight.
2. To start the secondary culture, add kanamycin to LB broth so that the final concentration of antibiotic in the media is 50 µg /ml. Pour the primary culture (20 mL) into 1 L LB broth. Incubate it at 37°C while shaking at 250 rpm. Check the OD (at 600 nm) at intervals of 30 min till the OD (at 600 nm) reaches 0.6–0.8. Once the desired OD is reached, induce protein expression by adding IPTG (1 mM). Further incubate for 4 h at 25°C with shaking at 250 rpm. Harvest the cells by centrifuging the culture at 5000 rpm for 10 min at 4°C.
3. Suspend the bacterial cell pellet in 5–10× volumes of lysis buffer. Add protease inhibitor tablet (1 tablet per 10 mL lysis buffer) and sonicate the cells.
4. Centrifuge the lysed cells at 10,000 rpm for 10 min at 4°C.
5. Collect the supernatant and put it in boiling water bath for 30 min. Again, centrifuge it at 10,000 rpm for 10 min and 4°C.
6. Nickel-NTA column chromatography and size exclusion chromatography were used to get pure TauM.
7. Purity was checked by SDS-PAGE (Fig. 1) for unboiled and boiled TauM. The TauM was quantified using BCA protein assay kit. Concentration of 1 mg/mL was set and stored at –80°C (*see* **Note 4**).

3.2 Characterization of TauM

SDS-PAGE

1. Take 15 µL (10 µg) of protein sample.
2. Add 5 µL of loading dye to the protein sample.

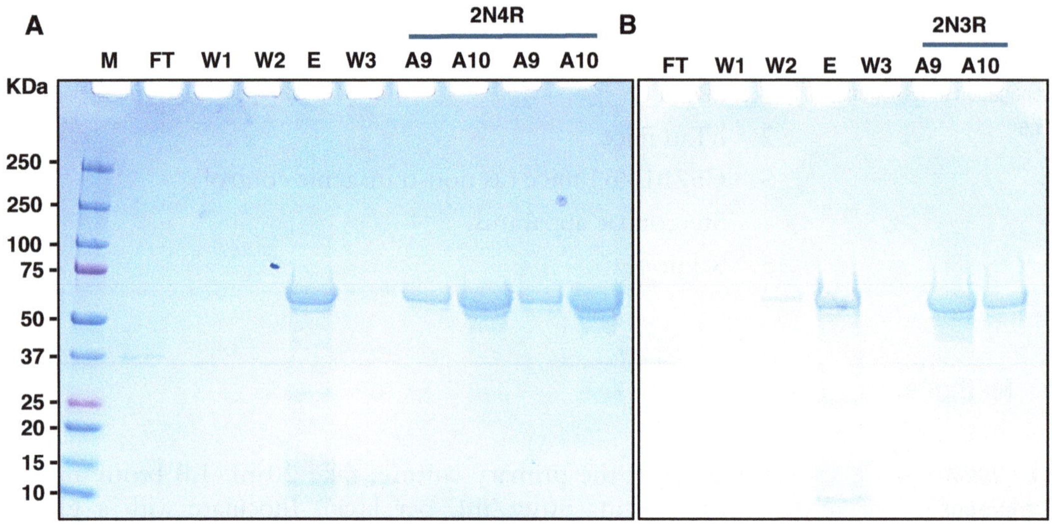

Fig. 1 SDS PAGE of recombinant tau monomer purification steps. (**A**) M, Protein marker; FT, flow through; W1, wash with 20 mM Tris, 500 mM NaCl and pH 8.0; W2, wash with 20 mM Tris, 500 mM NaCl, 10 mM imidazole and pH 8.0; E, eluate of tau protein from Ni-NTA beads (elution buffer-20 mM Tris, 500 mM NaCl, 100 mM imidazole and pH 8.0); (**B**) A9 and A10, fraction of tau after size exclusion chromatography

3. Mix and boil the sample at 95°C for 5–10 min. Bring it to room temperature.
4. Centrifuge at 14,000–16,000 × *g* for 5–15 min, to remove any particulate matter.
5. Load the sample into the gel.
6. Run the gel between 100–200 V and 20–50 mA for sufficient time till dye front reaches three fourth length of the gel.
7. After the run, remove the gel and place it in a container. Pierce Silver stain kit (Thermo Scientific) was used for the visualization of protein bands on SDS PAGE gel.

WB

This is done to confirm the presence of TauM present in the sample.

1. Thaw the Precision Plus Protein Dual Color Standard and samples on ice. Bring 4× loading buffer to RT.
2. Dilute samples in 4× loading buffer in the ratio 1:4 (*see* **Note 5**).
3. Set up the gel electrophoresis unit with the pre-cast gel and the 1× SDS-MES running buffer. Load marker and samples in the gel.
4. Run gel electrophoresis initially for 10 min at 80 V. Later increase the current to 120 V for 1 h or till the dye reaches the bottom of the gel.

5. Cut a piece of nitrocellulose membrane to gel size, take transfer pads and filter paper. Soak the three in 1× transfer buffer.
6. Then make a sandwich of all three with gel in the transfer cassette in order from black to red: transfer pad, filter paper, gel, nitrocellulose membrane, filter, transfer pad.
7. Run the transfer reaction at 97 V for 1 h at 4°C.
8. Following the transfer reaction, block the nitrocellulose membrane with 10% milk solution at RT for 2 h (*see* **Note 6**).
9. Use 1× TBST to wash the membrane.
10. Add the primary antibodies solution in 5% milk to the membrane and leave it at RT for 1 h.
11. Wash the membrane with 1X TBST.
12. Add the secondary antibodies solution in 5% milk, to the membrane and leave it at RT for 1 h.
13. Wash the membrane with 1× TBST.
14. Treat the membrane with the detection substrate and develop the membrane on X-ray autoradiography film (Fig. 2A).

HPLC

1. Load 10–15 μg of protein on to Superdex 200 10/300 column.
2. Set flow rate was 0.75 ml/min.

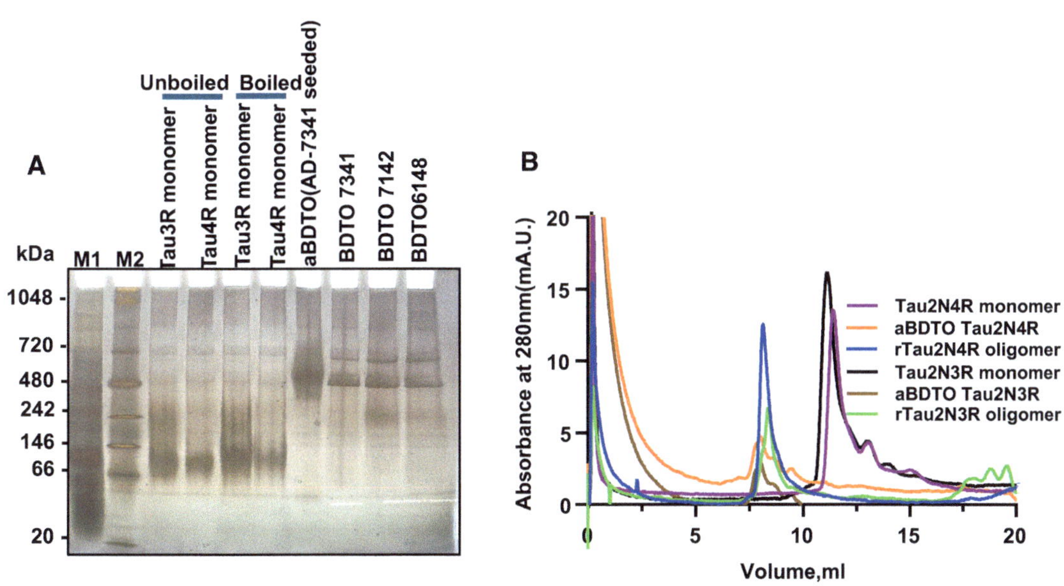

Fig. 2 Characterization of tau monomers and oligomers. (**A**) Shows Native-PAGE silver-stained gel of recombinant tau monomer, brain derived tau oligomers (BDTO), and amplified brain derived tau oligomer (aBDTO). (**B**) Size exclusion chromatography profile of tau monomer and oligomers: 15–20 μg of protein was loaded onto Superdex 200 10/300 column, flow rate was 0.75 mL/min). A couple of peaks were seen around 8–10 mL (500–600 kDa) when tau oligomer (TauO) was injected, and peaks were seen around 10–12 mL (50–60 kDa) when tau monomer (TauM) sample was injected

3. A couple of peaks were seen around 8–10 mL (500–600 kDa) when tau oligomer (TauO) sample was injected, and peaks were seen around 10–12 mL (50–60 kDa) when tau monomer (TauM) sample was injected (Fig. 2B).

Dot-Blot Assay

This technique allows for quick and easy semi-quantitative estimation of TauM for large sample size.

1. Take 0.45 μm nitrocellulose membrane. Draw a grid of 1 cm^2 boxes on it for desired number of samples.
2. On the individual squares pipette the sample.
3. Leave the membranes to dry for 30 min.
4. Leave the membrane in 10% milk solution for 1 h on slow rocker. This will block the strips.
5. Use 1× TBST to wash the membrane.
6. Dilute the primary antibodies in 5% milk solution. Add the primary antibody to the membrane and leave it on the rocker for 1–2 h at RT.
7. Use 1× TBST to wash the membrane.
8. Dilute the secondary antibodies in ratio of 1:3000 in 5% milk solution. Add the secondary antibodies to the membrane and leave it for 1 h at RT.
9. Use 1× TBST to wash the membrane.
10. Pipette the detection reagent (ECL solutions from Pierce) over the blot till it is evenly covered. Transfer it to a cassette and develop on X-ray autoradiography films.

The schematic overviews of the methods used for preparation and characterization of TauO (rTauO and aBDTO) have been summarized in Fig. 3.

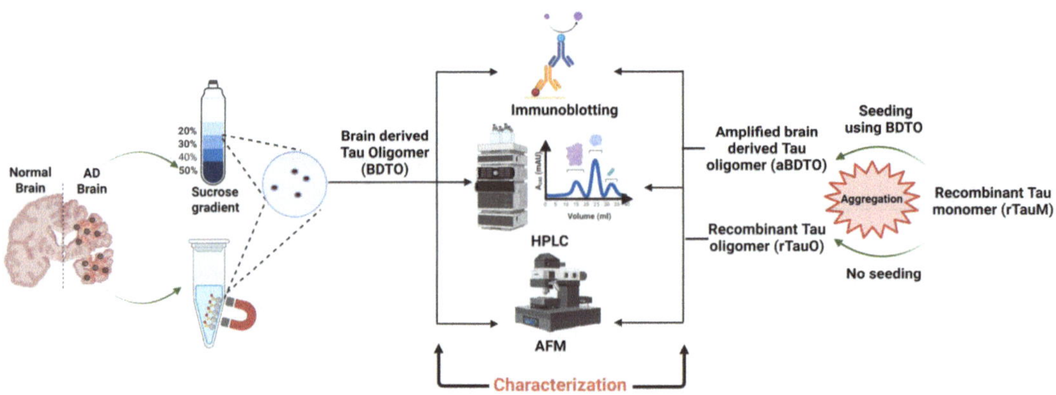

Fig. 3 Illustrative portrayal of sequential steps for preparation and characterization of tau oligomers (TauO)

DLS

1. Prepare 1 mg/mL sample (rTauM) in a microcentrifuge tube.
2. Pass through 0.22 micron syringe filtre or centrifuge to remove the dust particles.
3. 10–20 μL sample will be filled in clean and dried quartz Ultra-Micro cell.
4. Data were acquired in DLS (Zetasizer Malvern) using fixed angle of 90° at 20 ± 0.1 °C.
5. For each sample, three to five measurements were made for 20 s acquisition time.
6. Data was analyzed using software provided by Malvern Inc.

Mass Spectrometry [5]

1. Reconstitute the peptides in 20 μL 4% ACN/0.1% formic acid and separate them on online nanoflow Easy nLC1000 UHPLC system and analyse on Q Exactive Orbitrap mass spectrometer.
2. Inject 10 μL of sample into the capillary peptide trap column.
3. For the peptides to separate maintain a flow rate of 300 nL/min on the 25-cm UHPLC reversed-phase column.
4. Use a 2 h linear gradient from 2% solvent A (0.1% formic acid in H_2O) to 35% solvent B (0.1% formic acid in ACN) for each LC–MS/MS run.
5. Use Xcalibur 2.3 software in a positive ion mode at a spray voltage of 2.1 kV to perform data-dependent acquisition.
6. Use MaxQuant software (version 1.5.2.8) to analyze the data.

3.3 Generation of TauO

Production of rTauO

1. Take 0.6 mg/mL of 500 μL TauM in 1.5 mL micro centrifuge tube.
2. Incubate it for 72 h at 37 °C with continuous shaking at 300 rpm.
3. Characterize the oligomers formed (see Sect. 3.4).

Production of aBDTO

Homogenization of Brain Tissue and Extract Preparation

1. To 10 mL of 1× PBS add one protease inhibitor mini tablet and one phosphatase inhibitor cocktail. This is the homogenization buffer.
2. Add the frozen brain tissue to the homogenization tube (HT).
3. Add the homogenization buffer at 1/3 (w/v) ratio to the brain tissue in the HT.
4. Add steel beads (washed with 1× PBS) to each HT.
5. Use tissue lyser to homogenize the brain tissue (*see* **Note 7**).

6. Remove the steel beads from HT.
7. Centrifuge the HT at 10,000 rpm for 10 min at 4°C.
8. Remove the supernatant from HT and aliquot (*see* **Note 8**).

The supernatant obtained is further processed by two methods to obtain the BDTO.

Using Co-Immunoprecipitation Kit for BDTO Isolation

1. Prepare 1× coupling buffer from 20× coupling buffer provided in the kit using 1× PBS.
2. Pipette 100 μL of AmionoLink Plus Coupling Resin (provided in the kit) into the spin column. Centrifuge at 1000 rcf for 1 min.
3. Wash the column with 1× coupling buffer and couple the resin with purified T22 and T18 antibody.
4. Add sodium cyanoborohydride to the column and incubate at room temperature (RT) for 2 h.
5. Remove any unbound antibody from the column by washing it with 1× coupling buffer. Add quenching buffer provided in the kit.
6. Wash the column with wash solution followed by washing with phosphate buffer.
7. Add the supernatant of the brain homogenate in the column and put on rotation for overnight at 4°C.
8. Wash the column with wash lysis buffer next day.
9. Add 50 μL of 0.1 M glycine (pH 2.8) to the column. Centrifuge at 1000 rcf for 1 min.
10. Collect the fraction (the flow through) and screen for BDTO with WB (T22 and Tau13)

Using Sucrose Step Gradient to Isolate BDTO

1. Take the supernatant of the brain homogenate. Add 10% Sarkosyl to make the final concentration of the Sarkosyl in the sample to 1%.
2. Put the sample on an orbital shaker for 60 min at RT.
3. From a step gradient of sucrose in a clear Beckman tube. This is done by layering 0.8 mL of each concentration of sucrose (50%, 40%, 30%, 20%, 10%) in a Beckman tube.
   (i) Make use of 18 G 1 ½ needles to draw up sucrose.
   (ii) Make use of 30 G needles to release the sucrose, creating the gradient
4. Add 400 μL of supernatant of the brain homogenate to the top of the sucrose gradient formed.
5. Centrifuge the tube for 16 h at 107,000 × *g* at 4°C.

6. Five distinct layers will be formed on centrifuging the sample. Take 200 µL pipette to draw the sample from each of the five layers.
7. Aliquot each layer into separate tubes and screen for BDTO using WB (T22 and tau13)

Generation of aBDTO

1. Prepare 0.6 mg/mL of TauM in 1× PBS.
2. BDTO at 1:100 ratio of BDTO: TauM are added to the above. Use gentle pipetting to mix the proteins.
3. Keep the sample for 48 h at 37°C with continuous rotation.
4. Characterize the aBDTO (Fig. 2A) (see Sect. 3.4).

3.4 Characterization of TauO

Characterization of TauO is essential to differentiate between different strains (conformational variants) [5, 47, 56, 68]. Figure 4 is the summary of methods used for characterization of BDTO. The methods have been classified as biochemical, biophysical and bioactivity measurements based on the techniques involved.

Biochemical Analysis of TauO

Indirect ELISA

1. Take the 96-well ELISA plate. Into each well, pipette 50 µL of 0.05 M sodium bicarbonate (pH 9.6) (*see* **Note 9**).
2. Add BDTO to each well and ensure that final concentration is 0.5–1 µg BDTO per well
3. Keep for overnight incubation at 4°C.
4. Discard the ELISA plate contents next day. Wash it with 1× TBST.

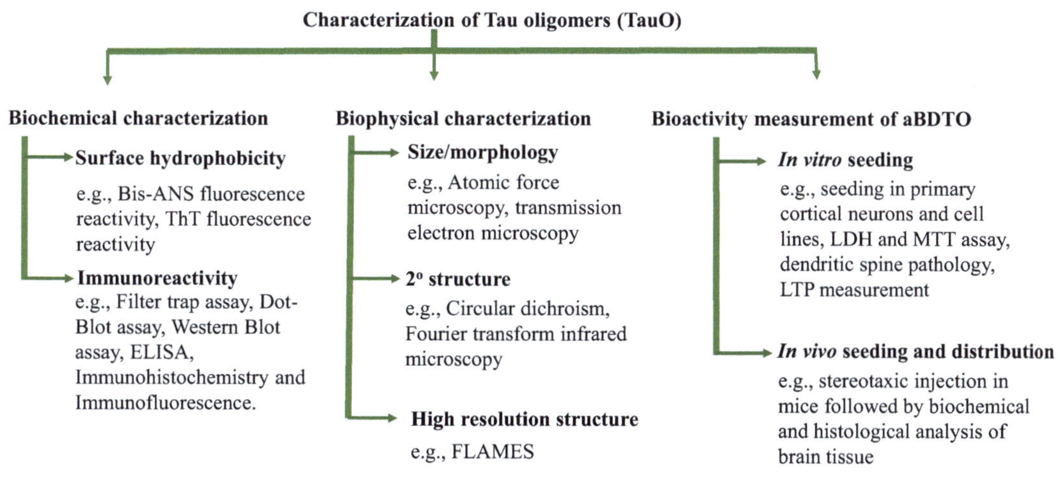

Fig. 4 Summary of methods used for characterization of tau oligomers (TauO)

5. Take 10% milk solution and pipette 150 μL into each well. Leave it at RT for 2 h.
6. Discard the contents of the ELISA plate and wash it with 1× TBST.
7. Dilute the primary antibody in 5% milk solution and pipette 100 μL into each well. Leave it for 1 h at RT.
8. Discard the contents of the ELISA plate and wash it with 1× TBST.
9. Dilute the secondary antibody in 5% milk solution in the ratio 1:10,000 and pipette 100 μL into each well. Leave it for 1 h at RT.
10. Discard the contents of the ELISA plate and wash it with 1× TBST.
11. Bring TMB to room temperature and add 100 μL into each well. Keep the plate in dark till blue color develops.
12. Once the blue color develops, pipette 100 μL of 2 M HCl in each well. The blue color turns to yellow indicating the reaction has stopped. Read it at 450 nm in the plate reader.

Sandwich ELISA

1. Take the 96-well plate. Coat the wells with T22 (anti-rabbit) diluted in 0.05 M sodium bicarbonate (pH 9.6) in 1:250 ratio.
2. Incubate the plate overnight at 4°C. Wash it with 1× TBST the next day.
3. To block the plate, add 10% non-fat dry milk solution into each well and leave it for 2 h at RT.
4. Discard the plate contents and wash it with 1× TBST.
5. Pipette 2 μg of rTauO or BDTO diluted in 1× PBS to each well and incubate it at 37°C for 90 min.
6. Discard the plate contents, wash it with 1× TBST.
7. Pipette Tau5 (anti-mouse) diluted in 1× PBS in the ratio of 1:1000. Leave for 1 h at RT.
8. Wash with 1× TBST and add an anti-mouse secondary IgG diluted in 1× PBS in the ratio 1:10,000, to each well and leave it for 1 h at RT.
9. Wash with 1× TBST. Bring TMB to RT and add 100 μL to each well and leave it till blue color develops.
10. Once the blue color develops, pipette 100 μL of 2 M HCl in each well. The blue color turns to yellow indicating the reaction has stopped.
11. Read it at 450 nm in the plate reader.

Dot-Blot Assay

This technique allows for quick and easy semi-quantitative estimation of TauO for large sample size. See Section: Dot-blot assay for the procedure and follow the same steps.

WB

This is done to identify the different strains of aBDTO present in the sample. See Section: WB for the procedure and follow the same steps.

FTA

This method allows for rapid detection of amyloid aggregates. Previous studies have successfully used this assay for characterization of tau oligomers [47, 65].

1. Take a piece of nitrocellulose membrane to fit in the slot blot chamber moisten it with 1× TBST.
2. Place the sample template on the top of the membrane and put between the two block and tighten the screw.
3. Connect the apparatus to a vacuum source. Gently remove the buffer from the well by vacuum.
4. Apply 1 µg of rTauO or aBDTO on the membrane in the apparatus.
5. Block the membrane with 10% milk solution after removing it from the apparatus. Leave it overnight at 4°C.
6. Following blocking, add the primary antibodies solution in milk to the membrane and leave it for 1 h at RT. Wash with 1× TBST.
7. Add the secondary antibodies solution in milk to the membrane and leave it for 1 h at RT. Wash with 1× TBST.
8. Apply the ECL detection solution from Pierce to the membrane and develop the membrane on X-ray autoradiography films.

PK Digestion

This assay is based on the resistance of amyloid aggregates to proteolysis [69]. The tau different conformations of tau will be differentially digested by the PK enzyme. Hence, this method can be used for the detection and characterization if different forms/strains of tau oligomers [47, 70].

1. Prepare a solution of 1 M Tris HCl and 50 mM NaCl in molecular grade water. Add rTauO or aBDTO to it and mix.
2. Add to it proteinase K enzyme (ensure that the concentration of the enzyme is 1–2.5 µg/mL in the sample).
3. Keep the samples for incubation for 1 h at 37°C.
4. Add 1× LDS buffer to the samples to stop the reaction.
5. Perform the WB with (see Section: WB and follow the same steps).

Immuno-Histochemistry in Frozen Tissue

With this method, we can observe the localization of tau oligomers in tissue or cell samples.

1. To fix brain sections keep them in chilled methanol at −20°C for 20 min.
2. Wash them in 1× PBS.
3. Leave them in 0.3% H_2O_2 for 15 min (*see* **Note 10**).
4. Wash them in 1× PBS.
5. Add 5% goat serum prepared in 1× PBS to the brain sections and leave it for 60 min. This is done to block non-specific sites.
6. Add primary antibodies diluted in antibody diluent to the brain sections. Leave them at 4°C in the humidity chamber overnight.
7. Wash them in 1× PBS.
8. Prepare the biotinylated goat secondary antibody, anti-mouse IgG (Vectastain ABC Kit) or biotinylated goat anti-rabbit IgG (Vectastin ABC Kit) by adding one drop (50 μL) in 10 mL of antibody diluent. Then leave the brain sections in this solution for 1 h at RT.
9. Wash them in 1× PBS. Prepare the ABC solution by mixing two drops (100 μL) of Reagent A and two drops (100 μL) of Reagent B to 10 mL of antibody diluent. Leave the brain sections with ABC solution for 1 h at RT.
10. Wash them in 1× PBS. Prepare DAB and apply the mixture to brain sections and wait until they turn brown.
11. Wash in 1× PBS.
12. Leave the brain section in hematoxylin solution for 1 min at RT.
13. Dip in lukewarm water for 5 min.
14. Perform dehydration of brain sections in different concentrations of ethanol and xylene in the order: 1 min in 80% ethanol, 2 min in 80% ethanol, 1 min in 95% ethanol, 2 min in 95% ethanol, 2 min in 100% ethanol, 3 min in 100% ethanol, 3 min in xylene, and 4 min in xylene.
15. Add mounting medium (Cytoseal 60). Put the coverslip over the sections (*see* **Note 11**).
16. Allow to dry and seal with nail polish.

Immuno-Fluorescence in Frozen Tissue

1. To fix brain sections keep them in chilled methanol at −20°C for 20 min. Wash in 1× PBS.
2. Wash in 70% ethanol. Circle with the brain section with PAP pen.

3. Apply 1× TrueBlack Lipofuscin Autofluorescence Quencher and keep it for 10 min.
4. Wash in 70% ethanol until the ethanol turns clear.
5. Wash in 1× PBS.
6. Circle the section using PAP pen.
7. Block the section in 5% goat serum diluted in 1× PBS. Leave it for 60 min.
8. Add the primary antibodies diluted in antibody diluent to the brain sections. Leave it overnight in the humidity chamber at 4°C.
9. Wash it with 1× PBS.
10. Add the secondary antibodies diluted in antibody diluent. Leave it in the humidity chamber for 1 h at RT.
11. Wash in 1× PBS.
12. Mount the coverslips using Invitrogen ProLong Gold Antifade Mountant with DAPI (15–20 μL) and seal it with clear nail polish.

Immuno-Histochemistry in Paraffin-Embedded Tissue

1. Heat the brain sections for 2 h and 30 min at 60°C.
2. Hydrate the brain tissue with xylene and different concentrations of ethanol in the order: Xylene (three times for 5 min), 100% ethanol (three times for 3 min), 95% ethanol (three times for 3 min), 80% ethanol (three times for 3 min), distilled water (30 s).
3. Put the 10% target retrieval solution diluted in sterile water on the brain sections. Microwave at power two for 1 min. Allow the sections to cool down for 1 min.
4. Wash with 1× PBS.
5. Add 3% H_2O_2 to the sections and leave it for 15 min at RT.
6. Wash with 1× PBS Block the section in 5% goat serum diluted in 1× PBS. Leave it for 60 min.
7. Add the primary antibodies diluted in antibody diluent to the brain sections. Leave it overnight in the humidity chamber at 4°C.
8. Wash it with 1× PBS Prepare the biotinylated goat secondary antibody, anti-mouse IgG (Vectastain ABC Kit) or biotinylated goat anti-rabbit IgG (Vectastin ABC Kit) by adding one drop (50 μL) in 10 mL of antibody diluent. Then leave the brain sections in this solution for 1 h at RT. Wash it with 1× PBS Prepare the ABC solution by mixing two drops (100 μL) of Reagent A and two drops (100 μL) of Reagent B to 10 mL of antibody diluent. Leave the brain sections with ABC solution for 1 h at RT. Wash it with 1× PBS Prepare DAB and

apply the mixture to brain sections and wait for them to turn brown. Wash it with 1× PBS Leave the brain section in hematoxylin solution for 1 min at RT.

9. Dip in lukewarm water for 5 min.
10. Perform dehydration of brain sections in different concentrations of ethanol and xylene in the order: 1 min in 80% ethanol, 2 min in 80% ethanol, 1 min in 95% ethanol, 2 min in 95% ethanol, 2 min in 100% ethanol, 3 min in 100% ethanol, 3 min in xylene, and 4 min in xylene.
11. Add mounting medium (Cytoseal 60). Put the coverslip over the sections.
12. Allow to dry and seal with nail polish.

Immuno-Fluorescence in Paraffin-Embedded Tissue

1. Heat the brain sections for 2 h and 30 min at 60°C.
2. Hydrate the brain tissue with xylene and different concentrations of ethanol in the order: Xylene (three times for 5 min), 100% ethanol (three times for 3 min), 95% ethanol (three times for 3 min), 80% ethanol (three times for 3 min), distilled water (30 s).
3. Put the 10% target retrieval solution diluted in water on the brain sections. Microwave at power two for 1 min. Allow the sections to cool down for 1 min.
4. Wash with 1× PBS Add 3% H_2O_2 to the sections and leave it for 15 min at RT. Wash with 1× PBS Leave the sections with 70% ethanol for 5 min.
5. Circle the sections with PAP pen. Apply 1× TrueBlack Lipofuscin Autofluorescence Quencher and keep it for 10 min. Wash in 70% ethanol until the ethanol turns clear.
6. Wash in 1× PBS.
7. Circle the section using PAP pen.
8. Block the section in 5% goat serum diluted in 1× PBS. Leave it for 60 min.
9. Add the primary antibodies diluted in antibody diluent to the brain sections. Leave it overnight in the humidity chamber at 4°C.
10. Wash it with 1× PBS.
11. Add the secondary antibodies diluted in antibody diluent. Leave it in the humidity chamber for 1 h at RT.
12. Wash in 1× PBS.
13. Mount the coverslips using Invitrogen ProLong Gold Antifade Mountant with DAPI (15–20 μL) and seal it with clear nail polish.

Biophysical Analysis of TauO

Bis-ANS Fluorescence Assay

This is based on the differences in the surface hydrophobicity of oligomers of tau and other proteins [47, 70].

1. Take a 96-well plate. Pipette 198 μL of bis-ANS solution into each well.
2. Pipette 2 μL of rTauO and BDTO to be tested, into each well.
3. Add 2 μL of glycine-NaOH buffer to the wells (*see* **Note 12**).
4. Cover the plate with aluminum foil and leave at RT for 10 min.
5. Read the plate at λ-emission 520 nm upon λ-excitation 380 nm.
6. Subtract the emission spectrum of bis-ANS alone.

ThT Fluorescence Assay

This method specifically detects and identifies fibrils formed in the sample [47, 70].

1. Take a 96-well plate. Pipette 198 μL of ThT solution in each well.
2. Pipette 2 μL of rTauO or BDTO to be tested into each well. Add 2 μL of glycine-NaOH buffer to the wells (*see* **Note 12**).
3. Cover the plate with aluminum foil and leave at RT for 10 min. Read the plate at λ-emission 490 nm upon λ-excitation 440 nm.
4. Subtract the emission spectrum of ThT alone.

Circular Dichroism Spectroscopy

The secondary structure of tau oligomers can be determined by circular dichroism spectroscopy [4, 70].

1. Rinse the cuvette with 70% ethanol, water and acetone, respectively. Leave it to dry for some time.
2. Prepare TauO in 1× PBS to a final concentration of 0.1 μg/μL. From this pipette 200 μL to the cuvette.
3. Measure the spectra at 0.20 nm intervals with spectral bandwidth of 1 nm, speed of 20 nm/min and response time of 8 s.
4. Measure the spectra from 195 to 250 nm with an average of three iterations.
5. Measure the spectra for 200 μL of buffer only (1× PBS) without any protein sample.
6. Wash the cuvette with 70% ethanol and water between every sample.
7. Deduce the secondary structure of protein from the spectra using K2D3 software [71].

AFM [68]

1. Prepare TauO samples in 1× PBS. The concentration of protein in the sample should be around 0.03–0.05 μg/mL.

2. Smooth out the surface of the mica disc using double-sided tape.
3. Add 3–5 μL of TauO sample on the freshly cleaved mica and leave it to dry for 1 h.
4. After the sample dries on the mica disc, pipette 200 μL of molecular grade water on the mica disc and rinse it by pipetting water over the entire disc.
5. Leave the mica disc to dry completely.
6. Scan the samples under AFM using ScanAssyst-Air (non-contact tapping method).

TEM [67]

1. Aliquot (2 μL) TauO. Add glutaraldehyde so that the final concentration remains 1% glutaraldehyde in the sample.
2. Take a strip of para Film and on it pipette the sample, the Milli-Q water (a drop) and 2% uranyl acetate (UA) (two drops), all apart from each other.
3. Place the carbon-coated copper grid on the surface of solution, ensuring the shiny side on the grid touching the solution, for 1 min.
4. Lift the copper grid and place its edge on the filter paper to remove excess solution. Place the shiny side of grid on the Milli-Q water drop. Repeat this process next with the first UA drop, then the second UA drop.
5. Then wash it by placing the shiny side on the water drop.
6. Use a transmission electron microscope to take images.

DLS

It is used to get an estimate of the distribution of size of oligomers in the sample. For procedure see Section: Dynamic Light Scattering (DLS) and follow the same steps for TauO.

Mass Spectrometry

It is used for characterizing oligomers, enabling the determination of their molecular weight, identification of oligomerization states, and even sequencing of unknown oligomers. See Section: Mass Spectrometry for procedure.

HPLC

It helps in identification of different TauO. See Section: High-performance liquid chromatography (HPLC) for the procedure.

FTIR

FTIR is another widely used method to detect secondary structures of protein and can be used for detecting secondary structures of tau oligomers [70].

1. Take alcohol wipes to clean the two zinc selenium windows.
2. Put them in a holder and place it NICOLET 6700 FT-IR machine. This is done to record the background.

3. Pipette 10 μL distilled water in the zinc selenium windows and record the spectrum at RT.
4. Pipette 10 μL of rTauO or aBDTO in zinc selenium windows and record the spectrum.
5. Subtract the background spectra.
6. Normalize the spectra for background spectrum of distilled water.
7. Plot normalized spectra focusing on amide I region from 1600 to 1700 cm^{-1}.

Fluorescent Amyloid Multi-Emission Spectra (FLAMES) Microscopy

This method is used to detect the conformationally variable strains or forms of amyloid aggregates [5, 63, 64].

1. Prepare a solution of 0.5% low melting point agarose in double-distilled water. Pipette mix the agarose with rTauO/aBDTO and control samples, separately.
2. Add this protein-gel mixture to the 96-well microplate ensuring the final protein concentration is 2 μg in each well.
3. After the gel solidifies, wash it with water.
4. Add the fluorescent dyes (50 μM ThT, 75 μM bis-ANS, 10 μ M FSB, 20 μM curcumin) to the wells and leave it overnight at RT.
5. Wash with water the next day.
6. Image to collect emission spectra using confocal microscope.

Bioactivity Measurement of aBDTO

Primary Cortical Neurons

1. Take the mouse primary cortical neurons and plate them on poly-d-lysine-coated glass coverslips (Corning, Inc.) in 24-well plate. Ensure the density is 80,000 cells/well (*see* **Note 13**).
2. Replace 50% of the media every three days.
3. Let the cells grow in vitro (DIV) for 10 days before performing the experiments.
4. Add 0.1–0.5 μM rTauO/aBDTO to the cells and leave them for 3 h.
5. Change the whole media after 3 h and leave the cells to grow for 3 days.
6. Perform live cell imaging or immunocytochemistry.

Cell Lines

1. Take human neuroblastoma SH-SY5Y cells and plate them on poly-d-lysine-coated glass coverslips (Corning, Inc.) in 24-well plate. Ensure the density of 60,000 cells/well. Leave them overnight.

2. Perform transient transfection of the cells with fluorophore-tagged tau plasmid using Lipofectamine 2000 (Invitrogen).
3. Take the transfected cells and grow them for 24 h. Expose the transfected cells 1 day old culture to 0.1–0.5 μM tau oligomers for 3 h.
4. Change the whole media after 3 h and leave the cells to grow for 3 days. Perform live cell imaging or immunocytochemistry.

LDH Assay [47, 70, 72]

1. Take human neuroblastoma SH-SY5Y cells and plate them in 96-well plate. Ensure the density is 10,000 cells/well. Grow them overnight.
2. Add (1–2 μM) rTauO/BDTO to the culture and leave them to grow for 24 h.
3. Evaluate cytotoxicity by performing LDH assay using Cytotoxicity Detection kit PLUS.
4. Read the plate at 490 nm wavelength using with Polar Star Omega plate reader (BMG Labtech).

MTT Assay [47, 70, 73]

1. Take human neuroblastoma SH-SY5Y cells and plate them in 96-well plate. Ensure the density is 10,000 cells/ well. Grow them overnight. Next day, add (1–2 μM) rTauO/BDTO to the culture and leave them to grow for 24 h.
2. Perform MTT assay using the kit.
3. Read the plate at 490 nm wavelength using Polar Star Omega plate reader (BMG Labtech).

Dendritic Spine Pathology [67, 70]

1. Take the mouse primary cortical neurons and plate them on poly-d-lysine-coated glass coverslips (Corning, Inc.) in 24-well plate. Ensure that density is 80,000 cells/well (*see* **Note 13**).
2. Replace 50% of the media every 3 days.
3. Let the cells grow in vitro (DIV) for 10 days before performing the experiments.
4. Add 0.1–0.5 μM rTauO/aBDTO to the cells and leave them for 6 h.
5. Wash the cells with 1× PBS and add 4% formaldehyde solution and leave them for 15 min to fix them. Wash them with 1× PBS. Add 0.25 % Triton X-100 in PBS for 10 min at RT. Take 1:1000 dilution for the primary antibodies. For post-synaptic density protein PSD95 antibody (Abcam, Ab18258) and neuronal marker βIII-tubulin antibody (Abcam, ab78078).
6. Count the PSD95 puncta after each treatment. Make sure to choose five different areas of 20 μm dendritic shafts (without any branches).

7. Subtract the background from the intensity of mean PSD95 using ImageJ software. Perform One-way ANOVA followed by Tukey's post hoc analysis to evaluate dendritic spine results.

Immuno-Cytochemistry [34, 47, 67, 70, 74]

1. Wash the cells with 1× PBS. Add 4% formaldehyde solution to the cells. Leave it for 15 min at RT.
2. Wash the cells with 1× PBS. Add 0.25 % Triton X-100 dissolved in PBS. Leave it for 10 min at RT.
3. Wash with 1× PBS. Add 5% goat serum and leave it for 30 min at RT.
4. Add the primary antibodies and incubate overnight at 4°C overnight.
5. Wash the cells with 1× PBS. Add Alexa-conjugated species-specific secondary antibodies (1:1000, Life Technologies) and leave them for 1 h at RT.
6. Wash the cells with 1× PBS. Mount the coverslip with Prolong Gold antifade reagent with DAPI.

LTP Measurement [5, 10, 66]

LTP is a component of memory function and is impacted by both recombinant and brain-derived tau oligomers [10, 66, 75, 76]. Therefore, LTP measurement in fresh brain slices following tau oligomers incubation provides insight into their effect on synaptic functions.

1. Take the 3-month-old hTau mice. Anaesthetize them by transcardially perfusion of carbogenated (mixture of 95% O_2 and 5% CO_2) NMDG-aCSF (25–30 ml), at RT.
2. Make thick (375 μm) transverse sections of the brain tissue (having Schaffer collateral synapses). Use Compresstome VF-300.
3. Take the sections and add NMDG solution (32–34°C). Leave them for 10 min. After the completion of 10 min, put them into carbogen bubbling HEPES holding aCSF solution, at RT. Store the sections in this solution till the experiments reach completion.
4. Add the rTauO/ aBDTO to the brain sections stored in carbogen bubbling HEPES holding aCSF. Leave them for 1 h at RT.
5. Wash the brain sections in fresh aCSF solution. Following the wash, put the brain sections on the recording stage. Perfuse them with carbogen bubbling aCSF solution at a flowrate of 3 mL/min.
6. Electrically stimulate the Schaffer collateral pathway in a hippocampal slice preparation.
7. Measure the resulting electrical field potential changes in the CA1 region of the hippocampus, specifically the excitatory postsynaptic potentials (EPSPs).

8. Analyze the data as percentage change from the initial average baseline EPSP slope. EPSP slope is defined as the average slope obtained for 10 min before HFS.

Stereotaxic Injection of aBDTO and Brain Tissue Analysis

This allows for the estimation of the propagation of aBDTO and development of the pathology [10, 45, 67, 77].

1. Anesthetize transgenic and wild-type animals and place them on heating pad to maintain the body temperature
2. Give stereotaxic injection of 1 μL of aBDTO (1 μg/μL) into the hippocampus of 3-month-old transgenic mice and non-transgenic C57BL/6 control animals. (Note: from bregma; −2.5 mm anterior/posterior, ±2.0 mm lateral, −1.8 mm dorsoventral) and overlying cerebral cortex (from bregma; −2.5 anterior/posterior, ±2.0 mm lateral, −0.8 mm dorsoventral) at a rate of 1.25 μL/min. Alternatively, the injection can also be given on both the hemispheres and the spread of pathology evaluated by biochemical analysis.
3. Isolate and collect brain tissues from animals at 3, 6, and 9 months after injection.
4. Collect both hemispheres and store at −80°C for biochemical analyses.
5. Prepare the brain homogenate as described in the section. Follow the steps described in Section: Biochemical analysis of TauO
6. Perform analyses of the brain extracts as described in Sect. 3.4.

4 Notes

1. The molecular weight of 2N4R and 2N3R is 45.9 kDa and 42.6 kDa, respectively.
2. The conditioning buffer helps with the wash and especially to provide the correct environment to enhance the efficiency of the low pH Elution Buffer. The overall result is more antigen recovered in the first volume of eluate (more concentrated antigen).
3. Tau conformation specific antibodies developed by Kayed lab. These antibodies target the oligomeric form of tau (e.g., T22 and TOMA 2) or the misfolded toxic tau (e.g., TOMA 1, TOMA 3, TOMA 4, TTC M1, TTC M2, T18, TTC 35, and TTC 99). All these anti-tau antibodies have been tested for use in tauopathies and other protein misfolding diseases. Many of these antibodies have been commercialized, including T22, ABN454; TOMA, MABN819, Millipore). All the

antibodies developed by Kayed lab have been predicted to react with mouse, rat, human and drosophila. They have applications in western blot, filter trap assay, dot-blot assay, immunofluorescence, and immunohistochemistry. These antibodies and the other commercially available general tau antibodies help in identification, characterization, and assessment of the bioactivity of the tau oligomers (Fig. 5).

4. TauM can be stored for 6 months at −20°C.
5. The samples are mixed with PBS and loaded into the gel. The samples should not be heated to 95°C before loading into the gel.
6. It can also be left overnight at 4°C.
7. To prevent heating up, keep the sample on ice.
8. The aliquots can be snap-frozen and stored at −80°C till further use.
9. 1× PBS can also be used as an alternative to sodium bicarbonate.
10. This is done to quench endogenous peroxidases.
11. If you find bubbles after mounting, soak the slide in xylene overnight and remount again.
12. The wells containing the buffer serve as blanks for the proteins.

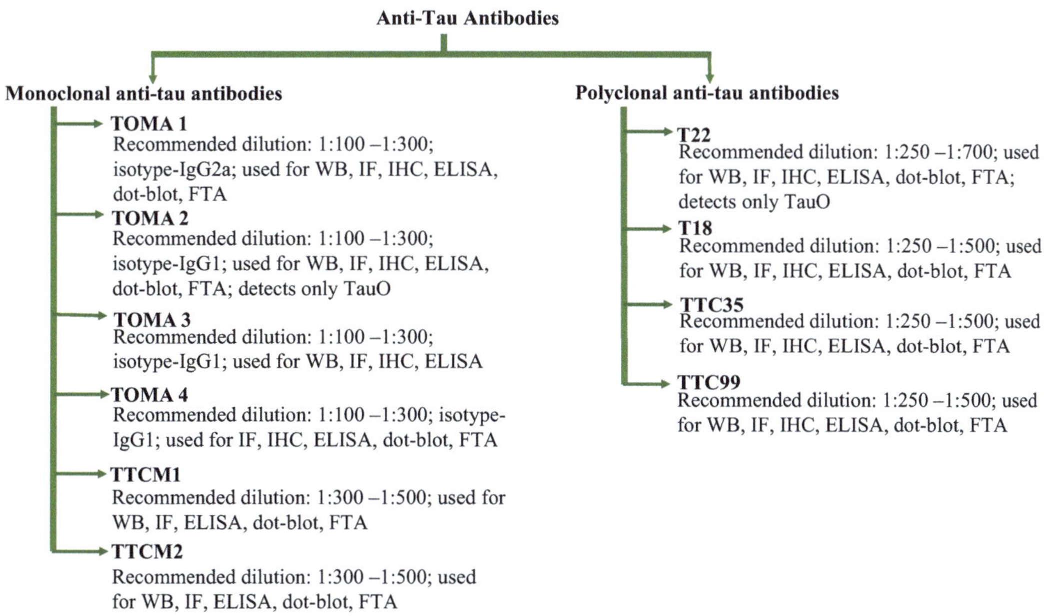

Fig. 5 Overview of anti-tau antibodies developed by Kayed lab targeting oligomeric/misfolded tau

13. Neurobasal medium supplemented with 2% B-27, 0.5 mM GlutaMax, 10,000 units/mL penicillin, 10,000 μg/mL streptomycin and 25 μg/mL amphotericin B supplement was used to plate the cells.

Acknowledgments

We thank lab members of Kayed lab for their contributions to the optimization of the protocols and useful suggestions. This work was supported by Mitchell Center for Neurodegenerative Diseases, and National Institute of Health grants: U24 AG072458, R01AG054025, RFA1AG055771 and R01AG060718.

Conflict of Interest The authors have no conflicts of interest to declare that are relevant to the content of this chapter.

References

1. Kovacs GG (2017) Tauopathies. Handb Clin Neurol 145:355–368
2. Gao YL et al (2018) Tau in neurodegenerative disease. Ann Transl Med 6(10):175
3. Daniel Estrella L et al (2025) Tau association with synaptic mitochondria coincides with energetic dysfunction and excitatory synapse loss in the P301S tauopathy mouse model. Neurobiol Aging 147:163–175
4. Bhopatkar AA et al (2024) MAPT mutations associated with familial tauopathies lead to formation of conformationally distinct oligomers that have cross-seeding ability. Protein Sci 33(9):e5099
5. Lo Cascio F et al (2025) Brain-derived tau oligomer polymorphs: distinct aggregations, stability profiles, and biological activities. Commun Biol 8(1):53
6. Zhang C et al (2025) Synergistic associations of amyloid-beta and phosphorylated tau with tau aggregation and cognitive decline in Alzheimer's disease. J Alzheimers Dis 13872877251322196
7. Abu-Rumeileh S et al (2025) Phosphorylated tau 181 and 217 are elevated in serum and muscle of patients with amyotrophic lateral sclerosis. Nat Commun 16(1):2019
8. Hsieh PF et al (2025) Plasma phosphorylated tau 217 as a discriminative biomarker for cerebral amyloid angiopathy. Eur J Neurol 32(2):e70066
9. Gerson JE et al (2014) Characterization of tau oligomeric seeds in progressive supranuclear palsy. Acta Neuropathol Commun 2:73
10. Lasagna-Reeves CA et al (2012) Alzheimer brain-derived tau oligomers propagate pathology from endogenous tau. Sci Rep 2: 700
11. Maeda S et al (2006) Increased levels of granular tau oligomers: an early sign of brain aging and Alzheimer's disease. Neurosci Res 54(3):197–201
12. Patterson KR et al (2011) Characterization of prefibrillar Tau oligomers *in vitro* and in Alzheimer disease. J Biol Chem 286(26):23063–23076
13. Cox K et al (2016) Analysis of isoform-specific tau aggregates suggests a common toxic mechanism involving similar pathological conformations and axonal transport inhibition. Neurobiol Aging 47:113–126
14. Takashima A (2013) Tauopathies and tau oligomers. J Alzheimers Dis 37(3):565–568
15. Pavlova A et al (2016) Protein structural and surface water rearrangement constitute major events in the earliest aggregation stages of tau. Proc Natl Acad Sci U S A 113(2):E127–E136
16. Mamun AA et al (2020) Toxic tau: structural origins of tau aggregation in Alzheimer's disease. Neural Regen Res 15(8):1417–1420
17. Ballatore C, Lee VM, Trojanowski JQ (2007) Tau-mediated neurodegeneration in Alzheimer's disease and related disorders. Nat Rev Neurosci 8(9):663–672
18. Gyparaki MT et al (2021) Tau forms oligomeric complexes on microtubules that are distinct from tau aggregates. Proc Natl Acad Sci U S A 118(19):e202146111

19. Mietelska-Porowska A et al (2014) Tau protein modifications and interactions: their role in function and dysfunction. Int J Mol Sci 15(3):4671–4713
20. Berger Z et al (2007) Accumulation of pathological tau species and memory loss in a conditional model of tauopathy. J Neurosci 27(14):3650–3662
21. Cowan CM, Quraishe S, Mudher A (2012) What is the pathological significance of tau oligomers?. Biochem Soc Trans 40(4):693–697
22. Kopeikina KJ et al (2011) Tau accumulation causes mitochondrial distribution deficits in neurons in a mouse model of tauopathy and in human Alzheimer's disease brain. Am J Pathol 179(4):2071–2082
23. Lasagna-Reeves CA et al (2011) Tau oligomers impair memory and induce synaptic and mitochondrial dysfunction in wild-type mice. Mol Neurodegener 6(1):39
24. Maeda S et al (2007) Granular tau oligomers as intermediates of tau filaments. Biochemistry 46(12):3856–3861
25. Sahara N et al (2013) Characteristics of TBS-extractable hyperphosphorylated Tau species: aggregation intermediates in rTg4510 mouse brain. J Alzheimers Dis 33(1):249–263
26. Spires TL et al (2006) Region-specific dissociation of neuronal loss and neurofibrillary pathology in a mouse model of tauopathy. Am J Pathol 168(5):1598–1607
27. Dujardin S et al (2020) Tau molecular diversity contributes to clinical heterogeneity in Alzheimer's disease. Nat Med 26(8):1256–1263
28. Niewiadomska G et al (2021) Tau oligomers neurotoxicity. Life (Basel) 11(1):28
29. Martinisi A et al (2021) Severe oligomeric tau toxicity can be reversed without long-term sequelae. Brain 144(3):963–974
30. Sengupta U, Kayed R (2022) Amyloid beta, tau, and alpha-synuclein aggregates in the pathogenesis, prognosis, and therapeutics for neurodegenerative diseases. Prog Neurobiol 214:102270
31. Ghag G et al (2018) Soluble tau aggregates, not large fibrils, are the toxic species that display seeding and cross-seeding behavior. Protein Sci 27(11):1901–1909
32. Lo CH, Sachs JN (2020) The role of wild-type tau in Alzheimer's disease and related tauopathies. J Life Sci (Westlake Village) 2(4):1–17
33. Nieznanska H et al (2021) Neurotoxicity of oligomers of phosphorylated tau protein carrying tauopathy-associated mutation is inhibited by prion protein. Biochim Biophys Acta Mol Basis Dis 1867(11):166209
34. Puangmalai N et al (2020) Internalization mechanisms of brain-derived tau oligomers from patients with Alzheimer's disease, progressive supranuclear palsy and dementia with Lewy bodies. Cell Death Dis 11(5):314
35. Alonso AdC et al (2001) Interaction of tau isoforms with Alzheimer's disease abnormally hyperphosphorylated tau and *in vitro* phosphorylation into the disease-like protein. J Biol Chem 276(41):37967–37973
36. Alonso AdC et al (2001) Hyperphosphorylation induces self-assembly of τ into tangles of paired helical filaments/straight filaments. Proc Natl Acad Sci U S A 98(12):6923–6928
37. Harada M, Isersky C, Cuatrecasas P, Page D, Bladen HA, Eanes ED, Keiser HR, Glenner GG (1971) Human amyloid protein: chemical variability and homogeneity. J Histochem Cytochem 19(1):1–15
38. King ME et al (1999) Ligand-dependent tau filament formation: implications for Alzheimer's disease progression†. Biochemistry 38(45):14851–14859
39. King ME et al (2000) Differential assembly of human tau isoforms in the presence of arachidonic acid. J Neurochem 74(4):1749–1757
40. von Bergen M et al (2001) Mutations of tau protein in frontotemporal dementia promote aggregation of paired helical filaments by enhancing local β-structure. J Biol Chem 276(51):48165–48174
41. Wille H et al (1992) Alzheimer-like paired helical filaments and antiparallel dimers formed from microtubule-associated protein tau *in vitro*. J Cell Biol 118(3):573–584
42. Wilson DM, Binder LI (1997) Free fatty acids stimulate the polymerization of tau and amyloid beta peptides. *In vitro* evidence for a common effector of pathogenesis in Alzheimer's disease. Am J Pathol 150(6):2181–2195
43. Combs B et al (2017) Production of recombinant tau oligomers *in vitro*. Methods Cell Biol 141:45–64
44. Gerson JE, Mudher A, Kayed R (2016) Potential mechanisms and implications for the formation of tau oligomeric strains. Crit Rev Biochem Mol Biol 51(6):482–496
45. Castillo-Carranza DL et al (2014) Specific targeting of tau oligomers in Htau mice prevents cognitive impairment and tau toxicity following injection with brain-derived tau oligomeric seeds. J Alzheimer's Dis 40(0):S97–S111
46. Lasagna-Reeves CA et al (2010) Preparation and characterization of neurotoxic tau oligomers. Biochemistry 49(47):10039–10041
47. Lo Cascio F et al (2020) Modulating disease-relevant tau oligomeric strains by small molecules. J Biol Chem 295(44):14807–14825
48. Sengupta U et al (2015) Pathological interface between oligomeric alpha-synuclein and tau in synucleinopathies. Biol Psychiatry 78(10):672–683

49. Goedert M et al (1996) Assembly of microtubule-associated protein tau into Alzheimer-like filaments induced by sulphated glycosaminoglycans. Nature 383(6600):550–553
50. Wilson DM, Binder LI (1997) Free fatty acids stimulate the polymerization of tau and amyloid beta peptides. *In vitro* evidence for a common effector of pathogenesis in Alzheimer's disease. Am J Pathol 150(6):2181–2195
51. Montgomery KM et al (2023) Chemical features of polyanions modulate tau aggregation and conformational states. J Am Chem Soc 145(7):3926–3936
52. Liu C et al (2024) PM2.5 triggers tau aggregation in a mouse model of tauopathy. JCI Insight 9(14):e176703
53. Zheng H et al (2024) The enigma of tau protein aggregation: mechanistic insights and future challenges. Int J Mol Sci 25(9):4969
54. Blair LJ et al (2013) Accelerated neurodegeneration through chaperone-mediated oligomerization of tau. J Clin Invest 123(10):4158–4169
55. Castillo-Carranza DL et al (2015) Tau immunotherapy modulates both pathological tau and upstream amyloid pathology in an Alzheimer's disease mouse model. J Neurosci 35(12):4857–4868
56. Castillo-Carranza DL et al (2014) Passive immunization with Tau oligomer monoclonal antibody reverses tauopathy phenotypes without affecting hyperphosphorylated neurofibrillary tangles. J Neurosci 34(12):4260–4272
57. Gerson J et al (2014) Characterization of tau oligomeric seeds in progressive supranuclear palsy. Acta Neuropathol Commun 2(1):73
58. Lasagna-Reeves CA et al (2012) Identification of oligomers at early stages of tau aggregation in Alzheimer's disease. FASEB J 26(5):1946–1959.
59. Vuono R et al (2015) The role of tau in the pathological process and clinical expression of Huntington's disease. Brain 138:1907–1918
60. Arar S, MA Haque, R Kayed (2025) Protein aggregation and neurodegenerative disease: structural outlook for the novel therapeutics. Proteins 93(8):1314–1329
61. Gaikwad S et al (2024) Nasal tau immunotherapy clears intracellular tau pathology and improves cognitive functions in aged tauopathy mice. Sci Transl Med 16(754):eadj5958
62. Montalbano M et al (2023) Pathological tau signatures and nuclear alterations in neurons, astrocytes and microglia in Alzheimer's disease, progressive supranuclear palsy, and dementia with Lewy bodies. Brain Pathol 33(1):e13112
63. Condello C et al (2018) Structural heterogeneity and intersubject variability of Abeta in familial and sporadic Alzheimer's disease. Proc Natl Acad Sci U S A 115(4):E782–E791
64. Strohaker T et al (2019) Structural heterogeneity of alpha-synuclein fibrils amplified from patient brain extracts. Nat Commun 10(1):5535
65. Montalbano M et al (2020) TDP-43 and tau oligomers in Alzheimer's disease, amyotrophic lateral sclerosis, and frontotemporal dementia. Neurobiol Dis 146:105130
66. Bittar A et al (2019) Neurotoxic tau oligomers after single versus repetitive mild traumatic brain injury. Brain Commun 1(1):fcz004
67. Castillo-Carranza DL et al (2018) Alpha-synuclein oligomers induce a unique toxic tau strain. Biol Psychiatry 84(7):499–508
68. Sengupta U, Carretero-Murillo M, Kayed R (2018) Preparation and characterization of tau oligomer strains. Methods Mol Biol 1779:113–146
69. Kraus A et al (2019) Seeding selectivity and ultrasensitive detection of tau aggregate conformers of Alzheimer disease. Acta Neuropathol 137(4):585–598
70. Sengupta U et al (2020) Polymorphic alpha-synuclein strains modified by dopamine and docosahexaenoic acid interact differentially with tau protein. Mol Neurobiol 57(6):2741–2765
71. Louis-Jeune C, Andrade-Navarro MA, Perez-Iratxeta C (2012) Prediction of protein secondary structure from circular dichroism using theoretically derived spectra. Proteins 80(2):374–381
72. Gaikwad S et al (2021) Tau oligomer induced HMGB1 release contributes to cellular senescence and neuropathology linked to Alzheimer's disease and frontotemporal dementia. Cell Rep 36(3):109419
73. Dasari AKR et al (2024) Characterization of alpha-synuclein oligomers formed in the presence of lipid vesicles. Biochem Biophys Rep 38:101687
74. Moreira GG et al (2021) Dynamic interactions and Ca(2+)-binding modulate the holdase-type chaperone activity of S100B preventing tau aggregation and seeding. Nat Commun 12(1):6292
75. Fa M et al (2016) Extracellular Tau oligomers produce an immediate impairment of LTP and memory. Sci Rep 6: 19393
76. Puzzo D et al (2017) LTP and memory impairment caused by extracellular abeta and tau oligomers is APP-dependent. Elife 6:e26991
77. Bittar A, Bhatt N, Kayed R (2020) Advances and considerations in AD tau-targeted immunotherapy. Neurobiol Dis 134:104707

Chapter 10

Capturing Tau and RNA Interactions

Sophie J. F. van der Spek and Benjamin L. Wolozin

Abstract

Tauopathies feature the intraneuronal aggregation of hyperphosphorylated tau, a protein that, despite not being a classical RNA-binding protein, interacts with RNA to promote aggregation and undergo liquid–liquid phase separation (LLPS). We present a streamlined in vitro method to study tau–RNA–driven LLPS under near-physiological conditions. The protocol includes isolation of RNA from mouse brain, fluorescent labeling of recombinant tau with minimal perturbation, and induction of LLPS using RNA scaffolds and a molecular crowding agent. This assay can be readily adapted and enables precise dissection of how tau–RNA interactions govern LLPS, offering insight into their contribution to tau pathology in neurodegenerative disease.

Keywords Tau, RNA, Liquid liquid phase separation, Biomolecular condensates

1 Introduction

Tauopathies are marked by the accumulation of intraneuronal fibrillary tangles containing hyperphosphorylated fibrillar Tau. Despite not being a classical RNA-binding protein (RBP), Tau has a well-documented ability to interact directly with RNA, which enhances Tau's tendency for aggregation [1]. More recent studies show that Tau, like RBPs, can undergo liquid–liquid phase separation (LLPS), which is also enhanced by the presence of RNA [2].

LLPS is a process by which molecules self-organize into membraneless organelles, which are essential for normal cellular functioning and RNA metabolism. For instance, stress granules are formed during stress for the temporal inhibition of RNA translation. Tau expression facilitates the formation of stress granules, where it interacts with RBPs such as T-cell intercellular antigen 1 (TIA1) [3]. In turn, coacervation with TIA1 not only promotes Tau's condensation but also greatly influences its aggregation pathway and toxicity: TIA1 promotes the formation of oligomeric tau species, which are more neurotoxic than fibrillar forms, both in cultured cells and mouse models [4].

Jessica S. Fortin (ed.), *Translational Research Methods for Tauopathies*, Neuromethods, vol. 233,
https://doi.org/10.1007/978-1-0716-5234-3_10,

The mechanism behind phase separation is grounded in the Flory–Huggins theory, which describes how temperature, pH, and the concentrations of RNA scaffolds, salt, and crowding agents generate energetically favorable molecular interactions [5]. Phase separation is triggered when the gain in enthalpy from molecular associations (like Tau-RNA interactions) overcomes the inherent loss of entropy accompanying the transition from dispersed molecules to condensed droplets [5]. Within neurons, physiological LLPS of Tau likely depends on Tau-RNA coacervation under crowded intracellular conditions, and this can be replicated in vitro to probe the underlying biophysics [4, 6, 7].

Previous work suggests aberrant tau phase separation—and its partnership with RBPs like TIA1—contributes to Tau pathology [4, 6, 7]. Understanding these molecular processes provides a clearer window into both normal cellular architecture and the development of neurodegenerative disease. This chapter outlines a standardized workflow for the preparation, execution, and quantitative analysis of Tau LLPS, which can be readily adapted to examine how proteins, RNA species, and defined physicochemical conditions influence Tau condensate formation.

2 Materials

2.1 Apparatus

1. 1.5 mL tube centrifuge
   i. Centrifuge 5417 R with a F45-30-11 rotor (Eppendorf).
2. Handheld homogenizer
   i. Pellet pestles (Sigma-Aldrich).
3. Glass slides
   i. Epredia™ Superfrost™ Plus Slides (Fisher Scientific), 25 × 75 × 1 mm.
4. Coverslips
   i. Fisherbrand™ Miscroscope Cover Glass (Fischer Scientific), 60 × 24 mm, #1.5.
5. Imaging spacers
   i. SecureSeal™ (Grace Bio Labs), 10-6.35 mm DIA × 0.12 mm Depth, 25 × 51 mm OD.
6. Inverted widefield microscope
   i. Axio Observer A1 (Zeiss).

2.2 Software

1. Fiji.
2. GraphPad Prism v.8 or higher.

2.3 Reagents

1. RNA isolation kit
   i. RNeasy plus mini kit (Qiagen) with adjusted protocol (see below); Trizol™ Reagent (Thermo Scientific), Chloroform (Sigma-Aldrich), and Ethanol (Sigma-Aldrich).
2. Fluorescent label
   i. DyLight® 488 Conjugation Kit (Fast) - Lightning-Link® (ab201799, Abcam).
3. Purified protein
   i. 0N4R Tau in 10 mM HEPES, pH 7.4 (>1 μg/μL).
4. Sodium chloride solution
   i. 5 M NaCl (Sigma Aldrich).
5. HEPES solution
   i. 1 M HEPES, pH 7.4 (Sigma Aldrich).
6. Crowding agent
   i. PEG-8000 (Polyethylene Glycol) (Bioworld).

2.4 Tissue

1. Mouse brain.

3 Methods

3.1 Isolate RNA

1. Homogenize mouse brain tissue in 1.5 mL tube with the motorized handheld pestle in 0.5 mL Trizol.
2. Add another 0.5 mL Trizol, vortex and incubate at RT for 5 min.
3. Add 200 μL Chloroform, shake until the samples have a pink milky color, and incubate at RT for 3 min.
4. Centrifuge samples at 12,000×*g*, 15 min, 4°C.
5. Transfer the upper aqueous layer to a DNA eliminator column and spin down 8000×*g*, 30 s, RT.
6. Add 1 volume of 70% Ethanol to a new tube, mix well.
7. Transfer 700 μL solution to an RNA column, spin down 8000×*g*, 30 s, RT and discard the flowthrough.
8. Add the remaining sample to the column and repeat the spin.
9. Wash column with 700 μL RW1 buffer, spin down 8000×*g*, 30 s, RT and discard the flowthrough.
10. Wash column with 500 μL PE buffer, spin down 8000×*g*, 30 s, RT and discard the flowthrough.
    - Make sure the appropriate volume of ethanol is added to the PE buffer before use.

11. Repeat Step 10 and transfer the column to a new 2 mL collection tube.
12. Dry spin for 2 min, 8000×*g*, RT and transfer column to a 1.5 mL clean collection tube.
13. Leave the cap open and air-dry samples for 4 min.
14. Add 30 μL nuclease- free water to the column and incubate for 10 min, RT.
15. Elute RNA by centrifugation for 2 min, full speed, RT.
16. Determine RNA concentration using a nanodrop, and nuclease-free water as a blank.
17. Adjust to 100 ng/μL end-concentration, aliquot and store at −80°C until further use.

3.2 Label Tau with Fluorophore

1. Take the correct volume of Tau for 200 μg and add 1 μL modifier/10 μL of Tau.
   - Keep the remainder of Tau to mix with labeled Tau just before the LLPS experiment (see Sect. 3.3.2. Prepare reaction mixes).
2. Transfer mix directly into the glass vial.
   - To prevent activation of the conjugate; do not pre-open the glass vial.
3. Mix quick and complete by pipetting and inverting and incubate at RT for 15 min in the dark.
4. Add 1 μL quencher/10 μL of mix and mix by pipetting.
5. Incubate at RT for 5 min in the dark.
6. Aliquot and store at −80°C until later use.

3.3 Perform In Vitro LLPS

Prepare Glass Slides

1. Cut plastic spacers to fit your glass slide and stick the plastic spacer onto the glass slide.
   - Make sure the sticker is attached tightly to prevent solution from leaking into neighboring wells.
2. Label the slots and keep slide ready for use.

Prepare Reaction Mixes and Induce LLPS

1. Dilute the Tau-DyLight-488 with unlabeled Tau in a 1:5 ratio.
   - Use unlabeled Tau from the same original aliquot as Tau-DyLight-488, so you can assume the same concentration.
2. Mix the appropriate volumes of listed reagents (Table 1) in 1.5 mL tubes to obtain a solution of 5 μM Tau, 10 mM HEPES, pH 7.4, 100 mM NaCl, 20 ng/μL RNA with or without 10% PEG.
   - Add PEG last as this will induce LLPS.

Table 1
Reagents used for Tau-LLPS experiments

			Uninduced		Induced	
Reagent	**Stock conc.**	**Unit**	**Final conc.**	**Vol (μL)**	**Final conc.**	**Vol (μL)**
Tau	40	μM	5	1.25	5	1.25
HEPES, pH 7.4	100	mM	10	0.875	10	0.88
NaCl	1000	mM	100	1	100	1
RNA	100	ng/μL	20	2	20	2.00
PEG	40	%	–	–	10	2.50
Water	–	–	–	4.88	–	2.38
Final Rxn Vol		μL		10		10

3. Mix by tapping the tubes.
4. Pipette 3 μL mix in the appropriate well on the glass slide, remove the upper plastic layer of the plastic spacer, and mount the coverslip by pressing it on the spacer.
5. Incubate the sample for 2 h at RT in the dark.
   - Invert your glass slide to let the droplets settle on the coverslips.

Image Acquisition

1. Use a 63× oil objective and focus on droplets in the LLPS-induced condition using the DIC channel first.
2. Set the parameters in the 488 channel to visualize Tau-DL-488 on the coverslips as bright droplets.
3. Take images of all conditions using the same laser settings, take ≥3 images/condition, and repeat each reaction three times.

3.4 Analysis of Droplets

1. To calculate Coverage of Field of View (% FOV), Number of Particles, and Particle Area (μm^2) in Fiji, binarize the image using a chosen threshold.
   - To pick a threshold, go to Image > Adjust > Auto Threshold > Try all, and repeat using the selected threshold.
2. Select droplets using the Analyze Particle function.
   - Analyze > Analyze Particles > "Add to Manager"; "Exclude on Edge"; "Display Results"; "Summarize," keep the rest at default.

3. From the summary section, collect the % Area (% FOV) and Count (Number of Particles); from the Results section, collect the Area (Particle Area (μm^2)).
4. When performing statistical analyses in Prism, use nested-data analysis.
   - Particle Area: First, determine the average Particle Area per image.
   - Treat each reaction as a biological replicate (column, N = x), and each image taken from the same reaction as a technical replicate (subcolumn, n = x).

4 Recommendations

4.1 Perform Quick LLPS Directly After Labeling

After labeling Tau, perform a "quick LLPS" reaction to test for the formation of droplets before you set up a full LLPS reaction.

1. Pipette 0.5 μL of the final Tau-DL-488 labelled reaction into two wells made from a slide covered with the Grace Biolabs spacers.
2. Pipette directly 4.5 μL of either 10 mM HEPES or 10 mM HEPES, 10% PEG on top.
3. Incubate 10 mins, at RT in the dark and image.

4.2 Titrations to Reaction Components on Tau LLPS

The effect of reaction components on Tau droplet formation is best observed using a range of concentrations. The example reaction described in this chapter mimics physiological conditions, and can be taken as a point of reference. We recommend making titration curves covering at least a tenfold range of, e.g., the concentration of a specific type of RNA or other protein that is added in.

4.3 Timing of Imaging after LLPs Induction

With increasing incubation time, phase-separated Tau exhibits reduced fluidity (vitrification), as demonstrated by previous FRAP experiments [4, 6], which may affect droplet size and shape. Since samples are not fixed and the duration of incubation influences Tau LLPS properties, where possible, one must include all conditions on each slide equally and limit imaging time per sample. When multiple slides are needed, start the LLPS reactions on each slide at different times to allow enough time to image each slide one after the other without delay.

4.4 Fluorescent Labeling

In order to study LLPS dynamics, proteins are preferably labeled with a fluorescent tag. However, the addition of fluorescent tags can modify the chemical and surface properties of proteins altering their biophysical behavior [8–11]. The inherent transient and weak interactions in condensates make the study of LLPS susceptible to labeling-induced artifacts.

To minimize the effect of labeling on LLPS properties, proteins of interest are preferably tagged with a fluorophore (like DyLight 488) and mixed with unlabeled protein, such that the labeled protein constitutes a small fraction of the total protein. Compared to larger fluorescent fusion proteins, small-molecule dyes are less likely to interfere with the protein's native structure and interactions. Diluting labeled protein with a larger excess of unlabeled protein further helps maintain the native behavior of condensates. Include an unlabeled control (images with DIC) where possible.

When fluorescent tags are fused to proteins, their placement is crucial. Since fusion proteins are large and tags are fixed at the N- or C-terminus, it's important to position the tag away from key LLPS-related regions (e.g., the proline-rich and microtubule-binding domains of Tau [4]).

5 Troubleshooting

5.1 Failure to Induce LLPs of Tau (or Other RBP)

(a) Ensure Tau protein is not already in aggregated form before us for LLPS studies, and Tau concentration is sufficiently high (5 μM). Validate protein purity by SDS-PAGE and aggregation state by Native-PAGE. Extended storage often leads to oxidation with the formation of disulfide cross-links and the formation of small protein aggregates. These aggregates will rapidly seed further aggregation, which interferes with the process of LLPS. To reduce aggregation, add reducing agents such as DTT (1 mM) during protein preparation to maintain cysteine residues in a reduced state and include glycerol (5%) to stabilize protein structure. Always use freshly purified or aliquoted Tau protein to avoid degradation or aggregation from repeated freeze-thaw cycles.

(b) Check critical parameters including salt (100 mM NaCl) and RNA (20 ng/μL) concentrations. The response of Tau LLPS on these parameters follows a U-curve. This means that both too little and too much salt or RNA lead to inefficient LLPS. With 5 μM 0N4R Tau protein, 10% PEG-8000, pH 7.4, and temperature of 24°C, salt concentrations between 25 and 100 mM NaCl, and total murine brain RNA concentrations between 20 and 70 ng/μL are most optimal.

(c) Consider alternative fluorescent tags or labeling strategies (e.g., amine reactive versus maleimide-based), as the conjugation strategy of a tag can preclude LLPS depending on the RBP. Also, be aware that attaching fluorescent proteins, such as eGFP or mCherry, can alter the phase separation properties of the cognate protein. Studies suggest that sfGFP has the lowest effect on phase separation [12].

(d) Try alternative crowding agents (e.g., Ficoll or Dextran-70) or vary their concentration to optimize crowding effects.

5.2 Inconsistent Microdomain Formation

When mixing Tau with other RBPs in an LLPS experiment, Tau may partition separately from the other RBPs or form microdomains within RBP droplets (or vice versa); this partitioning is normal and reflects the biophysical properties of the proteins used. However, consistency between experiments is improved by controlling RBP concentration and pH, as partitioning depends on molecular ratios and net charge, with Tau LLPS primarily driven by electrostatic interactions. It is important to note that some variability may still occur, partly due to the inherently stochastic nature of co-partitioning and the dynamic electrostatic landscape, and may therefore be unavoidable. Protein isoelectric point (pI) can be predicted here: https://web.expasy.org/protparam/

5.3 Low Fluorescence Signal

(a) Optimize the labeling ratio of Tau-488 to Tau; excessive labeling can impair phase separation, while too little reduces detectability. Start with the recommended 1:5 ratio, and adjust accordingly.

(b) Prepare labeling reactions fresh and protect from light to prevent fluorophore degradation.

Conflict of Interest

BW is CSO and Co-Founder of Aquinnah Pharmaceuticals Inc.

References

1. Kampers T, Friedhoff P, Biernat J, Mandelkow EM, Mandelkow E (1996) RNA stimulates aggregation of microtubule-associated protein tau into Alzheimer-like paired helical filaments. FEBS Lett 399(3):344–349. https://doi.org/10.1016/S0014-5793(96)01386-5
2. Hernández-Vega A, Braun M, Scharrel L et al (2017) Local nucleation of microtubule bundles through tubulin concentration into a condensed tau phase. Cell Rep 20(10):2304–2312. https://doi.org/10.1016/j.celrep.2017.08.042
3. Vanderweyde T, Apicco DJ, Youmans-Kidder K et al (2016) Interaction of tau with the RNA-binding protein TIA1 regulates tau pathophysiology and toxicity. Cell Rep 15(7):1455–1466. https://doi.org/10.1016/j.celrep.2016.04.045
4. Ash PEA, Lei S, Shattuck J et al (2021) TIA1 potentiates tau phase separation and promotes generation of toxic oligomeric tau. Proc Natl Acad Sci U S A 118(9):e2014188118. https://doi.org/10.1073/pnas.2014188118
5. Flory PJ (1942) Thermodynamics of high polymer solutions. J Chem Phys 10(1):51–61. https://doi.org/10.1063/1.1723621
6. Wegmann S, Eftekharzadeh B, Tepper K et al (2018) Tau protein liquid–liquid phase separation can initiate tau aggregation. EMBO J 37(7):e98049. https://doi.org/10.15252/embj.201798049

7. Kanaan NM, Hamel C, Grabinski T, Combs B (2020) Liquid-liquid phase separation induces pathogenic tau conformations in vitro. Nat Commun 11(1):2809. https://doi.org/10.1038/s41467-020-16580-3
8. Fatti E, Khawaja S, Weis K (2025) The dark side of fluorescent protein tagging—the impact of protein tags on biomolecular condensation. Mol Biol Cell 36(3):br10. https://doi.org/10.1091/mbc.E24-11-0521
9. Dörner K, Gut M, Overwijn D et al (2024) Tag with caution - how protein tagging influences the formation of condensates. bioRxiv 616694. https://doi.org/10.1101/2024.10.04.616694
10. Barkley RJR, Crowley JC, Brodrick AJ, Zipfel WR, Parker JSL (2024) Fluorescent protein tags affect the condensation properties of a phase-separating viral protein. Mol Biol Cell 35(7):ar100. https://doi.org/10.1091/mbc.E24-01-0013
11. Ziaunys M, Sulskis D, Veiveris D, Sakalauskas A, Mikalauskaite K, Smirnovas V (2024) Diverse effects of fluorescent labels on alpha-synuclein condensate formation during liquid-liquid phase separation. Int J Biol Macromol 283:137688. https://doi.org/10.1016/j.ijbiomac.2024.137688
12. Avecilla ARC, Thomas J, Quiroz FG (2025) Genetically-encoded phase separation sensors enable high-fidelity live-cell probing of biomolecular condensates. ACS Sens 10(3):1857–1869

Check for updates

Chapter 11

Generation of Stable Tau Biosensor Cell Lines

Jaime Vaquer-Alicea and Vaibhav Bommareddy

Abstract

Tau protein aggregation is a pathological hallmark of numerous neurodegenerative conditions. Cell-based tau biosensor assays provide a sensitive and quantitative readout of pathogenic tau aggregation activity by using reporter cell lines that stably express aggregation-prone tau fragments fused to fluorescent proteins. This approach enables highly sensitive detection of proteopathic tau seeds and has been instrumental in studying tau "prion-like" propagation, strain diversity, and the mechanisms of aggregation biology. In this chapter, we describe a detailed protocol for **generation of stable tau biosensor cell lines** and their validation with tau seeding assays. We also describe methods for assaying tau aggregation in these cell lines, including treatment with recombinant tau fibrils or pathogenic brain extracts and quantification of FRET-positive cells via flow cytometry or imaging. Troubleshooting guidance is provided to address common challenges.

Keywords Tau, Alzheimer's, Biosensors, HEK293T, Flow cytometry, FRET

1 Introduction

Tau is a microtubule-associated protein that abnormally aggregates in a class of neurodegenerative diseases known as tauopathies, which include Alzheimer's disease (AD), frontotemporal dementias, Pick's disease, progressive supranuclear palsy, and others. In AD, pathological tau accumulation follows a characteristic spatiotemporal pattern through connected brain networks, correlating with clinical disease progression [1]. This spread of tau pathology is hypothesized to occur via prion-like propagation, wherein misfolded tau "seeds" are released from an affected cell and taken up by neighboring cells to template the misfolding of native tau [2, 3]. This concept builds on Prusiner's original prion hypothesis, which proposed that certain proteinaceous infectious particles (prions) can self-propagate without nucleic acids [4]. By a similar principle, aggregated tau appears to act as a proteopathic seed that instigates the autocatalytic spread of tau amyloids throughout the brain [3]. Prion-like tau transfer has been demonstrated in vitro with cell models and in vivo *rodent*

Jessica S. Fortin (ed.), *Translational Research Methods for Tauopathies*, Neuromethods, vol. 233,
https://doi.org/10.1007/978-1-0716-5234-3_11,

models [5–9]. This propagation models is thought to underlie the stereotyped progression of tau pathology in AD and related disorders (see Fig. 1).

A major challenge in studying and targeting prion-like tau propagation is the sensitive detection of tau seeds. Traditional histopathological methods detect end-stage neurofibrillary tangles but do not readily capture soluble tau seeds. To address this, researchers have developed biosensor cell lines that report tau aggregation via Förster resonance energy transfer (FRET) [5]. The first-generation tau biosensor line, developed by the Diamond laboratory, stably expresses the repeat domain of tau with the P301S mutation tagged to CFP/YFP fluorophores. When pathogenic tau seeds enter these cells, they induce aggregation of the reporter tau proteins, bringing CFP and YFP into proximity and producing a quantifiable FRET signal [5]. This cell-based FRET assay can detect femtomolar concentrations of tau seeds in complex samples [5]. In fact, tau seeding activity was detected in tauopathy mouse brains 4–5 months before the appearance of overt histopathology, highlighting that tau seeds are an early and sensitive biomarker of incipient disease [5]. Subsequent optimizations have further improved the assay's sensitivity. For example, second-generation HEK293T tau biosensor lines (v2H) were engineered with enhanced fluorescent proteins and expression levels, yielding ~100-fold greater sensitivity than the original biosensor [10]. Coupled with techniques like immunoprecipitation to concentrate tau, these advanced biosensors detect seeding activity from as little as ~100 attomoles of synthetic tau fibrils or tau from 1 nanogram of AD brain tissue [10]. Such sensitivity approaches the threshold needed to probe biological fluids. Indeed, biosensor cells have been used to test for seeding activity in cerebrospinal fluid, though robust detection in living patients remains challenging [10]. Beyond FRET, newer

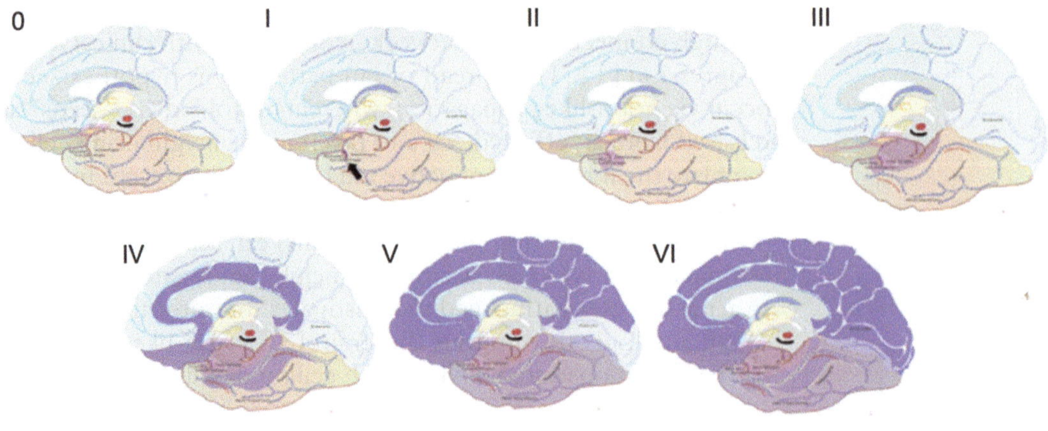

Fig. 1 Topographic representation of the spread of neurofibrillary tau tangles. Braak stage 0 (absence of tau accumulation) through Braak stage 6, illustrated above with affected brain regions are colored in different shades of purple. Adapted from Macedo AC et al (2023) *J Nucl Med* 64:1423–1431. CC BY 4.0

biosensor platforms (e.g., bimolecular luciferase or split fluorescence complementation) are being developed to monitor tau self-assembly in living cells [11, 12]. Together, these biosensor tools enable *quantitative, cell-based assays* for tau aggregation with greater sensitivity and specificity than conventional methods (see Figs. 2 and 3).

Importantly, tau biosensor assays have provided insights into tau "strains" and conformational diversity. It is now evident that misfolded tau is not a singular entity; instead, it can adopt multiple conformations that self-propagate, analogous to distinct prion strains. Different tau strains have been shown to cause unique patterns of pathology and clinical phenotypes. Sanders et al. (2014) first demonstrated this by isolating two distinct tau aggregate strains that propagated in cells and mice with markedly different neuropathological outcomes [13]. Subsequent studies have

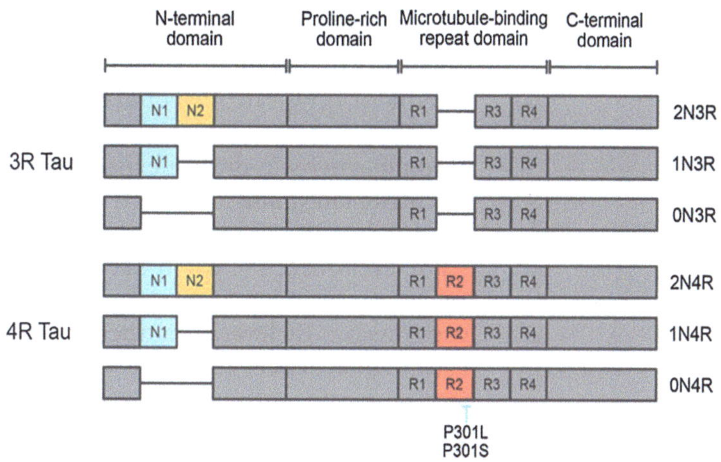

Fig. 2 Full-length tau isoforms (0N/1N/2N and 3R/4R) with exon structure and aggregation-prone segments. Adapted from Pinzi L et al (2023) *Molecules* 28(11):4544. CC BY 4.0. [PMID: 37299020]

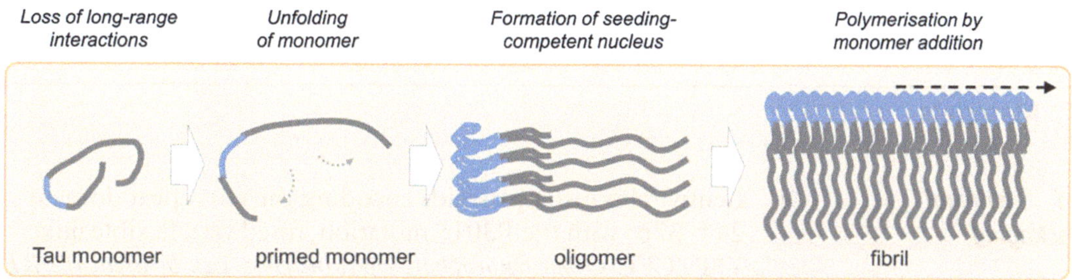

Fig. 3 Schematic representation of tau aggregation. Soluble tau monomers can assemble into oligomers, which ultimately organize into neurofibrillary tangles (NFTs). Pathological tau assemblies can act as proteopathic "seeds," promoting cell-to-cell propagation of tau pathology. [Adapted from Limorenko and Lashuel (2022) *Chem. Soc. Rev.* 51:513–565, licensed under CC BY-NC 3.0]

corroborated that tau exists as conformationally diverse assemblies which can *template* their shape onto natively expressed tau [7]. For instance, biochemical seeding assays can distinguish tau aggregates from Alzheimer's disease versus those from corticobasal degeneration [13, 14]. Such findings reinforce the concept of conformational templating, wherein a given tau strain templates its fold onto tau molecules in the recipient cell, perpetuating strain-specific structures. Notably, tau seeding activity in homogenates from human AD brains correlates with clinical heterogeneity: more aggressive cases of AD tend to harbor tau assemblies that seed more readily in vitro, and tau posttranslational modifications that enhance seeding are associated with faster cognitive decline [15, 16]. This implies that individual differences in tau strain composition may contribute to the variable progression rates of tauopathies.

Given the centrality of tau aggregation to neurodegeneration, the advent of tau biosensor cell lines has provided a powerful tool to investigate tau prion-like behavior with high sensitivity and throughput (Fig. 4). These cell models not only enable detection of minute quantities of pathological tau seeds but also allow controlled experiments of how various cellular perturbations affect aggregation [17, 18]. Biosensor assays have been successfully used to *quantify* seeding activity in diverse experimental and clinical samples, to differentiate tau strains associated with different diseases and to identify modulators of tau propagation in cells [10, 13, 14]. The insights gained reinforce the prion-like model of tauopathies, wherein misfolded tau spreads in a self-perpetuating manner and underlies disease progression. By leveraging these tau biosensor systems, researchers can further dissect the molecular mechanisms of tau aggregation and screen for interventions (e.g., small molecules or antibodies) that block seeding. Ultimately, such efforts may pave the way for tau-directed diagnostics and therapeutics, aimed at detecting pathogenic tau seeds early and halting the tau propagation associated with neurodegeneration. The following sections will detail the development and use of tau RD biosensor cell assays, offering practical guidance on harnessing this technology to study and potentially combat tau amyloids.

2 Materials

2.1 Lentiviral Packaging

- Lentiviral transfer plasmids encoding the tau repeat domain (aa 244–376) with the P301S mutation, fused via a flexible linker to YFP/CFP. *Other fluorophore combinations can be used, if FRET compatibility is confirmed.*

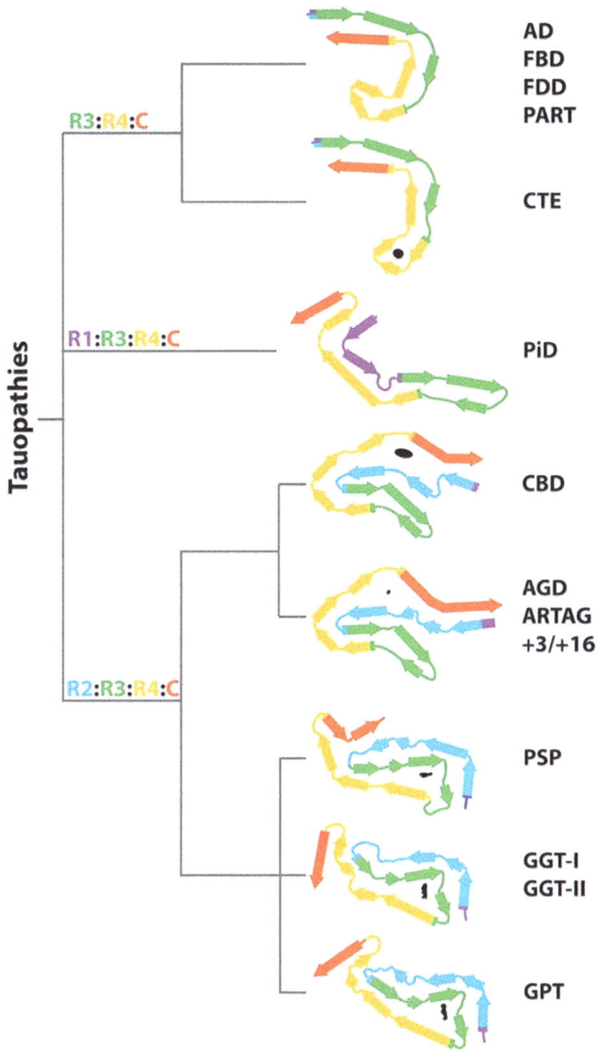

Fig. 4 Schematic representation of tau filament structures associated with major neurodegenerative diseases (e.g., AD, CTE, PiD, CBD, PSP, AGD, GGT, GPT), highlighting conformational diversity in fibril types. Adapted from Shi Y, Zhang W, Yang Y, Murzin AG, Falcon B, Kotecha A et al (2021) *Nature* 598:359–363. Licensed under CC BY 4.0. [PMC7611841]

- Plasmid with gene of interest will preferably have a constitutive mammalian promoter (e.g., CMV) and antibiotic-resistance cassette (e.g., ampicillin or kanamycin resistance).
- Bacterial mini-prep kit (QIAGEN Cat. # 27104).
- Second-generation packaging plasmids (psPAX2 and VSVg) are packaged with a modified FUGW plasmid (David Baltimore laboratory), which has a CMV promoter, Kozack sequence, and flexible polypeptide linker (GSAGSAAGSGEF) attached to a unique fluorophore: later referred to as the FM5 plasmid.

- Transfection reagent—TransIt293 (Mirus Bio MIR 2704).
- Lenti-X™ Concentrator kit in case viral yield is low (Takara Cat. # 631231).
- 6-well dish treated cell culture dish (Corning Cat. # 07-200-83).

2.2 Cell Culture

- Cells: HEK293T (ATCC-CRL-1268) cells maintained at 37°C, 5% CO_2, with 85%–95% humidified air.
- Media: Dulbecco's Modified Eagle's Medium (DMEM)—high glucose (Gibco Cat. # 11965118) + 10% fetal bovine serum (HyClone Cat. # SH30071.03) + 1% glutamax (Gibco Cat. # 35050061) + 1% penicillin/streptomycin (Gibco Cat. # 15140122).

2.3 Cell Sorting

- Fluorescence-activated cell sorting (FACS) with the following lasers:
  1. Violet: 405 nm: 50 mW
  2. Blue: 488 nm: 50 mW
  3. Yellow-Green: 561 nm: 50 mW
  4. Red: 640 nm: 40 mW
- Optical filters: eCFP (450/50 bandpass) and eYFP (525/40 bandpass) optical filters.
  - *Long pass filters may also be included in the FACS machine to allow for minimum laser length cutoffs to be analyzed.*
- Sorting media: 1× Hanks' Balanced Salt Solution (HBSS) (Gibco Cat. # 14-025-076) + 1% FBS + 1 mM EDTA (Invitrogen Cat. # AM9260G).

2.4 Tau Seed Transfection

- Recombinant human tau fibrils (purified tau monomer, heparin, DTT, HEPES, NaCl).
- Lipofectamine 2000/3000 (Invitrogen Cat. # 11668030/ L3000008) in Opti-MEM (Gibco Cat. # 31985070).

2.5 Cell Fixation and Analysis

- 0.05% trypsin–EDTA (Gibco Cat. # 25300054).
- 2% PFA in PBS (Gibco Cat. # 14190144).
- U-bottom 96-well plate (Thermo Cat. #143761).

2.6 Tau Aggregate Detection and Analysis

- Flow cytometer with FRET capability:
  - For FRET between CFP and YFP: You will need a channel where you are able to excite the fluorophore with the Violet laser (405 nm: 50 mW) and read the emission with the YFP optical filter set (530/30 bandpass).
  - *Optional: fluorescence microscope or high-content imager with CFP/YFP filters.*

- Western blot reagents, Triton X-100 lysis buffer, NuPAGE Bis–Tris Mini Protein Gels, 4%–12% (Thermo Cat. # NP0335BOX), anti-tau monoclonal antibody (Thermo Cat. # 13-6400), and anti-GFP monoclonal antibody (Thermo Cat. # MA5-15256), for fusion-protein verification.

2.7 General Lab Supplies

- Sterile pipette tips, tubes, PPE, CO_2 incubator, biosafety cabinet, cell counter, etc.

3 Methods

3.1 Overview

The process of generating a stable tau biosensor cell line involves several major steps. First, the tau reporter constructs are designed and cloned. Next, the constructs are introduced into the host cell line, typically via lentiviral transduction for high efficiency and propagation through cell divisions. Monoclonal cell lines are then isolated either by single-cell FACS sorting or limited dilution and expanded. Each clonal line is screened for robust expression of both reporter constructs and absence of spontaneous aggregation. Finally, the clones are tested in a tau seeding assay to validate that they produce a strong FRET signal upon exposure to tau aggregates, but not under basal conditions or with non-tau aggregates. Below, we detail each step.

3.2 Design and Cloning of Tau Biosensor Plasmids

1. **Select the tau sequence and mutation:** For a classic tau FRET biosensor, use the microtubule-binding repeat domain of tau (Tau RD) as the core aggregation-prone segment. In prior studies, the fragment spanning approximately residues 244–376 of the 4-repeat human tau isoform (2N4R) was used. Include the P301S mutation in the tau RD sequence, if a disease-associated variant is desired to enhance aggregation propensity. P301S is a familial mutation that promotes fibrillization; wild-type tau RD can also be used but has a higher threshold for seeding [5, 19, 20].
2. **Choose fluorescent reporters:** The standard approach uses a FRET donor–acceptor pair of fluorescent proteins attached to tau. eCFP and eYFP (or improved variants like mCerulean3 and mClover3) are common choices. Plan to create two versions of the tau construct: one fused to an eCFP and one fused to an eYFP. The fluorescent protein is typically fused to the C terminus of the tau RD fragment.
3. **Clone the constructs:** Synthesize or PCR-amplify the tau RD sequence with the desired mutation. Clone it in-frame with the fluorescent protein sequence in the FM5 vector. Ensure to include or maintain a Kozak consensus sequence before

the start codon of tau for efficient translation. Repeat for tauRD-YFP.

4. **Verify sequences:** Sequence the final plasmids to confirm the tau gene. Specifically, check for the presence of the tau gene, proper Kozak sequence, and fluorophores.
5. **Prepare DNA for transfection:** QIAGEN miniprep plasmid DNA for each construct to obtain high-quality, endotoxin-free DNA, which is important for lentiviral production yields.

3.3 Production of Lentiviral Particles Carrying Tau Constructs (see Fig. 5)

1. **Plate packaging cells:** One day before transfection, plate HEK293T packaging cells in a 6-well dish such that they are ~75% confluent on the day of transfection. These cells will produce the lentivirus.
2. **Transfection of packaging cells:** For each tau reporter plasmid (tauRD-CFP and tauRD-YFP), set up a separate transfection. In a sterile tube, combine the lentiviral transfer plasmid (500 ng), the packaging plasmid psPAX2 (1500 ng), and envelope plasmid pCMV-VSVG (500 ng). Add 7.5 μL TransIT 293 with the three plasmids and fill the volume to 100 μL with Opti-MEM. Incubate for 20–25 min before applying to cell culture. Gently rock the dish and incubate.
   - *These quantities are for a 6-well dish; scale accordingly.*
3. **Harvest viral supernatant:** 48–72 h post-transfection, collect the culture medium containing lentiviral particles. Spin down cell debris at 500 × *g*, keep viral supernatant on ice and aliquot

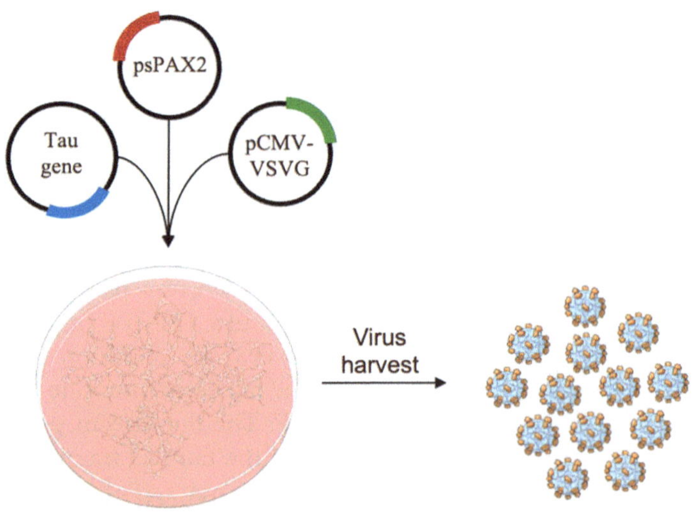

Fig. 5 Lentivirus production of tau biosensor constructs. Packaging plasmids (psPAX2), an envelope plasmid (pCMV-VSV-G), and the tau reporter gene are co-transfected into HEK293T cells. Transfected cells generate and release replication-deficient lentiviral particles. Viral supernatant is then harvested for downstream transduction of target cells

as needed. *Note: Lentivirus is sensitive to freeze–thaw; if not using immediately, snap-freeze aliquots in a −80°C freezer.*

4. **Titer:** If desired, determine the viral titer by transducing a test batch of cells with serial dilutions and measuring percentage and intensity of fluorescent cells after 48 h.
   - *Titrating the virus can allow you to estimate intracellular tau concentrations using fluorophore expression as a proxy.*
5. **Thaw or prepare target cells:** Make sure your target HEK293T cells are healthy and in log-phase growth.

3.4 Isolation of Clonal Dual-Expressing Cell Lines

To obtain a homogenous biosensor cell population, it is crucial to isolate **clonal cells** that express both CFP- and YFP-tagged tau reporters at appropriate levels. There are two main strategies:

3.5 Fluorescence-Activated Cell Sorting (preferred)

1. **Prepare cells for FACS:** Harvest the transduced cell population when it is in logarithmic growth (avoid over-confluence). Use trypsin to detach *cells*, neutralize with growth medium, spin cells down at 300 × *g* for 3 min and then resuspend cells in sorting buffer (e.g., HBBS + 1% FBS + 1 mM EDTA). Filter the cell suspension through a 40 μm cell strainer to remove clumps, as clumps can clog the sorter.
   - *Proper compensation controls are needed: use untreated HEK293T to set baseline CFP bleed-through into YFP detector and single-color controls, if available.*
2. **Sort for CFP and YFP positivity:** Using a flow cytometry sorter equipped with the proper lasers (e.g., a 405 nm for CFP and 488 nm for YFP), gate the cell population on live single cells, then on those cells that are positive for both CFP and YFP fluorescence (see Fig. 6). Set gates based on untransduced control cells to exclude autofluorescence. The goal is to sort the double-positive cells, which have both tauRD-CFP and tauRD-YFP integrated. If there is a broad range of intensities, you may sort a subset, for example, Diamond's group sometimes sorted cells into "low" and "high" expressors by gating on the lower vs higher ends of the fluorescence intensity distribution for the reporters [10]. This can yield clones with different expression levels, which might be useful for different sensitivity needs.
3. **Single-cell sorting:** Configure the sorter to deposit single cells into a 96-well plate (or one cell per well of a 96-well plate containing 200 μL of conditioned medium). You can also make 20% FBS media for sorting cells into. This allows for better single-cell survival.
4. **Post-sort culture:** After sorting, keep the plates in a 37°C incubator undisturbed for ~10–14 days. Do not change the

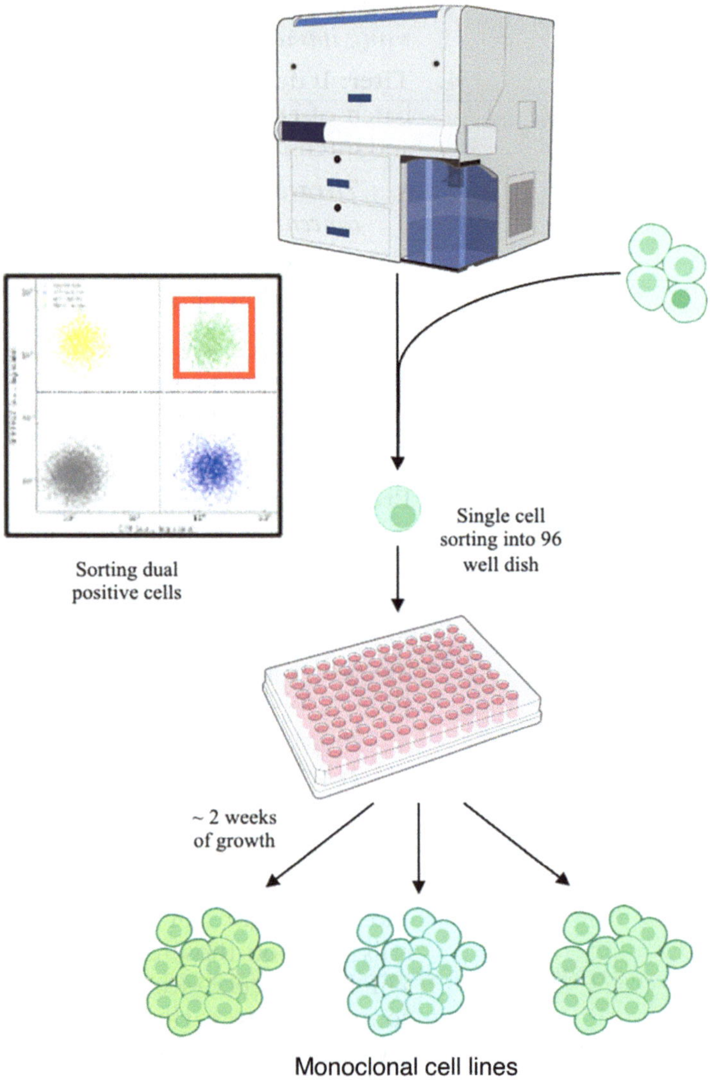

Fig. 6 Generation of monoclonal tau biosensor cell lines via flow cytometry. Dual-positive reporter cells are isolated by FACS and single cells are sorted into individual wells of a 96-well plate. After approximately 2 weeks of expansion, clonal populations are established, providing uniform and reproducible reporters for tau seeding assays

media for the first 5–7 days to avoid disturbing the single cells. Colonies will start becoming visible in wells after 1–2 weeks.

5. **Screen clones:** When clones reach sufficient size (20%–30% coverage of well), transfer them to larger wells for expansion. Note which well corresponds to which sorted gate (if you separated high vs low expressors). At this point, you should verify their fluorescence: check under a microscope for CFP and YFP

signals. Every clone should be positive for both, but the intensity may differ. Select ~10–20 clones for further analysis.

- *Watch for cell death in rapidly growing colonies of cells.*

3.6 Limit Dilution Cloning (If FACS Is Not Available)

1. **Dilution plating:** Take the mixed population of transduced cells (optionally after antibiotic selection) and perform a limiting dilution. For example, dilute the cells to ~1 cell per 200 μL and plate 100 μL per well in a 96-well plate. Another approach is to plate cells in a 10 cm dish at very low density (~100 cells) so that well-separated colonies form.
2. **Cultivate clones:** Allow diluted cells 1–2 weeks to form colonies. Under a microscope, mark wells that have a single colony or single initial cell. It's important to pick wells that clearly originated from one cell to ensure clonality. If using cloning rings on a dish, add trypsin inside the cylinder to lift that colony, then transfer to a well.
3. **Expansion:** Expand clones from 96-well to 24-well or 6-well as they grow. Be gentle in the early stages of expansion, since single-cell clones can be delicate.
4. **Initial clone screening:** Examine each clone for stable and homogenous CFP and YFP fluorescence. At this stage, without FACS enrichment, you will likely find some clones with only one fluorophore present. You should keep only those clones that show both CFP and YFP fluorescence, indicating both reporter genes are present. Clones with spontaneous aggregates (e.g., brightly fluorescent punctate structures indicating self-aggregation) should be discarded, as they likely overexpress the reporters to a degree causing aggregation artifacts or have some anomalous integration.

Regardless of the method, the outcome is a set of clonal cell lines. Each clone should be derived from a single cell and thus have a uniform integration pattern and expression level of the reporters. In practice, one usually obtains several candidate clones and then performs further testing to identify the best performer for the biosensor assay.

3.7 Screening and Expansion of Candidate Clones

1. **Western blot:** To validate tau expression, lyse cells from a 6-well dish with 70%–90% confluency of the tau biosensor clones. Run on NuPAGE 4%–12% Bis–Tris gel and transfer to western blot membrane to confirm the presence and size of tau-CFP and tau-YFP proteins. Probe with an anti-tau RD antibody or anti-GFP antibody.
   - *Bands corresponding to ~combined size of TauRD (~18 kDa) + FP (~28 kDa), roughly ~46 kDa. Clones with truncated products or low expression can be eliminated.*

2. **Growth characteristics:** Monitor the doubling time and morphology of each clone. A good biosensor line will grow robustly and tolerate many passages. If a clone grows unusually slow or tends to die off, it may not be ideal for experiments. Some very high expressor clones might impose stress on cells, slowing growth.
   - *1:10 splits of biosensor cells should be ready to split again in 2 days, with a doubling rate of ~16 h. This may vary slightly depending on expression, but a 1:10 split from an 80% confluent dish should not take more than 3 days. If you notice abnormally slower growth, then check for mycoplasma or thaw a new vial of biosensor cells, if at a high passage number.*
3. **Baseline FRET/background:** Use flow cytometry to measure baseline FRET signal in each clone *without any added tau seeds.* The baseline percentage of FRET-positive cells should be very low and is typically set up using the negative untreated control cells in your experiment. Any clone that shows a significant fraction of FRET-positive cells on its own is likely unstable or aggregating spontaneously (false positive). Exclude such clones from further consideration (see Fig. 7).
4. **Choose top candidates:** Based on the above (fluorescence intensity, lack of spontaneous aggregates, robust growth, low baseline FRET), select a few of the best clones for comprehensive validation.
5. **Cryopreserve clones:** It is wise to freeze down multiple aliquots of each selected clone at this stage in 90% FBS + 10% DMSO, then store in liquid nitrogen. This preserves the original early-passage state of the clones. The biosensor phenotype can drift with very long-term culture, so working from low-passage stocks is recommended.

3.8 Validation of Tau Biosensor Cell Lines in Seeding Assays

The final and most critical step is to demonstrate that the cloned cell lines indeed function as tau aggregation reporters. This is done by performing seeding assays: exposing the cells to exogenous tau aggregates and measuring the induced aggregation (FRET or fluorescent puncta). We describe a standard FRET flow cytometry assay, as established by Holmes et al. and used in subsequent studies [5].

3.9 Seeding With Recombinant Tau Fibrils (in 96-Well Format)

1. **Plating cells:** The day before treating with aggregates, plate biosensor cells in a 96-well plate at ~20,000 cells per well in 100 μL of complete DMEM. Aim for the cells to be ~50%–70% confluent at the time of treatment. *Seeding works best when cells are not overly confluent; high density can reduce uptake and increase background signaling.*

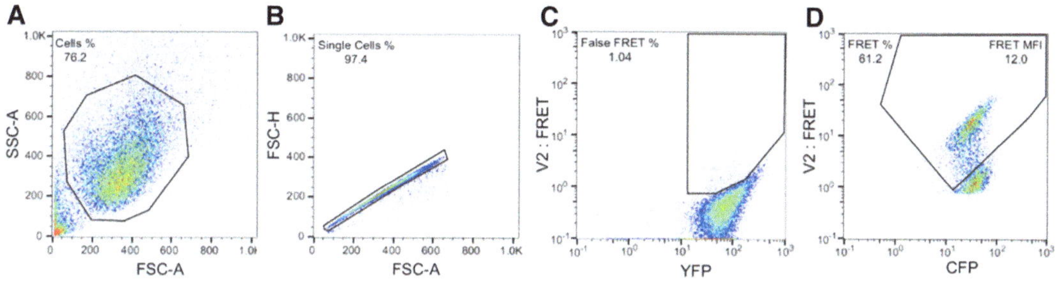

Fig. 7 Flow cytometry gating strategy for quantifying tau seeding activity by FRET. (**A**) Forward scatter (FSC-A) vs side scatter (SSC-A) plot to gate the main cell population and exclude debris. (**B**) FSC-A vs FSC-H plot to gate single cells, excluding doublets. (**C**) YFP vs FRET plot showing the "false FRET" gate established using YFP-single positive controls, to remove spectral bleed-through into the FRET channel. (**D**) CFP vs FRET plot showing the true FRET gate, derived from liposome-treated (no-seed) cells such that ≤1% of events fall within the FRET-positive region. Following fibril seed treatment, a clear population shift into the FRET gate is observed, enabling quantitation of seeding activity. Final readouts include %FRET-positive cells, median fluorescence intensity (MFI) of the FRET-positive population, and integrated FRET density (% positive × MFI). Adapted from Furman et al. 2015 [Tau RD(P301S)-CFP/YFP biosensor flow cytometry method]

2. **Prepare tau seeds:** Chose between using tau derived from brain homogenates or recombinant tau fibrils (Table 1). Preparations are outlined below:

3. **Mix with Lipofectamine:** In a separate tube for each sample, mix Lipofectamine 2000 (1 μL/well) with Opti-MEM (20 μL/well) and incubate ~5 min at room temperature. Then add the 30 μL of diluted tau fibril solution to the Lipofectamine mixture, yielding 20 μL + 30 μL = 50 μL total. Let this complex incubate for ~20 min at room temp. This allows liposome complexes to form, encapsulating tau seeds.
   - *Total volume added to cell wells can be decreased by changing Opti-MEM volume.*
4. **Apply to cells:** Add the 50 μL of Lipofectamine + tau seed mixture (Table 1) to each well of cells by gently dispensing it in.
5. **Incubation:** Return the plate to 37°C incubator and incubate for ~48 h. During this time, tau seeds will enter cells and induce aggregation of the reporter tau in some cells (see Fig. 8).
   - *Do not disturb the cells or change media during this period.*
6. **Harvest and fixation:** After 48 h, or the predetermined endpoint, carefully remove the media via aspiration and add 50 μL of 0.05% trypsin to each well. Incubate 5 min until cells detach, then neutralize and resuspend with 150 μL of complete media. Transfer the cell suspension to a U-bottom 96-well plate and spin down (500 ×*g*, 5 min). Remove trypsin and resuspend in

Table 1
Preparation of recombinant vs patient-derived tau fibrils

Recombinant 2N4R tau fibrils	Neurodegenerative brain homogenates
1. Start with purified recombinant human tau protein (2N4R full length or RD/K18 fragment).	1. Obtain frozen brain tissue confirmed by neuropathology.
2. Prepare buffer: 10 mM HEPES, 100 mM NaCl, pH 7.4.	2. Homogenize at 10% w/v (e.g., 100 mg tissue in 1 mL PBS or TBS) with protease/phosphatase inhibitors.
3. Mix: 10 μM tau + 8 μM heparin (~17 kDa) + 5 mM fresh DTT.	3. Clarify by centrifugation (21,000 ×*g*, 10 min) to remove debris.
4. Incubate at 37°C for 48–72 h in Eppendorf ThermoMixer at 650 rpm.	4. Collect supernatant as crude homogenate (contains tau seeds).
5. Fragment fibrils in water bath sonicator at 4 (~65% amplitude, 30–60 s pulses on/off).	5. Optional: sonicate briefly on ice to fragment and enhance seeding efficiency.
6. Aliquot and store: short term at 4°C or snap-freeze and store at −80°C.	6. Aliquot homogenate, store at −80°C; avoid repeated freeze–thaw cycles.

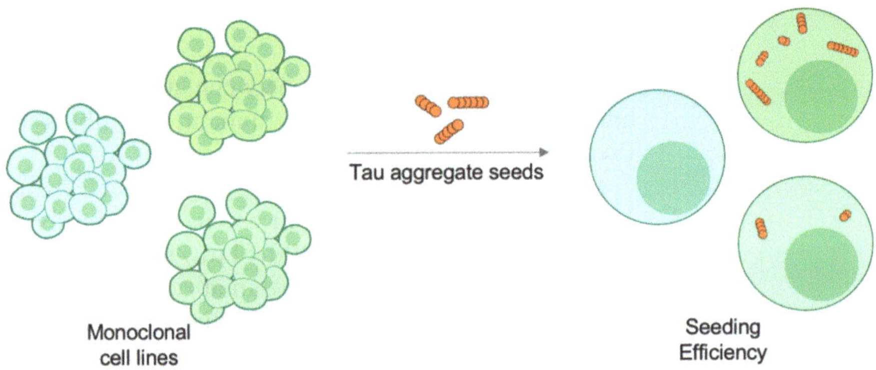

Fig. 8 Monoclonal reporter cell lines are exposed to exogenous tau aggregate seeds, which are taken up by individual cells. Internalized seeds template intracellular tau aggregation, producing a detectable reporter signal. The proportion of aggregate-positive cells reflects tau seeding efficiency

100 μL per well of 2% PFA in PBS to fix the cells (fixation preserves aggregates and FRET signals for flow analysis). Incubate 10 min at room temp. Spin (500 ×*g*, 5 min) and add finally resuspend cells in 200 μL FACS buffer for analysis.

7. **Flow cytometry analysis:** Using a flow cytometer, acquire data for CFP and YFP fluorescence of each sample. Detect FRET-positive cells by exciting CFP (e.g., with 405 nm laser) and detecting YFP emission (525/540 nm).

8. **Expected results:** Unseeded control wells should have <1% FRET-positive cells (often near 0.1%). Wells with tau seeds should show a dose-dependent increase in FRET-positive cells. Plot the dose-response to verify a sigmoidal curve and low background at zero seed. If a clone does not show any FRET

signal even at high seed concentrations, it may be nonfunctional. Conversely, if the "no seed" control shows significant FRET, the clone has too high background.

3.10 Alternative or Additional Validation

- **Seeding with brain-derived tau:** You can challenge the biosensor cells with diluted brain homogenates from tauopathy cases to test, if the line can produce templated aggregation. Prepare brain homogenates (10% w/v in TBS and protease inhibitor) and treat as above with Lipofectamine. AD brain seeds typically induce a robust FRET signal in tau RD P301S biosensors, whereas control (non-tauopathy) brains do not.
- **Fluorescence microscopy readout:** In addition to flow cytometry, examine seeded cells under a fluorescence microscope. In cells that received tau seeds, one should observe the formation of discrete **YFP-positive aggregates** (puncta or inclusions) in the cytoplasm or nucleus. Often these correspond to FRET-positive cells. In control cells, the fluorescence remains diffuse. Some labs also use microscopy as a quantitative measure to count cells with aggregates, By the end of these validation experiments, you should have identified one or more monoclonal cell lines that: (a) express tau-CFP and tau-YFP at levels sufficient for detection, (b) have low spontaneous FRET/aggregation, (c) show a strong, dose-dependent FRET response to tau seeds, and (d) do so specifically for tau aggregates. These stable tau biosensor cell lines are now ready to be used in various applications, such as comparing aggregation propensities of different tau mutants, testing patient samples for seeding activity, or screening for inhibitors of tau–tau interactions.

3.11 Troubleshooting

Problem	Solution
High background FRET from spontaneous aggregation in over- or under-confluent cultures	Passage cells before they exceed ~80% confluence and avoid seeding too sparsely; optimize seeding density (e.g., 2–3 × 10^4 cells/well in 96-well) so assays occur in the optimal growth window.
Expression level trade-offs: high noise vs low sensitivity	Screen and select clones balancing sensitivity and background adjust promoter strength or maintain both high- and low-expressing lines as needed.
Inefficient tau seed uptake	Titrate Lipofectamine starting with ~0.75–2uL/96-plate well. Follow reagent incubation times.

(continued)

Problem	Solution
False-positive FRET from dying or stressed cells	Include controls (no-seed, Lipofectamine-only, unrelated fibril); gate-out dead cells using DAPI/PI; set FRET-positive gate so ≤0.1% of control cells fall into it.
Loss of sensitivity over passages (gene silencing or aggregate accumulation)	Work with low-passage stocks, thaw fresh early-passage vials regularly, and re-validate seed response after thaw to ensure assay fidelity.
Lack of QC controls to verify assay performance	Keep aliquots of benchmark positive controls (e.g., K18 tau fibrils or AD brain extract); run them alongside samples to confirm biosensor responsiveness and reagent integrity.

By following this protocol and utilizing the troubleshooting guidelines above, you should be able to create stable tau biosensor cell lines that reliably report tau aggregate seeding. These cell models are powerful tools for dissecting the mechanisms of tau aggregation biology and for evaluating potential neurodegenerative therapeutics.

4 Conclusion

In this book chapter, we summarized practical methods to generate stable v2H FRET biosensor cell lines, validate isolated clones, and quantify seeding activity by flow cytometry. We outlined key biosensor parameters, essential controls, and a standard 96-well seeding workflow using recombinant or brain-derived aggregates. Last, we provided a guide for troubleshooting tau biosensors and a methodology for assay readouts to enable reproducible implementation across research settings. Together, these biosensors provide a reliable platform to detect tau seeds and support downstream mechanistic and exploratory studies.

Conflict of Interest

The authors have no conflicts of interest to declare that are relevant to the content of this chapter.

References

1. Braak H, Alafuzoff I, Arzberger T, Kretzschmar H, Del Tredici K (2006) Staging of Alzheimer disease-associated neurofibrillary pathology using paraffin sections and immunocytochemistry. Acta Neuropathol (Berl) 112(4):389–404. PMCID: PMC3906709
2. Goedert M, Eisenberg DS, Crowther RA (2017) Propagation of tau aggregates and

neurodegeneration. Annu Rev Neurosci 40(Volume 40, 2017):189–210
3. Rauch JN, Luna G, Guzman E, Audouard M, Challis C, Sibih YE, Leshuk C, Hernandez I, Wegmann S, Hyman BT, Gradinaru V, Kampmann M, Kosik KS (2020) LRP1 is a master regulator of tau uptake and spread. Nature 580(7803):381–385
4. Woerman AL, Aoyagi A, Patel S, Kazmi SA, Lobach I, Grinberg LT, McKee AC, Seeley WW, Olson SH, Prusiner SB (2016) Tau prions from Alzheimer's disease and chronic traumatic encephalopathy patients propagate in cultured cells. Proc Natl Acad Sci 113(50): E8187–E8196
5. Holmes BB, Furman JL, Mahan TE, Yamasaki TR, Mirbaha H, Eades WC, Belaygorod L, Cairns NJ, Holtzman DM, Diamond MI (2014) Proteopathic tau seeding predicts tauopathy in vivo. Proc Natl Acad Sci USA 111(41):E4376–E4385. PMCID: PMC4205609
6. Ficulle E, Kananathan S, Airey D, Gharbi SI, Humphryes-Kirilov N, Scherschel J, Dunbar C, Eastwood BJ, Laing E, Collier DA, Bose S (2022) A human tau seeded neuronal cell model recapitulates molecular responses associated with Alzheimer's disease. Sci Rep 12(1):2673
7. Narasimhan S, Guo JL, Changolkar L, Stieber A, McBride JD, Silva LV, He Z, Zhang B, Gathagan RJ, Trojanowski JQ, Lee VMY (2017) Pathological tau strains from human brains recapitulate the diversity of tauopathies in nontransgenic mouse brain. J Neurosci 37(47):11406–11423. PMCID: PMC5700423
8. Clavaguera F, Akatsu H, Fraser G, Crowther RA, Frank S, Hench J, Probst A, Winkler DT, Reichwald J, Staufenbiel M, Ghetti B, Goedert M, Tolnay M (2013) Brain homogenates from human tauopathies induce tau inclusions in mouse brain. Proc Natl Acad Sci 110(23):9535–9540
9. Clavaguera F, Bolmont T, Crowther RA, Abramowski D, Frank S, Probst A, Fraser G, Stalder AK, Beibel M, Staufenbiel M, Jucker M, Goedert M, Tolnay M (2009) Transmission and spreading of tauopathy in transgenic mouse brain. Nat Cell Biol 11(7):909–913. PMCID: PMC2726961
10. Hitt BD, Vaquer-Alicea J, Manon VA, Beaver JD, Kashmer OM, Garcia JN, Diamond MI (2021) Ultrasensitive tau biosensor cells detect no seeding in Alzheimer's disease CSF. Acta Neuropathol Commun 9(1):99
11. Lo CH (2021) Recent advances in cellular biosensor technology to investigate tau oligomerization. Bioeng Transl Med 6(3):e10231
12. Limorenko G, Lashuel HA (2022) Revisiting the grammar of tau aggregation and pathology formation: how new insights from brain pathology are shaping how we study and target tauopathies. Chem Soc Rev 51(2):513–565
13. Sanders DW, Kaufman SK, DeVos SL, Sharma AM, Mirbaha H, Li A, Barker SJ, Foley AC, Thorpe JR, Serpell LC, Miller TM, Grinberg LT, Seeley WW, Diamond MI (2014) Distinct tau prion strains propagate in cells and mice and define different tauopathies. Neuron 82(6):1271–1288. PMID: 24857020
14. Vaquer-Alicea J, Manon VA, Bommareddy V, Kunach P, Gupta A, Monistrol J, Perez VA, Tran HT, Saez-Calveras N, Du S, Batra S, Stoddard D, White CL, Joachimiak LA, Shahmoradian SH, Diamond MI (2025) Functional classification of tauopathy strains reveals the role of protofilament core residues. Sci Adv 11(4):eadp5978
15. Dujardin S, Commins C, Lathuiliere A, Beerepoot P, Fernandes AR, Kamath TV, De Los Santos MB, Klickstein N, Corjuc DL, Corjuc BT, Dooley PM, Viode A, Oakley DH, Moore BD, Mullin K, Jean-Gilles D, Clark R, Atchison K, Moore R, Chibnik LB, Tanzi RE, Frosch MP, Serrano-Pozo A, Elwood F, Steen JA, Kennedy ME, Hyman BT (2020) Tau molecular diversity contributes to clinical heterogeneity in Alzheimer's disease. Nat Med 26(8):1256–1263
16. Dickson JR, Sobolewski RGR, Fernandes AR, Cooper JM, Fan Z, Chung M, Donahue C, Oakley DH, Strickland DK, Hyman BT (2025) Alzheimer disease-associated tau post-translational modification mimics impact tau propagation and uptake. J Neuropathol Exp Neurol 84(6):459–470. PMCID: PMC12096005
17. Batra S, Vaquer-Alicea J, Valdez C, Taylor SP, Manon VA, Vega AR, Kashmer OM, Kolay S, Lemoff A, Cairns NJ, White CL, Diamond MI (2025) VCP regulates early tau seed amplification via specific cofactors. Mol Neurodegener 20(1):2
18. Pratt J, McCann K, Kuo J, Parker R (2025) Polyserine–tau interactions modulate tau fibrillization. J Biol Chem 301(9):110523. https://doi.org/10.1016/j.jbc.2025.110523
19. Yao QQ, Hong L, Wu S, Perrett S (2020) Distinct microscopic mechanisms for the accelerated aggregation of pathogenic tau mutants revealed by kinetic analysis. Phys Chem Chem Phys 22(14):7241–7249
20. Chen D, Drombosky KW, Hou Z, Sari L, Kashmer OM, Ryder BD, Perez VA, Woodard DR, Lin MM, Diamond MI, Joachimiak LA (2019) Tau local structure shields an amyloid-forming motif and controls aggregation propensity. Nat Commun 10(1):2493

Chapter 12

Cell-Based Tau Polymerization Assays

Natalie G. Horgan, Omnia M.H. Ibrahium, Taiwo A. Ademoye, Jessica S. Fortin and Raluca Ostafe

Abstract

Tau aggregation is a hallmark of tauopathies, such as Alzheimer's disease and related dementias, and the development of cell-based methods to monitor tau–tau interactions is critical for translational research and drug discovery. Here, we describe a split–green fluorescent protein (GFP) complementation assay optimized in Expi293 suspension cells to detect tau aggregation in living cells. In this system, tau is fused to complementary GFP fragments (GFP10C and GFP11C). When tau molecules oligomerize, the GFP fragments reconstitute fluorescence, which can be quantified by flow cytometry. Unlike most existing cell-based tau assays, which require seeding with preformed fibrils or cofactors, such as heparin, this approach allows spontaneous aggregation to be observed under conditions of high-level tau expression. The assay provides several advantages: it is sensitive, scalable, and compatible with both flow cytometry and fluorescence microscopy. Representative results demonstrate that a reference compound (**MO-19-10**) inhibits tau aggregation in a concentration-dependent manner, validating the assay's ability to detect pharmacological effects. The system can also be applied to screening panels of small molecules, providing a rapid means to discriminate between active and inactive compounds.

Keywords Aggregation, Cell-based assay, Flow cytometry, Neurodegeneration, Split-green fluorescent protein (GFP) complementation, Tau oligomers

1 Introduction

Tau oligomers are β-sheet rich pre-filament aggregates that can exist in both soluble and insoluble forms and may be either hyperphosphorylated or non-phosphorylated [1]. It is hypothesized that these dimers or oligomers form through a complementary electrostatic zipper mechanism, in which the positively charged proline domain of one tau molecule interacts with the negatively charged N-terminal tail of another [2]. When these oligomers are small, soluble, and hyperphosphorylated, they are considered toxic. In this state, they severely impact tau's synaptic and mitochondrial function, reduce its ability to bind and stabilize microtubules, and promote further aggregation into filaments and tangles [3–5]. On top of these functional

Jessica S. Fortin (ed.), *Translational Research Methods for Tauopathies*, Neuromethods, vol. 233,
https://doi.org/10.1007/978-1-0716-5234-3_12,

impairments, tau oligomers also contribute to neuronal loss, leading to learning and memory deficits [3, 6]. Moreover, tau oligomers are capable of interacting with alpha-synuclein, potentially promoting each other's fibrillization and accelerating neurodegeneration [7, 8].

To study tau aggregation in living cells, several cell-based fluorescence assays have been developed, including those based on Förster resonance energy transfer (FRET) [9] and bimolecular fluorescence complementation (BiFC) [10]. These approaches have provided important insights into tau–tau interactions and propagation, but in most cases robust aggregation was observed only when tau fibrils or other aggregation inducers were added to cells. Similarly, in vitro models based on truncated isoforms, such as K18 or K19, or disease-associated mutations like P301S or ΔK280, typically rely on polyanions, such as heparin or fatty acids, to accelerate aggregation [14]. These tools remain widely used in the field and have been foundational for understanding tau pathology.

In contrast, by adapting the split-GFP complementation system [15] to suspension-adapted HEK293 cells (Expi293), we observed spontaneous aggregation without the need for exogenous seeds or cofactors [16]. Expi293 cells are optimized for high-level protein expression and reach higher intracellular tau concentrations than adherent HEK293 cells. This enhanced protein expression likely increases the probability of tau–tau interactions, enabling GFP reconstitution upon oligomerization. Importantly, this spontaneous fluorescence can be readily quantified by flow cytometry, providing a robust, high-throughput readout of tau–tau interactions in live cells.

In this chapter, we present a detailed protocol for the split-GFP tau aggregation assay in Expi293 cells. In this system, tau is fused to complementary fragments of GFP (GFP10 and GFP11). When tau molecules interact, the GFP fragments reassemble into a functional fluorophore, producing a measurable signal by flow cytometry. This method, therefore, combines the physiological relevance of a mammalian expression system with the scalability of flow cytometry, offering a practical platform for evaluating compounds that inhibit tau oligomerization.

2 Materials

2.1 Cell Culture and Transfection

- Expi293F cells (Gibco, Cat. A14527), stored in liquid nitrogen.
- Expi293 Expression Medium (Gibco, Cat. A1435101), 2–8°C, protect from light.
- ExpiFectamine™ 293 Transfection Kit (Gibco, Cat. A14525), store at 2–8°C.
- Opti-MEM™ Reduced Serum Medium (Gibco, Cat. 31985070), 2–8°C

2.2 Plasmids and Molecular Biology

- pmGFP10C–0N4R tau plasmid (Addgene, #71433) [15].
- pmGFP11C–0N4R tau plasmid (Addgene, #71434).
- *Escherichia coli* competent cells suitable for plasmid amplification (e.g., XL10-Gold ultracompetent cells, Agilent Technologies, Cat. 200314).
- Endotoxin-free plasmid purification kit (e.g., PureLink™ HiPure Plasmid Maxiprep Kit, Invitrogen, Cat. K210006).
- LB medium supplemented with antibiotics: ampicillin (100 μg/mL) and kanamycin (50 μg/mL).

2.3 Assay and Detection Reagents

- Ethidium homodimer-1 (Invitrogen, Cat. E1169), –20°C, protect from light.
- Test compound(s) (e.g., urea derivative MO-19-10 as validation control).
- Protease inhibitor cocktail (Thermo Scientific, Cat. 78429).
- RIPA Lysis and Extraction buffer (Thermo Scientific, Cat. 89900).

2.4 Equipment

- 125 mL polycarbonate vented Erlenmeyer shake flasks (sterile, disposable).
- Orbital shaker incubator (37°C, 8% CO_2, ≥80% humidity).
- Automated cell counter (e.g., Invitrogen Countess II).
- BD LSRFortessa™ or equivalent flow cytometer.
- Flow cytometry tubes.
- Refrigerated centrifuge.
- Standard laboratory equipment: biosafety cabinet, incubator, pipettes, centrifuge tubes.

2.5 Software (Optional)

- FlowJo software (BD Biosciences).
- GraphPad Prism (GraphPad Software).

3 Methods

This protocol describes a split-GFP complementation assay to monitor tau aggregation in Expi293 suspension cells. 0N4R Tau is fused to complementary GFP fragments (Tau-GFP10C and Tau-GFP11C) (*see* **Note 1**). When tau molecules interact, GFP fragments reconstitute fluorescence, which can be quantified by flow cytometry. The assay enables detection of tau aggregation in the absence of exogenous seeds and provides a platform for testing molecules that inhibit oligomerization.

3.1 Culturing Expi293F Cells (see Notes 2–5)

1. Thaw a vial of Expi293F cells rapidly in a 37°C water bath (1–2 min until mostly thawed). Transfer to a biosafety cabinet immediately after thawing.
2. Add 10 mL fresh culture media and centrifuge at 300 ×*g* for 10 min to remove freezing medium, then resuspend in fresh Expi293 Expression Medium.
3. Transfer cells into a sterile 125-mL vented Erlenmeyer flask containing 30 mL pre-warmed Expi293 Expression Medium.
4. Incubate in a shaking incubator at 37°C, 8% CO_2, ≥80% humidity, 125 rpm.
5. After 72 h, confirm cell viability ≥90% and expand until the culture reaches 4–6 × 10^6 cells/mL. Subculture as needed to maintain exponential growth. Cells should be passaged at least three times post-thaw before being used for transfection (*see* **Notes 3–5**).

3.2 Plasmid Amplification in E. coli (see Notes 6 and 7)

Tau-GFP10C and Tau-GFP11C plasmids (Addgene #71433, #71434) are typically supplied in a cloning strain, such as *E. coli* DH5α. These can be streaked onto LB agar plates containing the appropriate antibiotic (ampicillin for Tau-GFP10C; kanamycin for Tau-GFP11C) and reinoculated into liquid LB medium for plasmid preparation (see Sect. 3.3).

If retransformation is required, plasmids can be introduced into competent *E. coli* as follows:

1. Transform competent cells (e.g., DH5α, XL10-Gold *E. coli*) with 10–100 ng plasmid DNA by heat shock at 42°C for 45 s.
2. Recover in 950 μLS.O.C. medium at 37°C with shaking (225 rpm) for 1 h.
3. Plate 50–1000 μL (depending on the cell competency) onto LB agar plates containing the appropriate antibiotic.
4. Incubate overnight at 37°C and select single colonies.
5. Inoculate selected colonies into 5 mL LB broth with the corresponding antibiotic and grow overnight at 37°C, 225 rpm.

3.3 Plasmid Purification for Transfection (see Notes 8–12)

1. Expand single colonies into larger culture volumes (100–200 mL LB broth with the appropriate antibiotic).
2. Harvest cells by centrifugation and purify plasmid DNA using an endotoxin-free maxiprep kit, following the manufacturer's instructions.
3. Elute plasmid DNA in nuclease-free water (*see* **Note 10**).
4. Measure plasmid concentration and purity using a NanoDrop spectrophotometer or equivalent system (*see* **Note 11**).
5. Store DNA at −20°C for long-term use (*see* **Note 12**).

3.4 Transfection of Expi293F Cells (see Notes 13–15)

1. One day prior to transfection, adjust Expi293F cells to 2.5–3.0 × 10^6 viable cells/mL in Expi293 Expression Medium and incubate overnight.
2. On the day of transfection, confirm culture density at 4.5–5.5 × 10^6 cells/mL with ≥95% viability. Dilute to 3.0 × 10^6 cells/mL in fresh medium.
3. Prepare plasmid DNA by combining Tau-GFP10C and Tau-GFP11C at equal concentrations (final 1 μg/mL of cell suspension each) in Opti-MEM reduced serum medium.
4. Dilute ExpiFectamine 293 reagent in Opti-MEM according to the manufacturer's instructions. Mix with plasmid DNA and incubate for 10–20 min at room temperature to allow complex formation.
5. Add the DNA–lipid complex dropwise to the Expi293F culture with gentle swirling.
6. After 3 h, add test compounds at desired concentrations (e.g., 0.3–40 μM) together with a vehicle control (0.1% DMSO) and a nontreated control (*see* **Note 13**).
7. At 18–22 h post-transfection, add ExpiFectamine enhancers 1 and 2 as recommended by the manufacturer. Continue incubation for an additional 24–48 h with shaking at 37°C, 8% CO_2 (*see* **Notes 14 and 15**).

3.5 Flow Cytometry Analysis

1. Harvest cells 48 h post-transfection and transfer 1.0 mL per sample to flow cytometry tubes (*see* **Note 16**).
2. Acquire samples data on a flow cytometer equipped with a 488 nm laser and 530/30 nm emission filter for GFP detection.
3. Record a minimum of 10,000 live single-cell events per sample. At least three independent replicates are recommended for each condition (*see* **Note 17**).
4. Apply the following gating strategy (see Fig. 1):
   a. FSC vs SSC to exclude debris.
   b. FSC-A vs FSC-H to identify singlets and remove doublets.
   c. FSC vs SSC to distinguish live from dead cells (dead cells typically show increased side scatter and reduced forward scatter; *see* **Note 18**).
   d. Within the live cell population, gate GFP+ cells.
   e. Quantify the percentage of GFP+ live cells for each condition.
5. Perform statistical analysis using GraphPad Prism. Data should be analyzed using ordinary one-way ANOVA with Dunnett's post hoc test to compare compound-treated groups to the vehicle (0.1% DMSO) control.

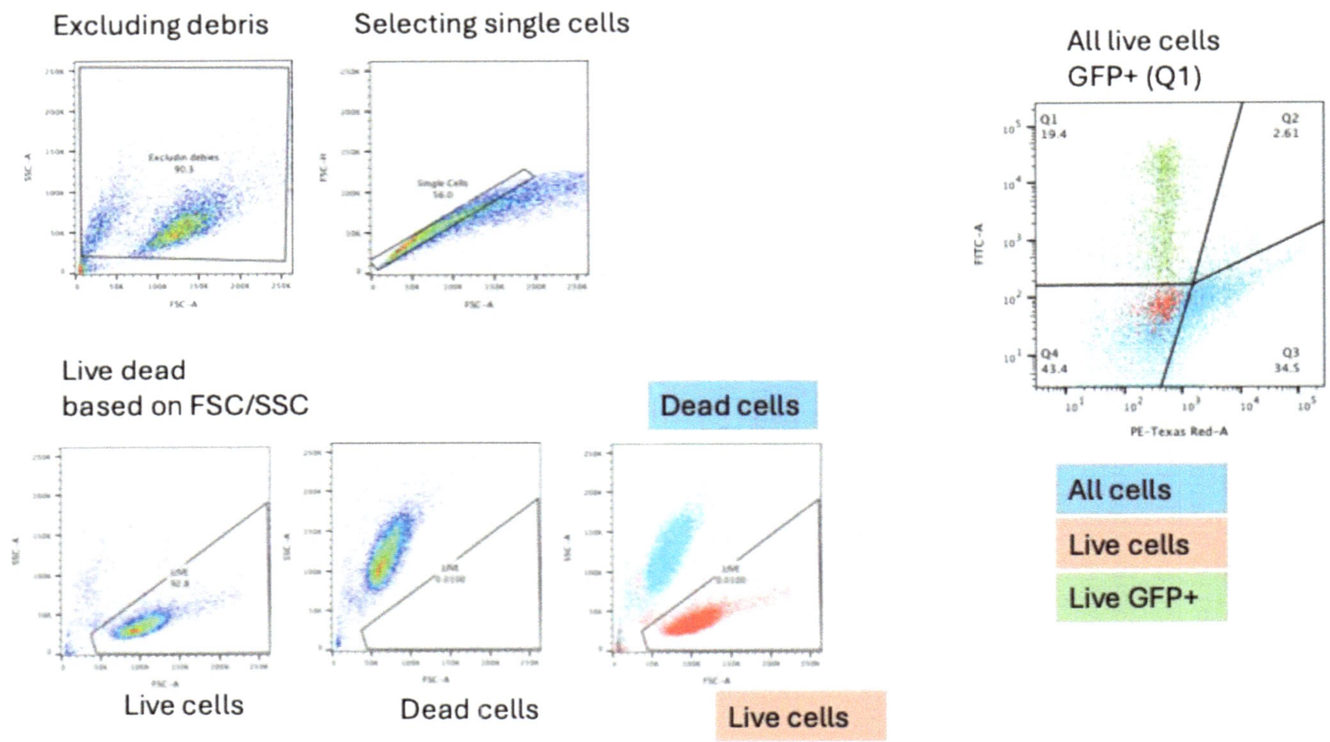

Fig. 1 Gating strategy for flow cytometry analysis of tau aggregation in Expi293 cells. Sequential gating was applied to identify GFP+ cells following split-GFP complementation. (**A**) FSC vs SSC gating to exclude debris. (**B**) FSC-A vs FSC-H gating to select singlets. (**C**) FSC vs SSC to separate live from dead cells, with dead cells typically showing increased side scatter and reduced forward scatter. (**D**) Representative dot plots illustrating exclusion of dead cells and retention of live cells. (**E**) Final gating of live GFP+ cells. The percentage of GFP+ cells within the live-cell population provides a direct readout of tau–tau interactions. Data shown were collected 48 h post-transfection; analysis can also be performed at 72 h, which increases GFP signal but may reduce viability. A minimum of 10,000 live single-cell events and at least three independent experiments are recommended for statistical robustness. In the absence of a viability dye, dead cells were excluded based on FSC/SSC properties, although viability stains (e.g., propidium iodide or ethidium homodimer-1) can provide greater accuracy

4 Representative Results

Using the 0N4R tau-split-GFP complementation assay, robust GFP signals were detected in Expi293 cells transfected with Tau-GFP10C and Tau-GFP11C, with approximately 40% GFP+ cells in the vehicle-treated group (Fig. 2).

Treatment with the reference compound **MO-19-10** resulted in a significant reduction in GFP+ cells, confirming the assay's ability to detect inhibitors of tau aggregation in a concentration dependent manner. In contrast, most of the other tested compounds did not alter GFP signal relative to vehicle, consistent with a lack of inhibitory activity. These results demonstrate that the assay can reliably distinguish active from inactive compounds and provide a quantitative readout of tau–tau interactions suitable for compound screening.

In addition to flow cytometry, fluorescence microscopy provided independent confirmation of tau aggregation and its inhibition by small molecules. In untreated Expi293 cells co-transfected with Tau-GFP10C and Tau-GFP11C, bright GFP fluorescence was observed in a subset of cells, indicating spontaneous tau–tau interactions leading to GFP complementation. By contrast, cells treated with **MO-19-10** displayed markedly reduced GFP fluorescence, consistent with the reduction in GFP+ populations quantified by flow cytometry (Fig. 3C). These microscopy observations support the conclusion that the split-GFP assay faithfully reports tau aggregation in live cells and that compound treatment reduces aggregate formation.

To further assess the assay's utility in screening, we tested a panel of small molecules with diverse chemical scaffolds. While most compounds showed little or no effect on the percentage of GFP+ cells compared with the vehicle control, a subset of compounds significantly reduced GFP signal (Fig. 3D). These findings demonstrate that the assay can discriminate between active and inactive compounds, capture concentration-dependent effects, and be adapted to medium-throughput screening. Together, the flow cytometry, microscopy, and compound screening data confirm that the split-GFP system provides a robust, versatile, and reproducible readout of tau aggregation in mammalian cells.

5 Advantages and Limitations

The split-GFP tau aggregation assay described here provides a number of advantages over previously reported cell-based systems:

- Tau aggregation without seeds. Unlike most cell-based tau assays that require preformed fibrils or cofactors, such as heparin to initiate aggregation, the Expi293 suspension system exhibits robust GFP reconstitution in the absence of exogenous seeds. This feature simplifies the workflow and enhances physiological relevance.

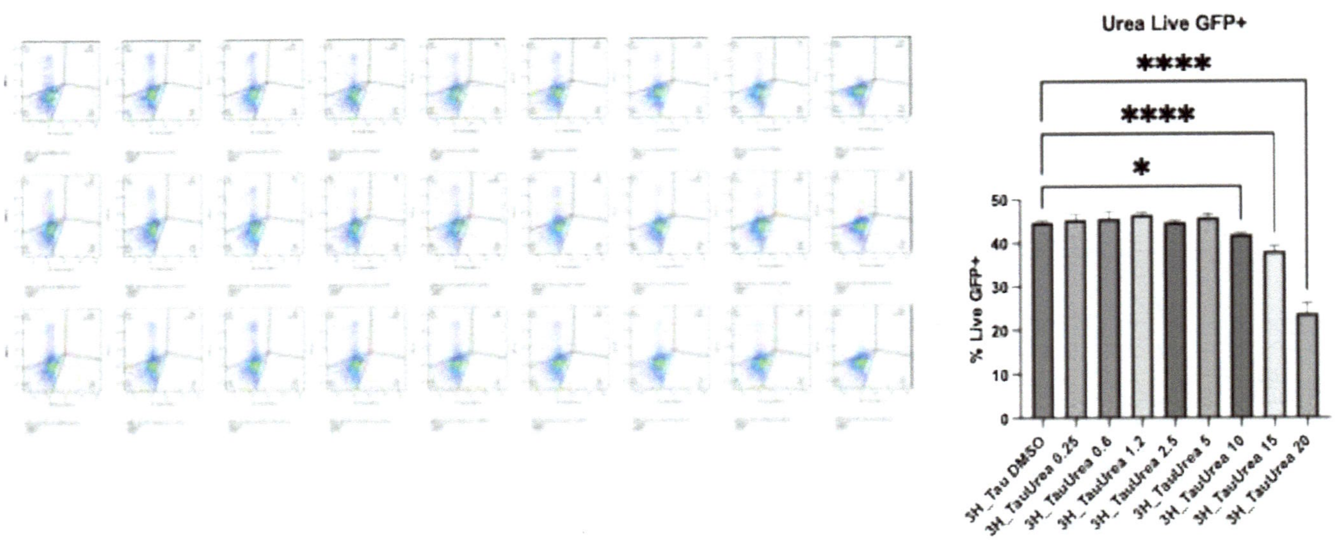

Fig. 2 Dose-dependent inhibition of tau aggregation by **MO-19-10** in Expi293 cells. Expi293 cells were transfected with Tau-GFP10C and Tau-GFP11C constructs and treated with vehicle (0.1% DMSO) or increasing concentrations of **MO-19-10**. Representative dot plots show GFP fluorescence within the live-cell population after 48 h. The summary bar graph (right) displays the percentage of GFP+ live cells as a function of compound concentration. Treatment with **MO-19-10** reduced the proportion of GFP+ cells in a concentration-dependent manner compared with vehicle. Data represent mean ± SEM from three independent experiments; statistical analysis was performed using ordinary one-way ANOVA with Dunnett's post hoc test versus vehicle

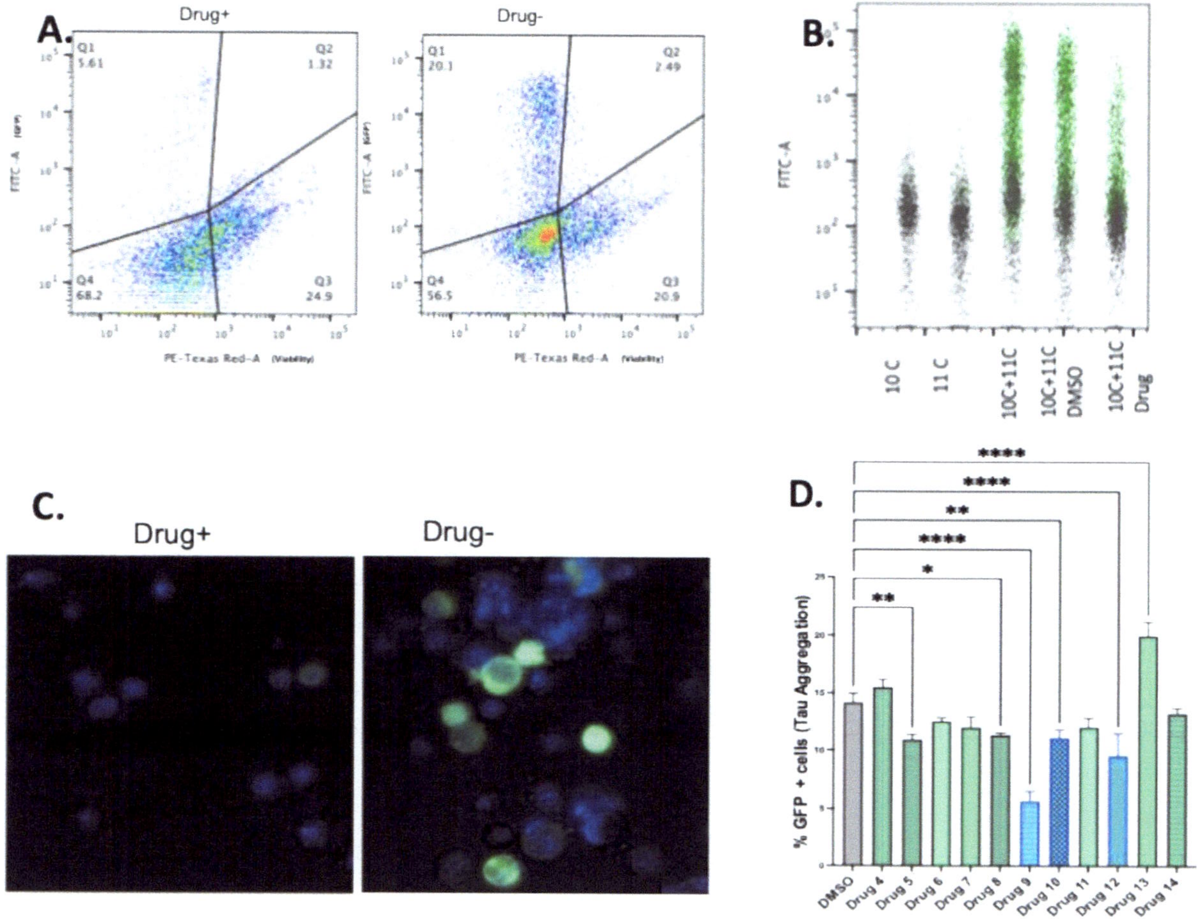

Fig. 3 (**A**) Flow cytometry dot plots showing GFP signal in transfected Expi293 cells under aggregating conditions (no drug) vs drug-treated cells. Drug candidate treatment reduces GFP signal, indicating inhibition of tau aggregation. (**B**) Concatenated histogram of GFP fluorescence in cells transfected with single constructs, both constructs, vehicle, or drug-treated. Drug candidate treatment decreases GFP signal compared with controls. (**C**) Representative fluorescence microscopy images of transfected cells with (left) and without (right) drug candidate treatment. Green: GFP (aggregated tau); blue: DAPI. Drug candidate reduces GFP signal. (**D**) Quantification of GFP+ cells by flow cytometry across a panel of drug candidate treatments. Bars represent normalized means ± SD (*$p<0.05$, **$p<0.001$, ****$p<0.0001$)

- **High sensitivity and scalability.** The use of flow cytometry enables the detection of tau–tau interactions at the single-cell level across thousands of events per sample. This provides both sensitivity and the throughput required for compound screening.
- **Quantitative and versatile readout.** GFP complementation yields a binary readout (GFP+ vs GFP–) that can be expressed as a percentage of positive cells, facilitating statistical comparisons. The assay can be adapted for dose–response studies, medium-throughput compound testing, and validation of candidate inhibitors.
- **Compatibility with microscopy.** GFP fluorescence can also be monitored by live-cell or fixed-cell imaging, providing complementary qualitative evidence of compound effects on tau aggregation.

Despite these strengths, there are limitations that should be considered:

- **Irreversibility of GFP complementation.** Once the GFP10 and GFP11 fragments reconstitute, the signal is stable and does not reflect dynamic disaggregation events. Thus, the assay is best suited to evaluate inhibitors of aggregation rather than enhancers of clearance.
- **Cell line specificity.** Robust spontaneous aggregation was observed in Expi293 suspension cells but may not occur in all mammalian cell lines. Optimization may be required, if other cell backgrounds are used.
- **Dependence on protein overexpression.** The assay relies on high levels of tau expression, which may not fully replicate endogenous tau biology. Findings should be validated in complementary systems.
- **Potential interference with GFP complementation.** Certain compounds may reduce GFP signal not by inhibiting tau–tau interactions but by interfering with the β-hairpin structure required for GFP fragment reconstitution. Because this interaction is mechanistically similar to tau's own β-hairpin aggregation motif, distinguishing between true tau inhibition and GFP complementation artifacts can be challenging. For this reason, hits from this assay should always be validated in orthogonal assays (e.g., thioflavin binding, biochemical aggregation assays, or other cell-based models).

Overall, the split-GFP assay offers a robust and practical method for detecting tau aggregation in mammalian cells and for evaluating potential therapeutic compounds. It complements existing in vitro biochemical and in vivo animal models by providing a quantitative, cell-based platform for translational studies.

6 Conclusions and Future Directions

The split-GFP complementation assay in Expi293 suspension cells provides a robust and scalable method for detecting tau aggregation in a cell-based context. By enabling spontaneous aggregation without the need for exogenous seeds, and by offering quantitative readouts through flow cytometry, the assay fills an important methodological gap between biochemical assays and in vivo models.

Future work should expand the assay to include additional tau isoforms and disease-associated mutants to capture the diversity of tau pathologies. Moreover, given the potential for compounds to interfere directly with GFP complementation, orthogonal assays—such as thioflavin-based aggregation assays, biochemical seeding assays, or alternative cell-based reporters—are essential for confirming specificity. Finally, the platform can be adapted for medium-throughput or automated workflows, making it suitable for early-phase compound screening and drug discovery pipelines.

Overall, this method provides a flexible foundation for studying tau aggregation and for identifying small molecules with potential therapeutic relevance in tauopathies.

7 Notes

1. We have used the 0N4R Tau split GFP constructs, but other forms of tau can be cloned as a fusion protein to split GFP.
2. Although this protocol was optimized for Expi293F cells, other high-density suspension lines may be suitable. We have also tested ExpiCHO cells, which showed tau aggregation under similar conditions.
3. Always consult the manufacturer's recommendations for culture medium, agitation speed, and CO_2/humidity settings, as these can vary between instruments and lots.
4. Expi293F cells should be used at ≤20 passages after thawing; higher passage reduces transfection efficiency and may affect reproducibility of tau aggregation.
5. Perform mycoplasma test regularly on all cell culture.
6. Plasmids received from Addgene are generally supplied in *E. coli* DH5α and can be expanded directly. Transformation is only necessary if plasmids are received as DNA or if a different host strain is preferred.
7. Maintain antibiotic selection throughout culture to avoid plasmid loss.
8. Add RNase A to the resuspension buffer prior to use to prevent RNA contamination.

9. During the lysis step, mix gently by inversion; vortexing may shear genomic DNA and reduce plasmid quality. Endotoxin-free preparations yield better transfection efficiency.
10. Elution in TE buffer may improve long-term stability but can interfere with some transfection reagents. Use water for immediate utilization in transfection.
11. High-quality plasmid DNA should have a 260/280 ratio of ~1.8 (acceptable range 1.8–1.9). Ratios lower than 1.8 indicate protein contamination; ratios significantly higher than 2.0 may indicate RNA contamination.
12. DNA preparations are generally stable for ≥6 months at −20°C. Always recheck concentration and purity before use, and periodically confirm integrity by agarose gel electrophoresis.
13. Always include a vehicle control in drug candidate assays to normalize for solvent effects (the MO-19-10) 0.1% DMSO.
14. The optimal shaking speed depends on flask size and fill volume. Consult the Expi293 manufacturer's guidelines to ensure sufficient aeration without excessive shear stress.
15. The assay can be performed at 48 h or extended to 72 h post-transfection; longer incubations may increase GFP signal but can reduce viability.
16. Collect at least 10,000 live single-cell events per sample to ensure robust statistics. A minimum of three independent experiments is recommended.
17. In the absence of a viability dye, dead cells can be excluded based on FSC/SSC properties (increased side scatter, reduced forward scatter). For greater accuracy, include a viability stain (e.g., propidium iodide or ethidium homodimer-1).
18. GFP complementation is irreversible; this assay is most suitable for inhibitors of aggregation, not clearance.

Conflict of Interest

The authors have no conflicts of interest to declare that are relevant to the content of this chapter.

References

1. Cowan CM, Quraishe S, and Mudher A (2012) What is the pathological significance of tau oligomers? Biochem Soc Trans 40:693–697
2. Rosenberg KJ, Ross JL, Feinstein HE et al (2008) Complementary dimerization of microtubule-associated tau protein: implications for microtubule bundling and tau-mediated pathogenesis. Proc Natl Acad Sci USA 105:7445–7450
3. Lasagna-Reeves CA, Castillo-Carranza DL, Sengupta U et al (2011) Tau oligomers impair memory and induce synaptic and

mitochondrial dysfunction in wild-type mice. Mol Neurodegener 6:39
4. Cowan CM, Bossing T, Page A et al (2010) Soluble hyper-phosphorylated tau causes microtubule breakdown and functionally compromises normal tau in vivo. Acta Neuropathol 120:593–604
5. Bramblett GT, Goedert M, Jakes R et al (1993) Abnormal tau phosphorylation at Ser396 in Alzheimer's disease recapitulates development and contributes to reduced microtubule binding. Neuron 10:1089–1099
6. Kim Y, Choi H, Lee W et al (2016) Caspase-cleaved tau exhibits rapid memory impairment associated with tau oligomers in a transgenic mouse model. Neurobiol Dis 87:19–28
7. Gerson JE, Farmer KM, Henson N et al (2018) Tau oligomers mediate α-synuclein toxicity and can be targeted by immunotherapy. Mol Neurodegener 13:13
8. Moussaud S, Jones DR, Moussaud-Lamodière EL et al (2014) Alpha-synuclein and tau: teammates in neurodegeneration?. Mol Neurodegener 9:43
9. Holmes BB, Furman JL, Mahan TE et al (2014) Proteopathic tau seeding predicts tauopathy in vivo. Proc Natl Acad Sci USA 111(41):E4376–E4385
10. Tak H, Haque MdM, Kim MJ et al (2013) Bimolecular fluorescence complementation; lighting-up tau-tau interaction in living cells. PLoS One 8:e81682
11. Hunke S, Müller VS (2012) Approaches to analyze protein-protein interactions of membrane proteins. In: Cai J (ed) Protein interactions. InTechOpen. https://doi.org/10.5772/38069
12. Masi A, Cicchi R, Carloni A et al (2010) Optical methods in the study of protein-protein interactions. In: Becchetti A, Arcangeli A (Eds.) Integrins and ion channels. Springer New York, New York, NY, pp 33–42
13. Schäferling M, Nagl S (2011) Förster resonance energy transfer methods for quantification of protein–protein interactions on microarrays. In: Wu CJ (Ed.) Protein microarray for disease analysis. Humana Press, Totowa, NJ, pp 303–320
14. Aggidis A, Devitt G, Zhang Y et al (2024) A novel peptide-based tau aggregation inhibitor as a potential therapeutic for Alzheimer's disease and other tauopathies. Alzheimers Dement 20:7788–7804
15. Brunello CA, Yan X, Huttunen HJ (2016) Internalized tau sensitizes cells to stress by promoting formation and stability of stress granules. Sci Rep 6:30498
16. Ibrahium OMH, Ademoye TA, Fortin JS, et al. (2025) A high-throughput drug screening assay for anti-tau aggregation using split GFP and flow cytometry. Sci Rep 15:37866

Part IV

In Vivo Models

Check for updates

Chapter 13

Modeling Functions and Pathology of Tau Protein in *C. elegans*

Ghulam Jeelani Pir, Bikash Choudhary and Eckhard Mandelkow

Abstract

Aggregation of Tau, a microtubule-associated protein, is a hallmark of multiple neurodegenerative disorders collectively termed tauopathies, which include dementia-related diseases like Alzheimer's disease. Technological advances in genetic engineering have transformed the way we study biology, offering powerful tools to explore fundamental processes and to recapitulate aspects of human disease in animal models in remarkably short time frames. Among these, *Caenorhabditis elegans*, a tiny, nonparasitic nematode, stands out for its fully mapped neuronal connectome, optical transparency, making it suitable for high-resolution behavioral and cellular analyses. These features offer an excellent platform for investigating mechanisms underlying tauopathies. In this chapter, we present methods for generating *C. elegans* tauopathy models and detail strategies to assess disease-related changes at behavioral, cellular, and biochemical levels, with the aim of advancing our understanding of Tau-driven neurodegeneration.

Keywords *C. elegans*, Tau, Posttranslational modifications, GABAergic neurons, Glutamate-responsive neurons, Thrashing assay, Mechanosensory neurons, Aging, Neurofibrillary tangles/paired helical filaments

1 Introduction

Tau, a microtubule-binding protein formally known as microtubule-associated protein Tau (MAPT), is highly soluble under physiological conditions and mostly localized to the axons, where it plays a critical role in neuronal development and maintenance [1]. Under pathological conditions—collectively termed tauopathies—Tau misfolds and aggregates into insoluble neurofibrillary tangles within neurons. These aggregates may arise from familial mutations in the endogenous *MAPT* gene or occur sporadically when wild-type Tau undergoes posttranslational modifications or interactions triggered by alterations in the cellular environment that lead to its loss of function, toxic gain of function or mislocalization [1].

The link between neurodegenerative disease and abnormal aggregates in brain cells was first noticed by Alois Alzheimer [2]

Jessica S. Fortin (ed.), *Translational Research Methods for Tauopathies*, Neuromethods, vol. 233,
https://doi.org/10.1007/978-1-0716-5234-3_13,

using light microscopic studies of patient brain tissue obtained post-mortem. They include "amyloid plaques" (made up of aggregated amyloid-beta peptide) and "neurofibrillary tangles" (NFT). The NFTs are now known to contain Tau protein in a filamentous state dominated by beta structure [3, 4], plus a variety of attached proteins [5]. This is in contrast to Tau in healthy neurons, which is highly soluble because of its hydrophilic composition. Tau's structure and interactions are regulated by multiple posttranslational modifications (PTMs) with phosphorylation being the most extensively studied. Cellular Tau is normally in a state of high phosphorylation; however, postmortem it is quickly dephosphorylated by phosphatases. In non-AD subjects, this leads to the artefact of "low" phosphorylation, whereas in AD-patients many phosphorylation sites are protected within Tau aggregates, reminiscent of the in vivo state [6, 7]. The current use of certain hyperphosphorylated Tau residues (e.g., 181, 217, etc.) may be valid as early biomarkers of AD, but derived from CSF which is different from neuronal Tau, and serve primarily as diagnostic indicators rather than evidence of functional roles. The appearance of "abnormal" Tau in the brain progresses in a characteristic pattern that correlates with the clinical observations on dementia [8]. However, since the earliest hallmarks precede dementia by years, the causative factor(s) are still unknown, and therefore, the available drugs (mainly acting on neuronal receptors) can only slow down but not cure the disease.

In research labs, a large number of animal models of Tau pathology have been generated to model AD and observe the effects by cell biological or behavioral methods (e.g., in mice, flies, zebra fish, or worms). Most of the models, specifically the rodent models, were mostly based on rare genetic mutations of Tau that promote its aggregation and proved invaluable for studying tauopathies. How ever, these models present limitations, including high costs, lengthy timelines for pathology development, complex nervous systems, and challenges in conducting large-scale genetic or pharmacological screens [9].

To address these challenges, smaller genetic model organisms such as *C. elegans* and *Drosophila melanogaster* offer clear advantages: rapid generation of transgenic lines, minimal genetic redundancy, short lifespans, and suitability for high-throughput screening [10, 11]. Here, we focus on studies in *C. elegans* that was first established as a genetic model by Sydney Brenner to study developmental biology, and is now widely used to investigate neurodevelopment, aging, and neurodegeneration [12].

C. elegans model offers several advantages. It shares many conserved neurobiological features with higher vertebrates, including excitatory and inhibitory neurons, comparable neurotransmitters, and similar signaling pathways. Its fully mapped neuronal connectome, transparent body, ease of transgenesis, and capacity for single-neuron resolution studies make it an ideal platform for modeling neurodegenerative diseases. The simplicity and cost-effectiveness of *C. elegans* husbandry—maintained in large numbers on bacteria-seeded plates, with self-fertilizing hermaphrodites producing genetically identical offspring—further enhance its utility. Its short life cycle, with development from egg to adult in just ~3 days (see Fig. 1), allows rapid experimental turnover.

These advantages have led to the generation of *C. elegans* models for amyotrophic lateral sclerosis (ALS), Parkinson's disease, Huntington's disease, and tauopathies [13]. Kraemer et al. [14] were the first to lay the foundation for tauopathy modeling in *C. elegans*, which have since been carried forward with remarkable research that ultimately translated to the mammalian systems. *C. elegans* involved engineering wild-type human Tau (Tau^{WT}) in the nervous system and its mutant variants that are either prone for aggregation such as V337M, P301L & deletion of K280 (ΔK280), or novel mutations such as A152T whose pathological consequences were not known at that time, or mutations that mimicked posttranslationally modified Tau, e.g., pseudohyperphosphorylated Tau with glutamate substitutions of serine or threonine [14–18]. Aggregation-prone Tau variants (i.e., V337M, P301L, or ΔK280) caused pathology partly through the buildup of insoluble Tau aggregates [14, 17].

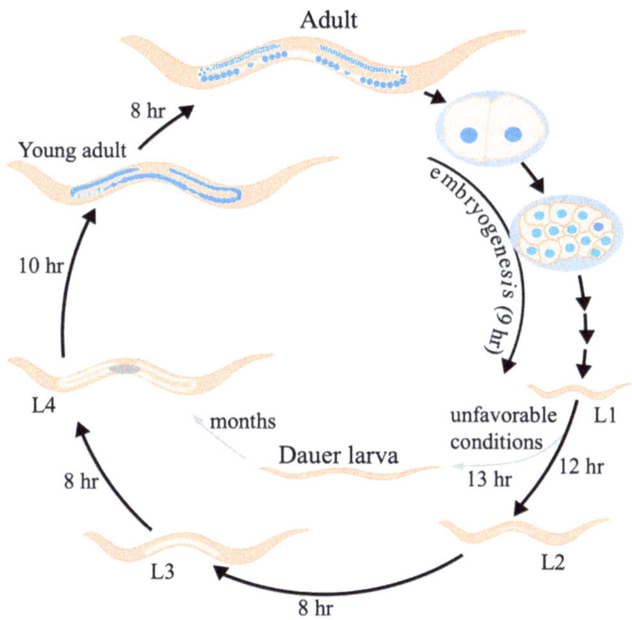

Fig. 1 Overview of the life cycle of a hermaphrodite worm. Oocytes are self-fertilized and embryos laid at ~30 cell stage. Embryonic development progresses through four larval stages (L1–L4) before reaching a fully mature adult and the entire cycle (egg to egg-laying adult) is completed in roughly 3 days under normal laboratory conditions. Under unfavorable conditions (e.g., crowding, food scarcity, or high temperature), the L1 larva enters an alternative developmental stage called dauer, which can live for months before metamorphosing into L4 larva once conditions improve

In contrast, Tau^{A152T} variant produced severe pathology without insoluble Tau accumulation, reinforcing the emerging notion that preinsoluble Tau species may be equally, if not more, toxic [18]. Several aspects of neurodegenerative diseases were recapitulated in these models, among which behavioral phenotypes like uncoordinated locomotion and paralysis were thoroughly exploited to identify genetic, peptide, or pharmacological modifiers, approaches that are either highly costly or impossible in mammalian models. An example is the genetic modifier "suppressor of Tau pathology" *sut-2* [19], whose homolog (MSUT-2) in the subsequent work was found to share the functional aspects in the mammalian brain, influencing the structure of RNAs [20]. Similarly, a peptide that reduced Tau aggregation and associated pathology was first identified in *C. elegans* and later validated in a cellular model [21]. The nearly transparent appearance of *C. elegans* has since enabled in vivo studies of Tau biology, including the selective vulnerability of glutamate-responsive versus GABAergic motor neurons [22], accelerated turnover of a Dendra2::Tau construct compared to Dendra2 alone [23], increased diffusion kinetics of the Tau^{A152T} variant (attributed to an extra phosphorylation site in the form of Thr152) relative to Tau^{WT} [24], and the progressive onset of Tau pathology across brain cells in humans affected by AD [25–27]. A subsequent genetic screen revealed modifiers of endolysosomal integrity, which was compromised in this model, and these findings were again validated in human cells [25]. For more, we refer our readers to articles where the authors have compiled *C. elegans* models of neurodegenerative diseases based on Tau and other human disease-associated proteins [11, 28–32], since covering all these models is out of scope of this article. In this chapter, we outline the methodology for:

(a) generating transgenic lines expressing wild-type human Tau (Tau^{WT}, 2N4R) and its mutant variants (e.g., Tau$^{\Delta K280}$, Tau^{A152T}) associated with tauopathies [17, 18],

(b) behavioral characterization of these models,

(c) biochemical assays to assess Tau aggregation, and

(d) neurodegeneration analysis, including single-neuron pathophysiology.

2 Materials

2.1 Reagents for the Generation of Tau-Transgenic Animals

Reagents for Generating Extrachromosomal Transgenic Animals

- Well growing L4 to young adult wild-type (N2) animals.
- NGM (Nematode growth medium) pouring material (NGM powder, AGscientific, 23 g + 1 mL of 1 M $MgSO_4$ + 1 mL of 1 M

$CaCl_2$ + 25 mL Phosphate buffer, make the volume to 1 L by adding H_2O).

- Phosphate buffer (108.3 g KH_2PO_4 + 35.6 g K_2HPO_4, make the final volume 1 L, and adjust pH to 6).
- NGM petri dishes seeded with concentrated OP50.
- Microinjection capillaries/needles, Microscope slides, Microloader pipette tips.
- Halocarbon oil (HC-700", 8 oz. bottle (Lab Scientific Inc.)).
- Plasmid DNA for injection were wild-type human Tau (Tau^{WT}, 2N4R) and its genetic mutant variants expressed under a *C. elegans* pan-neuronal promoter (e.g., *prab-3*, *psnb-1*, *prgef-1*) and a reporter plasmid (we generally use a plasmid that labels the coelomocyte or pharynx with GFP or RFP, which can be easily identifiable under stereo-fluorescence microscope, e.g., *punc-122::*GFP for GFP labeled coelomocyte, *pmyo-2::GFP* for labeled pharynx) [18, 33].
- M9 buffer (3 g KH_2PO_4 + 6 g Na_2HPO_4 + 5 g NaCl + 1 mL of 1 M $MgSO_4$, make final volume 1 L with H_2O, pH should be 6.8–7.0).
- 2% agarose pads (experimental grade agarose in melted condition, 2 g agarose in 100 mL M9 buffer).
- Calcium chloride (1 M) and Magnesium chloride (1 M).
- Platinum wire-based worm-picker, Bunsen burner/spirit lamp and Heat plate.
- Tabletop microcentrifuge.
- Dissecting stereomicroscope with bottom illumination and fluorescence stereomicroscope (for screening the fluorescently labeled transgenic animals) (this is common equipment for other material sections, so we are not mentioning in each section, unless there is some variant).
- Bright field Injection Microscope (should be equipped with DIC prisms and the objectives 10× and 40×) assembled with the micromanipulator.
- Worm lysis buffer (10 mM Tris, pH 8.2; 50 mM KCl; 2.5 mM $MgCl_2$; 0.45% NP40; 0.45% Tween 20; 0.01% gelatine; 0.5 mg/mL Proteinase K).

Generating the Integrated Transgenic Lines from the Extrachromosomal Lines

- 12-well cell culture plates with each well poured with NGM and seeded with OP50 bacterial strain.
- Well growing L4 or young adult extrachromosomal transgenic animals.
- Trioxsalen and DMSO.

- Plastic pasture pipette.
- UV strata linker.
- 20°C incubator.

2.2 Behavioral Analysis with the Isolated Integrated Tau-Transgenic Lines

Thrashing Assay (a Read-Out of Motor Activity of The Animal)

- Microslide.
- M9 buffer.
- Dissecting stereomicroscope with bottom illumination and equipped with camera (Olympus DP17).

2.3 Biochemical Analysis of the Total and Insoluble Tau in the C. elegans Tauopathies Model

- Four to five well-grown plates of worms of following genotypes: N2, transgenic worms overexpressing Tau^{WT} and its mutant variants.
- 2× protein sample buffer (0.5 M Tris-HCl pH 6.8, Glycerol, 10% SDS, 0.1% Bromophenol Blue).
- 5× protein sample buffer (1.25 M Tris-HCl pH 6.8, Glycerol, 50% SDS, 0.25% Bromophenol Blue).
- β-mercaptoethanol.
- 30% sucrose solution.
- High salt RAB buffer (100 mM 2-(N-morpholino) ethanesulfonic acid (MES), 1 mM EGTA, 0.5 mM MgSO4, 20 mM NaF).
- Protease Inhibitors (Sigma-Aldrich P8340, Hamburg, Germany).
- Primary and secondary antibodies (check in Methods section).
- RIPA buffer (150 mM NaCl, 1% Nonidet P-40, 0.5% deoxycholate, 0.1% SDS, 50 mMTris, pH 8.0).
- Urea buffer (30 mM Tris, 7 M urea, 2 Mthiourea, 4% CHAPS (3-[(3-cholamidopropyl)dimethylammonio]-1-propanesulfonate).
- Sonicator: Sonorex RK102 (Bandelin), Sonoplus Sonotrode 73 (Bandelin).
- Centrifuges: Centrifuge 5415 D (Eppendorf), MIKRO 200 R (Hettich), Centrifuge 5810 R (Eppendorf), TL-100 ultracentrifuge (Beckman Coulter).
- Western blot apparatus (Biorad and nitrocellulose membrane from Millipore).
- Blotting buffer (75 mM Tris-Cl, 20% (v/v) methanol, pH 7.4).

2.4 Neurodegeneration and Neuronal Aging Analysis of Tau Transgenic Worms

- Synchronized worm populations of various genotypes at various life stages such as L4, Day 1 (D1), Day 2 (D2).
- *C. elegans* transgenic strains with GABAergic nervous system (*juIs76*) and mechanosensory neurons (*zdIs5*) labelled with green fluorescent protein (GFP).
- Genetic crosses in cross plates.
- 2% Agarose pads, Microslides, Coverslips, and Immersion oil.
- 20 mM sodium azide.
- Epifluorescence microscope equipped with 20× and 60× objectives and CMOS type camera for image capturing.
- ImageJ software for image analysis and image processing.

3 Methods

3.1 Generation of Tau-Transgenic Worms Carrying Extrachromosomal Arrays

This protocol outlines the fundamental steps for generating transgenic animals. In *C. elegans*, plasmids injected into individuals are not stably inherited by all progeny across generations. Within a population, only a subset of animals will carry the injected plasmids [34]. When plasmids are transmitted to subsequent generations with high frequency, the resulting transgenic lines are described as exhibiting high-frequency transmission; conversely, if the transmission frequency is low, the lines are termed low-frequency transmission. These transgenic strains are referred to as extrachromosomal arrays. The following steps detail the procedure for generating such extrachromosomal arrays:

1. Prepare 2% agarose in water by bringing it to a boil, mixing thoroughly, and keeping it warm in a heat block. With a cut-off P200 tip, dispense a drop (~100 μL) of hot agarose onto a glass coverslip (50 × 22 mm), avoid air bubbles. Quickly place a second coverslip on top of the drop and gently tap to spread it. Once the agarose solidifies, carefully peel or slide the coverslips apart. Bake the coverslip–pad in a vacuum oven at 80°C for 1 h to overnight.
2. Prepare the DNA injection mixture consisting of Tau plasmids (*psnb-1::Tau*^{WT}or its mutant variants *psnb-1::Tau*$^{\Delta K280}$, *psnb-1::Tau*^{A152T}) and a reporter plasmid (*pmyo-2*::GFP) for injection. The recommended concentration is 20 ng/μL for reporter plasmids and 75 ng/μL for Tau constructs (*see* **Note 1**).
3. Centrifuge the DNA mixture for few minutes to settle down the particulate matter that might clog the microinjection needle. Pipette upper 5-10 μL of the mixture and transfer it to a

new microfuge tube which will be used for injection. Prepare DNA mixture for each Tau variants.

4. Pick 10–15 young adult animals (here, young adult refers to 3–4 h post L4) (*see* **Note 2**) in an OP50-seeded NGM plate.
5. Switch on the microscope and microinjector manipulator. The setup includes an Olympus IX51 microscope with 10× and 40× objectives, equipped with Narishige MO-202 and MMN-1 micromanipulators. Pressure for injections is supplied by an Eppendorf FemtoJet. Turn on the microscope (switch on the right side) and the FemtoJet (switch on the back right). Pressure settings are automatic: P1 = 5200 hPa (clean), P2 = 1200 hPa (injection), and P3 = 180 hPa (resting). The foot pedal controls injection—press to inject, release to stop. Use the clean button to clear needle clogs.
6. Load the DNA injection mixture in the microloader pipette tips and finally into the microneedle (2-3 μL). We use the prepulled microinjection needles. Carefully insert the new needle from the back end to avoid damaging the tip, ensuring approximately 1 cm of the needle extends beyond the rear of the holder. Secure the needle by tightening the assembly onto the manipulator arm. Adjust the fine and coarse control knobs on the micromanipulator to their midpoint positions. Using the coarse controls on the stage-mounted part of the micromanipulator, position the needle tip left/right and forward/back until it is just above the objective lens, visible as a glow from the condenser's light.
7. A drop of halocarbon oil (30–50 μL) is placed on the prepared agarose slide from Step 1. Gently lower the needle into the oil using the coarse Z control, stopping as soon as you notice slight oil displacement—avoid pushing too far down. At 10× magnification, center both the breaking bar (explained below) and the needle tip within the field of view. Switch to 40× magnification and repeat the centering process.
8. Once the injection needle is centered and loaded, it must be carefully broken to allow proper flow. We do it directly on the agarose pads. Using the fine adjustment controls, slowly lower the injection needle toward the stage until the needle tip is precisely in the same focal plane as the agarose pad—ensuring the tip is clearly visible. Then, gently move the needle back and forth, lightly rubbing it against the surface of the agarose pad. Next, test the needle by pressing the foot pedal to apply pressure and observe the flow. At pressure setting P2, a rapid laminar flow outward from the needle tip, followed by a return flow toward the shaft, should be visible. At resting pressure, no flow should be detected. If no flow is observed at P2, the needle is likely

still plugged; repeat the breaking procedure until proper flow is achieved. Once the needle is successfully broken and flow is confirmed, carefully lift the needle away from the stage to allow placement of the injection pad containing worms onto the floating stage.

9. Transfer a single animal to the edge of the halocarbon oil, ensuring it remains on the agarose pad. This step should be performed under a dissecting stereomicroscope. Place the slide with the immobilized worm onto the stage of the injection microscope. Using 10× magnification, locate the animal and ensure it is oriented with the vulva away from the needle. Adjust the coverslip to position the worm at around 45° angle to the needle, maximizing needle path length within the gonad and reducing the risk of piercing through. Lower the needle into the focal plane using the fine adjuster, moving only the stage to position the worm. Switch to 40× and focus on the syncytial gonad arm (clear, sausage-shaped with nuclei along the sides). Align the gonad against the body wall so the needle will enter it. Penetrate either by gently vibrating the needle or by moving the stage forward. Once inside, inject until the gonad visibly fills and liquid flows into the proximal region, then stop.
10. After successful injection, gently withdraw the needle from the injected animal and lift it out of the oil using the fine controls, then remove the pad. Using 10× objective on the dissecting stereomicroscope, locate the worm and place a drop of M9 buffer on top of the injecting worm to break it free, and move the worm onto freshly OP50-seeded NGM plates (*see* **Note 3**).
11. Inject 10–15 N2 worms for each plasmid mixture. Ensure that a different microneedle is used for each distinct plasmid to avoid cross-contamination.
12. Injected animals (parent, P0) will lay eggs, producing F1 progeny (filial 1). Screen F1 under a stereo fluorescence microscope for fluorescence (GFP in this case), and transfer each F1 to a different plate to get independent lines. Allow plates to grow for one or two more generations, checking the frequency of fluorescent (transgenic) animals. If F2 or later generations lack transgenic animals, discard the line as it does not stably inherit the extrachromosomal array.
13. The identified independent lines can be categorized into two major classes: (a) high frequency transmission (can range from 40% to 80%), (b) low frequency transmission (5%–10%). Maintain both classes of independent lines.

3.2 Transgene Confirmation by—Single-Worm PCR

1. Identify worms carrying the transgenic fluorescent marker under a fluorescence stereomicroscope.
2. Transfer a single worm into a PCR reaction tube containing 4 μ L of worm lysis buffer.
3. Incubate the samples at –80°C for 40 min.
4. Place the tubes in a thermal cycler and run using the settings: 65°C for 40 min, 95°C for 10 min, Hold at 4°C.
5. Use the lysate directly as the template for PCR. Optimize cycling parameters according to the specific primer set and the expected fragment size (confirmed using agarose gel electrophoresis).
6. Once confirmed, assign unique identifiers to transgenic strains carrying extrachromosomal arrays to facilitate accurate tracking of each generated line.

3.3 Generation of Stable Lines from Animals Carrying Extrachromosomal Array

The extrachromosomal array identified in Sect. 3.2 generally exhibits variable expression of the protein of interest in the nervous system across the population, making homogeneous scientific analysis challenging. Therefore, establishing a stable integrated transgenic line is crucial. In such a line, the extrachromosomal DNA becomes stably integrated into the genome, functioning like an endogenous gene to ensure stable inheritance across generations [35]. The following steps are performed to generate the integrated stable Tau^{WT} line and its mutant variants.

1. Prepare the Trioxsalen solution in DMSO at a concentration of 2 mg/mL (*see* **Note 4**).
2. Select extrachromosomal transgenic animals (young adults from a low frequency line are preferred) from well growing plates, approximately 80 animals of each genotype, and transfer them to separate OP50-seeded NGM plates (*see* **Note 5**).
3. Wash the worms with M9 two to three times to remove bacteria adhered to their bodies.
4. Retain 200 μL of M9 containing the worms in a microfuge tube, add 15 μL of Trioxsalen stock solution, and place the tube in the dark on a rotating microfuge mixer for 15 min.
5. While the worms are incubated with Trioxsalen, prepare concentrated OP50 NGM plates. "Concentrated OP50 refers to a well-grown OP50 bacterial culture (2 mL) that is centrifuged in a microfuge tube, with the resulting pellet resuspended in 200–300 μL of the remaining liquid culture."
6. After incubation with Trioxsalen (from step 4), use a glass pipette to place a drop of the worm-containing solution onto unseeded NGM plates (3–4 plates).
7. Place each plate (without lid) into the UV stratalinker and expose to 300 J/m² [35].

8. Immediately after UV treatment, add 100 μL of concentrated OP50 directly onto the worms.
9. Incubate plates at 20°C for 7–8 h.
10. Pick surviving adult worms onto 10–15 fresh OP50-seeded NGM plates, with 10–12 worms per plate per genotype.
11. The following day, transfer treated worms to fresh plates. The original plates containing laid eggs are designated as Day 0 (D0).
12. In the meantime, prepare 12-well cell culture plates containing NGM and seed each well with OP-50 bacteria. This format facilitates easier screening and allows a greater number of transgenic worms to be examined. One can also use regular NGM-poured petri dishes commonly used for *C. elegans* culture.
13. After 24 h, transfer worms to fresh OP50-seeded NGM plates, the parent plate is now termed as Day 1/D1.
14. Allow the F1 generation to develop, then transfer single transgenic worms from D0 plates (200 worms) and from D1 plates (200 worms) into individual wells of 12-well plates under a stereo-fluorescence microscope (*see* **Note 6**).
15. Let singled worms grow in the wells and check their F2 progeny for the fluorescence reporter under the stereo-fluorescence microscope.
16. Mark wells with the highest number of fluorescent animals and select 10–15 such wells (*see* **Note** 7).
17. From each selected well, single out worms into new 12-well plates (each plate corresponds to one selected well) and allow them to grow and produce progeny.
18. Examine the plates from Step 17 under a stereo-fluorescence microscope for the reporter fluorescence. If a strain is integrated, at least one or two wells will show reporter fluorescence in all animals, indicating homozygosity for the transgene and, therefore, an integrated line.
19. Mark wells originating from different 12-well plates, as each represents an independent integrated transgenic line. Pick worms from these wells, transfer them to standard NGM plates seeded with OP50, and allow them to grow.
20. Typically, three to four independent integrated lines are retained for future characterization.
21. These independent integrated lines are then backcrossed seven times with N2 to eliminate background mutations that may have arisen from the Trioxsalen and UV treatment.
22. Freeze the resulting strains for long-term storage and future use.

3.4 Analyzing Tau Expression Levels in the Integrated Transgenic Lines

At this stage, we have multiple stably integrated transgenic lines of Tau^{WT} and its mutant variants (e.g., $Tau^{\Delta K280}$, Tau^{A152T}). These independently isolated stable lines may exhibit variation in total Tau expression levels. Based on our experience, Tau expression exerts a dosage-dependent effect on various aspects of the animals, ranging from behavior to neurodegeneration. Therefore, we select isolated variants that show comparable expression levels of Tau^{WT} and the mutant variants [18]. To assess total Tau expression, we extract total protein from worms of the different isolated lines and perform standard western blotting using a Tau-specific antibody (we mostly use K9JA pan-tau antibody, Dako A-0024). We follow a basic Western blot protocol, briefly outlined below:

1. Grow worms from the designated independent integrated transgenic lines expressing Tau^{WT} and the mutant variants.
2. Wash the worms with M9 and collect the worm pellet for total protein extraction (see Sect. 3.6 in detail).
3. Store the final lysate at –80° for long-term storage.
4. Run SDS PAGE with the prepared lysates by loading equal amounts of protein. Transfer the proteins to a nitrocellulose membrane and probe with an anti-Tau-specific antibody (making sure a loading control is also included at this stage).
5. The probed blot is developed, imaged, and densitometry analysis is performed on the acquired blot image to compare the Tau expression levels.
6. Select isolated integrated transgenic lines exhibiting comparable Tau expression between Tau^{WT} and the mutant variants for future characterization or analysis (*see* **Note 8**).

3.5 Quantitative Scoring of Motor Activity in Tau Transgenic Lines

Expression of Tau, a microtubule-binding protein in the nervous system, can influence neuronal physiology in a neuron subtype-specific manner. One such effect is on worm locomotion (see Fig. 2), where the characteristic sinusoidal movement depends on the balance between excitatory and inhibitory neurons. Previous studies have shown that genetic variants of human Tau linked to tauopathies, when expressed in the *C. elegans* nervous system, lead to the loss of GABAergic motor neurons, thereby impairing the animals' motor activity [14, 17]. Consequently, the motor activity assay is a primary and routinely performed test to evaluate the pathological consequences of novel Tau mutations identified in human patients. The following steps outline the locomotion/thrashing assay performed on *C. elegans* tauopathy lines:

1. Grow a synchronized population of the selected Tau transgenic lines with comparable expression of Tau^{WT} and the mutant Tau variants.

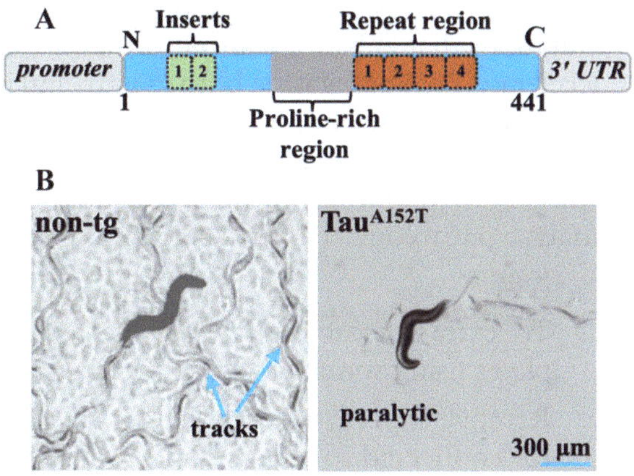

Fig. 2 (**A**) Schematic representation of a transgene construct showing the main elements needed for stable transgene expression: promoter, transgene cDNA and 3' UTR. A variety of *C. elegans* promoters are available to drive tissue-specific expression (e.g., neuronal, muscle, hypodermal, and intestinal). Multiple human Tau isoforms can be expressed in worms, depending on the number of N-terminal inserts (one or both) and C-terminal repeat regions (three or four), in either wild-type or mutant forms, to model tauopathies in transgenic worms. (**B**) An example of a nontransgenic healthy worm and a paralytic worm expressing mutant human Tau (Tau^{A152T}) is shown

2. Place a 20–30 μL drop of M9 on a microscope slide, transfer a single worm of the genotype to be analyzed, and allow it to acclimatize for 1–2 min before recording.
3. Start live imaging of the worms in the M9 drop using the dissecting stereoscope equipped with a monochrome camera.
4. Analyze the recorded videos by counting the number of body bends (thrashes) per unit time, typically over a 30–60 s interval. Quantify locomotor activity by averaging thrashing rates across multiple worms per genotype and perform statistical comparisons to assess differences between Tau^{WT} and mutant Tau variant lines.

3.6 Biochemical Analysis (Tau Extraction from Transgenic C. elegans)

Tau aggregation is a key feature of several neurodegenerative disorders but is also observed in cognitively normal elderly individuals with primary age-related tauopathy (PART) [36], a condition whose distinction from early Alzheimer's disease is still debated. Although Tau is natively unfolded, it can assemble into highly ordered fibrils, a process largely driven by two short hexapeptide motifs—VQIINK and VQIVYK—located in the repeat domain [37]. These sequences have a strong propensity to form β-sheet structures and are crucial for Tau fibrillization. The VQIVYK motif alone can form steric zippers, tightly interdigitated β-sheet that drive fibril assembly. Tau aggregation is thought to begin with exposure of these zipper-forming motifs, facilitating β-sheet stacking [38]. Disruption of these motifs, such as by introducing proline mutations, inhibits aggregation, whereas mutations that enhance β-sheet formation (e.g., ΔK280 or P301L) accelerate Tau aggregation both in vitro and in vivo [39], underscoring their functional importance. Worm models have been generated to mimic this pathological accumulation of insoluble Tau found in human tauopathies, and the following methods are carried out to analyze pathological accumulation of insoluble Tau in transgenic *C. elegans.*

Total Tau Extract

1. To extract total Tau, wash synchronized worms off NGM plates using water and leave on ice for 5 min to settle.
2. Remove supernatant containing *E. coli* OP50 using an aspirator connected to a glass trap.
3. Wash the worm pellets further with water and centrifuge at 1000 rpm to remove the residual bacteria—repeating at least three times until the worm suspension appears clear.
4. Resuspend the clean pellets in 1× protein sample buffer (2× w/v) containing 355 mM 2-mercaptoethanol, 1× Complete Protease Inhibitor Mixture (Sigma-Aldrich P8340, Hamburg, Germany), 1 μM okadaic acid, and 0.5 mM PMSF.
5. Boil the samples at 96°C for 10 min with continuous agitation at 14,000 rpm.
6. The lysates are then centrifuged briefly to pellet the debris, and the supernatant containing total Tau protein is collected.

Insoluble Tau by Sequential Extraction

To quantify insoluble Tau aggregates, a sequential extraction is performed using buffers of increasing stringency—first extracting soluble cytoplasmic, nuclear, and membrane-associated proteins, and finally isolating the insoluble fraction.

1. Harvest worms from 10 cm NGM plates using cold 0.1 M NaCl solution and transfer to a 15 mL conical tube.
2. Briefly vortex the tube; agar chunks will adhere to the tube walls while worms remain suspended, transfer the suspension to a fresh tube.
3. Centrifuge at 1000 rpm for 1 min and carefully discard the supernatant.
4. Resuspend the pellet in 10 mL cold (4°C) 0.1 M NaCl, centrifuge at 1000 rpm for 1 min, and remove the supernatant.
5. Resuspend the washed pellet in 5 mL cold 0.1 M NaCl. Add an equal volume (5 mL) of ice-cold 60% sucrose solution and mix gently but quickly.
6. Carefully overlay the mixture with 2 mL ice-cold 0.1 M NaCl, and centrifuge at 3000 rpm for 5 min.

7. Live worms will accumulate at the interface between the 0.1 M NaCl and sucrose layers.
8. Collect the interface (~2.5 mL) and transfer to a new 15 mL tube containing 10 mL M9 buffer.
9. Pellet the cleaned worms at 3000 rpm for 1 min, discard the supernatant till 1 mL mark.
10. Transfer the 1 mL worm suspension into a Beckman ultracentrifuge tube and pellet down.
11. Resuspend the worm pellets in twice the volume (w/v) of high-salt RAB buffer [100 mM 2-(N-morpholino) ethanesulfonic acid (MES), 1 mM EGTA, 0.5 mM $MgSO_4$, 20 mM NaF] containing 1× Complete Protease Inhibitor Mixture (Sigma-Aldrich P8340, Hamburg, Germany), 1 μM okadaic acid, and 0.5 mM PMSF, and sonicate (6 × 10 s bursts, 10 s intervals) on ice.
12. Centrifuge the lysates at 40,000 × g for 40 min at 4°C.
13. Collect the supernatant (RAB-soluble Tau fraction) and snap-freeze in liquid nitrogen.
14. Resuspend the RAB pellet in RAB buffer containing 1 M sucrose, 1× Complete Protease Inhibitor Mixture (Sigma-Aldrich P8340, Hamburg, Germany), 1 μM okadaic acid, and 0.5 mM PMSF, and sonicate briefly (~10 s).
15. Centrifuged at 40,000 × g for 20 min at 4°C and discard the supernatant.
16. Resuspended the pellet in RIPA buffer (150 mM NaCl, 1% Nonidet P-40, 0.5% deoxycholate, 0.1% SDS, 50 mM Tris, pH 8.0) containing 1× Complete Protease Inhibitor Mixture (Sigma-Aldrich P8340, Hamburg, Germany), 1 μM okadaic acid, and 0.5 mM PMSF, mix thoroughly by pipetting, and centrifuge at 40,000 × g for 20 min at 4°C to collect the remaining soluble nuclear and membrane-associated protein fraction in the supernatant (RIPA-soluble Tau fraction).
17. Before insoluble protein extraction, gently wash the pellet with 500 μL RIPA buffer without disturbing it.
18. Finally, extract the pellet with urea buffer [30 mM Tris, 7 M urea, 2 M thiourea, 4% (3-[(3-cholamidopropyl) dimethylammonio]-1-propanesulfonate) (CHAPS), pH 8.5] containing 1× Complete Protease Inhibitor Mixture (Sigma-Aldrich P8340, Hamburg, Germany), 1 μM okadaic acid, and 0.5 mM PMSF.
19. Centrifuge at 13,000 × g for 15 min. The supernatant represents the detergent-insoluble Tau fraction.

Immunoblotting

1. Subject the worm fractions (total or sequentially extracted–soluble, detergent-soluble, and detergent-insoluble) to SDS-PAGE using 10% polyacrylamide gels.
2. After electrophoresis, carefully remove the SDS gel and equilibrate in 1× western blotting buffer for 10 min at room temperature.
3. Activate a PVDF membrane in methanol, then transfer proteins from the gel to the membrane at 1 mA/cm^2 for 60 min using an electroblotting apparatus.
4. Incubate the membrane in 4% (w/v) milk in 1× TBST to block nonspecific binding sites.
5. Incubate with primary antibody (e.g., K9JA pan-Tau, diluted in 4% milk/TBST) at 37°C for 1 h and wash the membrane three times with 1× TBST.
6. Incubate with HRP-conjugated secondary antibody (diluted in 4% milk/TBST) at 37°C for 1 h and wash the membrane three times with 1× TBST.
7. Visualize the signal using an ECL-based detection system to detect the luminescent product of the secondary antibody.

Native PAGE for Higher-Order Tau Species

1. For native PAGE analysis, prepare the worm pellets from NGM plates as described above.
2. Resuspend the worm pellets in buffer C (20 mM Hepes, pH 7.9, 25% glycerol, 0.42 M NaCl, 1.5 mM $MgCl_2$, 0.2 mM EDTA, 0.5 mM DTT) containing Complete Protease Inhibitor Mixture (Sigma-Aldrich P8340, Hamburg, Germany), 1 μM okadaic acid, and 0.5 mM PMSF.
3. Sonicate the worm suspension (3 × 10 s bursts, 10 s breaks) on ice.
4. Centrifuge at 40,000 × g for 5 min at 4°C and collect the supernatant.
5. Analyze the supernatants on 4%–12% native PAGE gels and analyze by western blot.

3.7 Assaying Neurodegeneration in Defined Neuronal Subtypes and Aging Phenotypes at Single-Neuron Resolution

One of the hallmarks of the tauopathies is reported to be the loss of specific class of neurons with the aging of the individuals. As mentioned earlier, there is conservation in terms of neuronal subtypes, and the neurotransmitters are well conserved between *C. elegans* and higher species. Therefore, we can assess the pathophysiology of the Tau mutant variants in a neuron-specific manner in the *C. elegans* nervous system. The ease of the transgenesis and transparent body plan in *C. elegans* helps in visualizing individual neurons or neuronal-subtypes by expressing a fluorescent protein under a neuronal-subtype specific promoter. In this chapter, we will use two examples to demonstrate how we can understand the pathophysiology of the Tau mutations in a neuronal-subtype specific manner: (a) using a transgenic strain that marks the GABAergic nervous system of the *C. elegans* with GFP (juIs76) [14, 17], (b) a strain which specifically marks the mechanosensory neurons with GFP (zdIs5) [18] (see Fig. 3 for an example).

We have utilized these two strains in Tau^{WT} and Tau mutant variants background and analyzed the neuronal changes across different time points in the *C. elegans* life span, varying from L4 to Day 3 adult or sometime even Day 10 adults. These are the steps to be followed for the analysis of neuropathology in the Tau^{WT} and the mutant Tau variants at the neuronal subtype or at single neuron level:

1. Genetic crosses were performed to generate strains where in the Tau variants-expressing backgrounds also carried either the *juIs46* or *zdIs5* transgene, which label specific neuronal populations with GFP, to allow visualization under a epifluorescence microscope equipped with 10× and 60× (oil immersion) objectives.
2. Worms were synchronized for a uniform population and imaged at various stages of their life, such as L4, D1, D2, or even D7, depending upon the requirement.

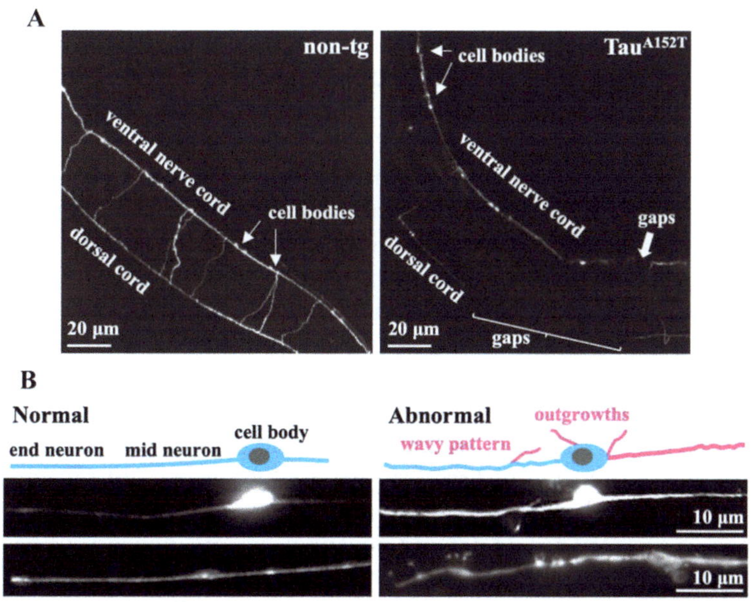

Fig. 3 (**A**) GABAergic motor neurons visualized with the *Punc-25::gfp* reporter (labeling these neurons with GFP) under fluorescence microscope are shown in a healthy non-transgenic worm with intact connectivity, and a Tau^{A152T} transgenic worm with abnormalities in the form of gaps. (**B**) An example of Touch neurons visualized with the *Pmec-4::gfp* reporter (labeling mechanosensory neurons with GFP) in a healthy worm with a normal, straight neuronal process, and in a Tau^{A152T} transgenic worm exhibiting aging-like phenotypes like wavy neuronal processes, sprouting and nonspecific outgrowths, as illustrated in the accompanying cartoon

3. Worms were scored for neurodegeneration by scoring the frequency discontinuities in the ventral cord (a bunch of neuronal fibers running along the body from anterior to posterior in the ventral region) in the N2, Tau^{WT} and the mutant variants of the Tau. For neuronal aging phenotype, mechanosensory neurons were used as a model, which has a characteristic morphological change with respect to neuronal aging, as established by researchers in the field (see Fig. 3).
4. Worms of respective genotypes are mounted on 2% agarose pads on microscope slides, after anesthetization with sodium azide.
5. Images are generally captured at 60× magnification to achieve high resolution, although 20× is occasionally used. For aging characterization, however, the images are acquired at 60× to cover the full spectrum of aging-related features (*see* **Note 9**).
6. Acquired images are qualitatively processed through ImageJ for representational purposes.

4 Conclusions

Engineering small genetic model organisms to study neurological disorders has become an increasingly common strategy to complement pathophysiological insights obtained from rodent models. The limited genetic redundancy in invertebrate genomes provides a clearer background for dissecting disease mechanisms and mapping the signaling pathways involved. *C. elegans* shares approximately 60%–80% of its genes with human orthologs, ensuring broad conservation of key biological pathways relevant to neurodegeneration. The ease of genetic manipulation, rapid generation time, low maintenance cost, and suitability for high-throughput screening make *C. elegans* an excellent platform for identifying genetic and pharmacological modifiers of disease phenotypes. Finally, the field can move to the CRISPR/Cas9-based genome-wide screens for genetic modifiers in the existing tauopathy models.

The fully mapped neuronal connectome of *C. elegans* provides unique opportunities to explore aspects of modern Tau biology, including the spatiotemporal progression of Tau pathological effects [25] across defined neuronal circuits, neuron subtype–specific vulnerability to distinct mutant Tau variants [22], and the interplay between cell-autonomous and non–cell-autonomous mechanisms of neurodegeneration. Collectively, these features position *C. elegans* as a powerful and complementary system for elucidating disease pathways and accelerating the discovery of therapeutic strategies for human neurological disorders.

5 Notes

1. DNA concentration varies with the construct and gene of interest and strongly influences transformation outcomes. Plasmids carrying human Tau cDNA are toxic at high concentrations, whereas at low concentrations transgenic worms exhibit minimal Tau expression and weak phenotypes. Optimal results are obtained with Tau constructs at ~75 ng/μL, together with coinjection markers (e.g., myo-2::RFP or myo-2::GFP for pharyngeal expression) at 20 ng/μL to aid transformant identification.
2. In our experience, young adults (4–5 h post-L4 stage) have a higher survival rate following injection and are more likely to yield transgenic animals carrying extrachromosomal arrays.
3. For beginners, immobilize one animal per agarose pad, positioning it with the vulva facing one direction to prevent the syncytial gonad from being on top or beneath. In oil, the syncytial gonad appears as two clear regions at the anterior and posterior. Wait until the animal floats with sideways body flexures, then gently press it onto the pad using a pick—avoiding the head to prevent injury. The animal should be fully immobilized except for the head, which remains free to move. Animals adhere best on first contact; repeated attempts at the same spot are ineffective. For streamlining injections, practice arranging multiple animals in a uniform line. Complete the procedure within 10 min to prevent dehydration.
4. Trioxsalen is highly carcinogenic and light sensitive. When making the stock solution or treating worms, make sure that the microfuge tube is wrapped with aluminum foil or work in a dark space. To fully dissolve Trioxsalen in DMSO, use frequent vortexing and intermittent heating at 45°C. Store the stock at −80°C and consider aliquoting it to avoid repeated freeze-thaw cycles. Since Trioxsalen is highly carcinogenic, any leftover solution after worm treatment should be discarded in a designated hazardous chemical container.
5. We have observed that Tau transgenic animals carrying extrachromosomal arrays grow more slowly then non-transgenic animals on the same plate. Therefore, when selecting transgenic animals under a stereo-fluorescence microscope, ensure that the plates are not overcrowded.
6. In our experience, a higher number of transgenic animals can be obtained from the D1 plate compared to D0. Although equal number of worms can be transferred from D0 and D1, we transfer a higher number from D1 (let us say ~100 from D0,

and ~300 from D1). We had a higher success of obtaining integrated lines from D1 plates.

7. The wells with the highest number of transgenics as scored (more than 75% of the population) have the maximum probability of being an integrated strain; however, there is still a possibility that it is not integrated but has acquired a high frequency-transmission of the extrachromosomal array, which can be dissected out in the next step.
8. Always select Tau transgenic lines with a comparable levels of Tau expression (wild-type verses mutant Tau) for subsequent analyses to ensure that any observed phenotypic differences arise from the mutation itself rather than differences in expression levels.
9. Individual neurons in *C. elegans* are very small, with cell body sizes of approximately 2–4 μm. Therefore, to see the minute details of aging-related morphological changes in neuronal processes or cell bodies, imaging at 60× magnification is required. Whereas for the ventral cord processes (a thick bundle of fibers), the discontinuity in the neuronal fibers can be observed at 20× magnification.

Acknowledgments

The authors acknowledge research support by the Academic Health Systems (Hamad Medical Corporation), Qatar, Department of Biology (The University of the South), TN, USA, German Center for Neurodegenerative Diseases, Bonn and the Max-Planck Society, Germany.

Conflicts of interest The authors have no conflicts of interest to declare that are relevant to the content of this chapter.

References

1. Wang Y, Mandelkow E (2016) Tau in physiology and pathology. Nat Rev Neurosci 17(1):5–21
2. Alzheimer A (1907) Über eine eigenartige Erkrankung der Hirnrinde. Allg Z Psychiatr Psych-Gerichtl Med 64:146–148
3. Goedert M, Crowther RA, Scheres SHW, Spillantini MG (2024) Tau and neurodegeneration. Cytoskeleton (Hoboken) 81(1):95–102
4. Sawaya MR, Hughes MP, Rodriguez JA, Riek R, Eisenberg DS (2021) The expanding amyloid family: structure, stability, function, and pathogenesis. Cell 184(19):4857–73
5. Spires-Jones TL, Attems J, Thal DR (2017) Interactions of pathological proteins in neurodegenerative diseases. Acta Neuropathol 134(2):187–205
6. Drepper F, Biernat J, Kaniyappan S, Meyer HE, Mandelkow EM, Warscheid B et al (2020) A combinatorial native MS and LC-MS/MS approach reveals high intrinsic phosphorylation of human Tau but minimal levels of other key modifications. J Biol Chem 295(52):18213–25
7. Garver TD, Harris KA, Lehman RA, Lee VM, Trojanowski JQ, Billingsley ML (1994) Tau phosphorylation in human, primate, and rat brain: evidence that a pool of tau is highly phosphorylated *in vivo* and is rapidly dephosphorylated *in vitro*. J Neurochem 63(6):2279–2287

8. Braak H, Braak E (1991) Neuropathological stageing of Alzheimer-related changes. Acta Neuropathol 82(4):239–259
9. Szegvari E, Holec SAM, Woerman AL (2025) Limitations and applications of rodent models in tauopathy and synucleinopathy research. J Neurochem 169(3):e70021
10. Giong HK, Subramanian M, Yu K, Lee JS (2021) Non-rodent genetic animal models for studying tauopathy: review of drosophila, zebrafish, and *C. elegans* models. Int J Mol Sci 22(16):8465
11. Pir GJ, Choudhary B, Mandelkow E (2017) *Caenorhabditis elegans* models of tauopathy. FASEB J 31(12):5137–48
12. Alcedo J, Jin Y, Portman DS, Prahlad V, Raizen D, Rapti G et al (2020) Nature's gift to neuroscience. J Neurogenet 34(3–4):223–224
13. van Ham TJ, Breitling R, Swertz MA, Nollen EA (2009) Neurodegenerative diseases: lessons from genome-wide screens in small model organisms. EMBO Mol Med 1(8-9):360–370
14. Kraemer BC, Zhang B, Leverenz JB, Thomas JH, Trojanowski JQ, Schellenberg GD (2003) Neurodegeneration and defective neurotransmission in a *Caenorhabditis elegans* model of tauopathy. Proc Natl Acad Sci U S A 100(17):9980–9985
15. Miyasaka T, Ding Z, Gengyo-Ando K, Oue M, Yamaguchi H, Mitani S et al (2005) Progressive neurodegeneration in *C. elegans* model of tauopathy. Neurobiol Dis 20(2):372–383
16. Brandt R, Gergou A, Wacker I, Fath T, Hutter H (2009) A *Caenorhabditis elegans* model of tau hyperphosphorylation: induction of developmental defects by transgenic overexpression of Alzheimer's disease-like modified tau. Neurobiol Aging 30(1):22–33
17. Fatouros C, Pir GJ, Biernat J, Koushika SP, Mandelkow E, Mandelkow EM et al (2012) Inhibition of tau aggregation in a novel *Caenorhabditis elegans* model of tauopathy mitigates proteotoxicity. Hum Mol Genet 21(16):3587–3603
18. Pir GJ, Choudhary B, Mandelkow E, Mandelkow EM (2016) Tau mutant A152T, a risk factor for FTD/PSP, induces neuronal dysfunction and reduced lifespan independently of aggregation in a *C. elegans* Tauopathy model. Mol Neurodegener 11:33
19. Guthrie CR, Schellenberg GD, Kraemer BC (2009) SUT-2 potentiates tau-induced neurotoxicity in *Caenorhabditis elegans*. Hum Mol Genet 18(10):1825–1838
20. Wheeler JM, McMillan P, Strovas TJ, Liachko NF, Amlie-Wolf A, Kow RL, et al (2019) Activity of the poly(A) binding protein MSUT2 determines susceptibility to pathological tau in the mammalian brain. Sci Transl Med 11(523):eaao6545
21. Pir GJ, Choudhary B, Kaniyappan S, Chandupatla RR, Mandelkow E, Mandelkow EM et al (2019) Suppressing tau aggregation and toxicity by an anti-aggregant tau fragment. Mol Neurobiol 56(5):3751–67
22. Choudhary B, Mandelkow E, Mandelkow EM, Pir GJ (2018) Glutamatergic nervous system degeneration in a *C. elegans* Tau(A152T) tauopathy model involves pathways of excitotoxicity and Ca(2+) dysregulation. Neurobiol Dis 117:189–202
23. Han M, Saxton A, Currey H, Waldherr SM, Liachko NF, Kraemer BC (2024) Transgenic Dendra2::tau expression allows *in vivo* monitoring of tau proteostasis in Caenorhabditis elegans. Dis Model Mech 17(3):dmm050473
24. Butler VJ, Salazar DA, Soriano-Castell D, Alves-Ferreira M, Dennissen FJA, Vohra M et al (2019) Tau/MAPT disease-associated variant A152T alters tau function and toxicity via impaired retrograde axonal transport. Hum Mol Genet 28(9):1498–514
25. Sandhof CA, Martin N, Tittelmeier J, Schlueter A, Pezzali M, Schoendorf DC, et al (2025) A novel *C. elegans* model for MAPT/Tau spreading reveals genes critical for endolysosomal integrity and seeded MAPT/Tau aggregation. Autophagy 21(12):2963–2981
26. Schoonhoven DN, Coomans EM, Millan AP, van Nifterick AM, Visser D, Ossenkoppele R et al (2023) Tau protein spreads through functionally connected neurons in Alzheimer's disease: a combined MEG/PET study. Brain 146(10):4040–4054
27. Goedert M, Eisenberg DS, Crowther RA (2017) Propagation of tau aggregates and neurodegeneration. Annu Rev Neurosci 40:189–210
28. Natale C, Barzago MM, Diomede L (2020) *Caenorhabditis elegans* models to investigate the mechanisms underlying tau toxicity in tauopathies. Brain Sci 10(11):838
29. Rani N, Alam MM, Jamal A, Bin Ghaffar U, Parvez S (2023) *Caenorhabditis elegans*: a transgenic model for studying age-associated neurodegenerative diseases. Ageing Res Rev 91:102036
30. Rodriguez P, Blakely RD (2024) Sink or swim: Does a worm paralysis phenotype hold clues to neurodegenerative disease?. J Cell Physiol 239(6):e31125
31. Wu Y, Chen Y, Yu X, Zhang M, Li Z (2023) Towards understanding neurodegenerative diseases: insights from *Caenorhabditis elegans*. Int J Mol Sci 25(1):443
32. Pir GJ, Buddenkotte J, Alam MA, Own A, Eck RJ, Kraemer BC et al (2026) TDP-43 proteinopathies and neurodegeneration: insights from *Caenorhabditis elegans* models. FEBS J 293(2):348–384

33. Choudhary B, Napier-Jameson R, Norris A (2025) Regulated microexon alternative splicing in single neurons tunes synaptic function. EMBO Rep 26(14):3640–3662
34. Stinchcomb DT, Shaw JE, Carr SH, Hirsh D (1985) Extrachromosomal DNA transformation of *Caenorhabditis elegans*. Mol Cell Biol 5(12):3484–3496
35. Kage-Nakadai E, Imae R, Yoshina S, Mitani S (2014) Methods for single/low-copy integration by ultraviolet and trimethylpsoralen treatment in *Caenorhabditis elegans*. Methods 68(3):397–402
36. Crary JF, Trojanowski JQ, Schneider JA, Abisambra JF, Abner EL, Alafuzoff I et al (2014) Primary age-related tauopathy (PART): a common pathology associated with human aging. Acta Neuropathol 128(6):755–766
37. von Bergen M, Friedhoff P, Biernat J, Heberle J, Mandelkow EM, Mandelkow E (2000) Assembly of tau protein into Alzheimer paired helical filaments depends on a local sequence motif ((306)VQIVYK(311)) forming beta structure. Proc Natl Acad Sci U S A 97(10):5129–5134
38. Sawaya MR, Sambashivan S, Nelson R, Ivanova MI, Sievers SA, Apostol MI et al (2007) Atomic structures of amyloid cross-beta spines reveal varied steric zippers. Nature 447(7143):453–457
39. Khlistunova I, Biernat J, Wang Y, Pickhardt M, von Bergen M, Gazova Z et al (2006) Inducible expression of Tau repeat domain in cell models of tauopathy: aggregation is toxic to cells but can be reversed by inhibitor drugs. J Biol Chem 281(2):1205–1214

Chapter 14

Tracking Tau Multimerization in the Adult *Drosophila* Brain

Meghna N. Chinchankar and Bess Frost

Abstract

Tauopathies are a distinct group of neurodegenerative disorders neuropathologically defined by the deposition of aggregated forms of the tau protein. We have previously leveraged *Drosophila* models of tauopathy to discover novel biological mechanisms that drive tau-induced neurotoxicity and neuroinflammation, and have developed a new model to monitor tau multimerization in the brain of living flies. In this chapter, we describe a suite of biochemical, in situ and in vivo tools that we utilize to study transgenic tau protein levels and multimerization in the adult *Drosophila* brain. Our experimental protocols include use of: (i) proximity ligation assay (PLA) to detect and visualize tau–tau protein interactions in the fly brain, (ii) native PAGE-Western blot to resolve monomeric and high molecular weight tau multimers, and (iii) in vivo luminescence to assess tau protein multimerization in live flies using the Tau^{LUM} model.

Keywords *Drosophila*, Tau, Tauopathy, Tau^{LUM}, Split-luciferase reporter, Luminescence assay, Proximity ligation assay (PLA), Native PAGE, Brain dissection

1 Introduction

Drosophila melanogaster has become a prominent model organism due to its small size, short life cycle, large brood sizes, genetic amenability, and high level of evolutionary conservation of mechanisms relevant to human biology. Though the nervous system physiology of *Drosophila* is much simpler than that of humans, *Drosophila* models of tauopathy feature age-dependent neurodegeneration and recapitulate many cellular features of human Alzheimer's disease and related tauopathies. Multiple comprehensive reviews highlight the varied mechanistic insights into tau protein biology that have arisen from studies in *Drosophila* [1–5]. In this chapter, we describe three main techniques that we implement to study tau–tau protein interactions in the adult fly brain.

We have optimized a proximity ligation assay (PLA) to study tau–tau interactions in fixed brain tissue. First established over 20 years ago, the PLA has been used in innumerable applications to directly assess protein–protein interactions with high specificity

Jessica S. Fortin (ed.), *Translational Research Methods for Tauopathies*, Neuromethods, vol. 233,
https://doi.org/10.1007/978-1-0716-5234-3_14,

at molecular resolution [6–8]. This assay involves the coupling of antibodies specific to interacting proteins of interest to complementary oligonucleotides that hybridize when they are within <40 nm. Under fluorescence or brightfield assay reaction conditions, these oligonucleotides ligate to form a closed DNA circle which then undergoes rolling circle amplification to generate antibody-tethered concatemeric copies. Fluorescently-labeled or horseradish peroxidase (HRP)-labeled complementary probes then bind to the amplified DNA to produce visually-detectable signal that can be easily quantified (Fig. 1A). PLA has been adapted in several cell- and tissue-based systems, including in *Drosophila*, where it has be utilized to address biological questions in development and neurobiology [9–13]. Furthermore, PLA has supported molecular pathology studies of tauopathies and Alzheimer's disease, especially with regard to protein interactions and post-translational modifications (PTMs) of tau in human patient samples, as well as in cellular and mouse models [14–19]. In this chapter, we describe an original PLA protocol to study tau–tau interactions in the fixed adult *Drosophila* brain.

To biochemically corroborate the PLA-based assessment of tau-tau interactions, we have included an adaptation of the conventional native PAGE-Western method to study tau protein multimerization in the heads of adult flies, modified from our previous report [20]. This technique allows for the electrophoretic migration of sample proteins based on their native size, charge and conformation.

A novel technical advancement from our group leverages a split-luciferase reporter-based system to detect tau–tau interactions in the brains of active, awake *Drosophila*, which we term "Tau^{LUM}" [20]. This system is based on split-luciferase assays and similar reporters of tau aggregation in vitro and in cellular models of tauopathy [21–25]. Briefly, this model relies on N terminal or C-terminal luciferase fragments (referred to as NLuc and CLuc, respectively) fused to the aggregation-prone human Tau^{P301L} mutant; a catalytically active luciferase enzyme is generated when monomeric tau fused to NLuc interacts with monomeric tau fused to CLuc. A complementary EGFP^{LUM} model serves as a control for overexpression of a transgenic protein. The Tau^{LUM} model emits a quantifiable luminescence signal in the presence of its substrate D-luciferin (Fig. 1B). The elegant design of this assay allows for easy plate reader-based measurement of luminescence emanating from live flies.

In the following sections, we detail the suite of assays that we perform to quantify tau protein levels and multimerization in *Drosophila* models of tauopathy. We begin with a brief overview of our genetic experimental setup (Protocol 1), followed by our standard Western blotting protocol to quantify tau protein in fly heads (Protocol 2). We then describe the method by which we dissect and fix whole brains from adult *Drosophila* flies (Protocol 3) followed by an immunofluorescence protocol to visualize tau protein in the *Drosophila* brain (Protocol 4). Next, we introduce our Proximity Ligation Assay (PLA)

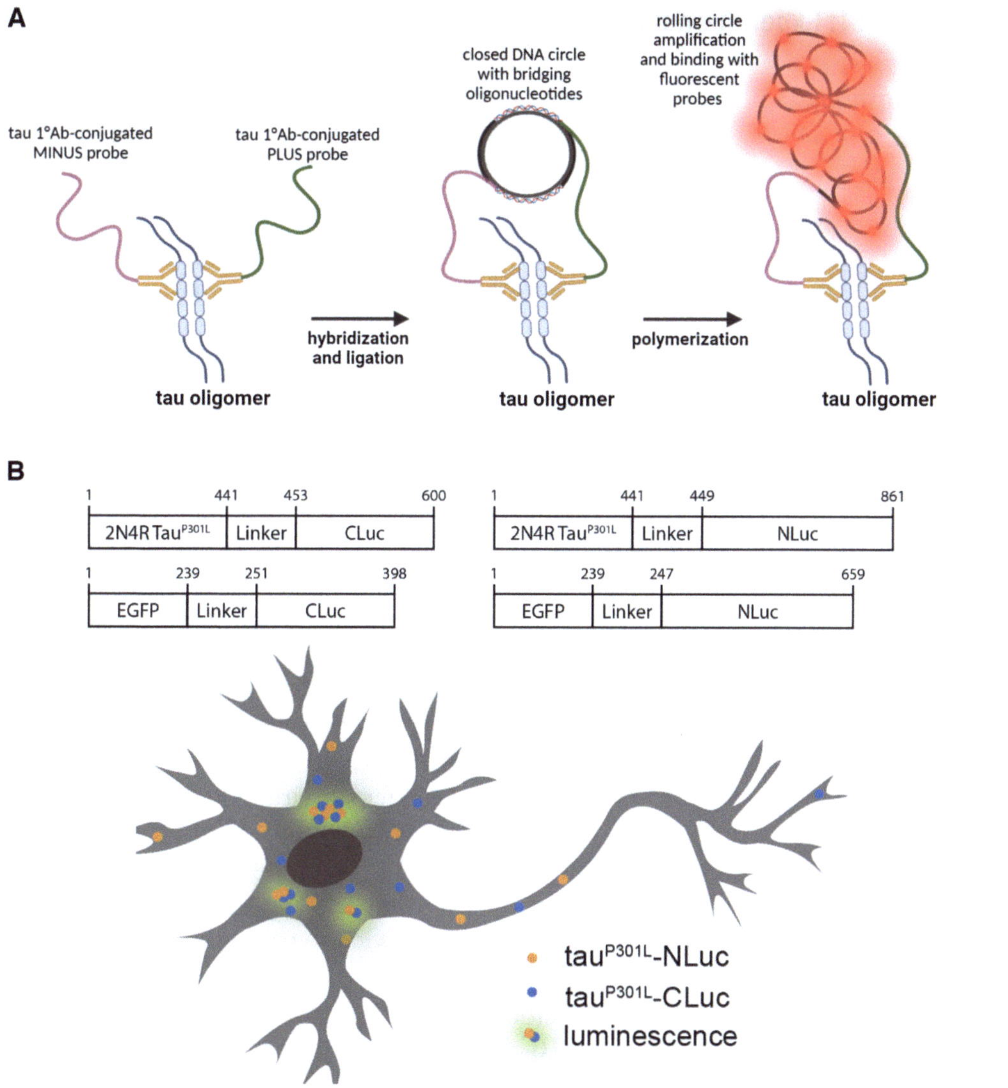

Fig. 1 Principles of proximity ligation assay (PLA) and in vivo luminescence assay to study tau multimerization in *Drosophila*. (**A**) Strategy of the PLA to study in situ tau–tau protein interactions: When tau-specific antibodies conjugated with complementary oligonucleotides (PLUS and MINUS probes) bind to interacting tau proteins, oligonucleotide probes hybridize and ligate to form a closed DNA circle, which is amplified; fluorescent probes enhance this signal for visual detection. (**B**) Schematic overview of the Tau^{LUM} model: The split luciferase reporter-based Tau^{P301L} and EGFP constructs of the transgenes used in Tau^{LUM} and EGFP^{LUM} fly models are shown. When pan-neuronally expressed Tau-NLuc and Tau-CLuc constructs multimerize in vivo, D-luciferin-fed Tau^{LUM} emit detectable luminescence signal as a function of these interactions. (Adapted from Levy et al. [20])

to detect tau–tau interactions in situ in the *Drosophila* brain (Protocol 5) and outline a native PAGE-Western blot protocol to biochemically detect multimeric tau in fly heads (Protocol 6). Finally, we describe a straightforward protocol for the in vivo luminescence assay to detect tau multimerization in the Tau^{LUM} model (Protocol 7).

2 Protocol 1: Experimental Setup

2.1 Requirements

- *Drosophila* stocks: elav-Gal4 (BDSC #458), w1118 (BDSC #3605), UAS-tau^{WT} [26], UAS-tau^{R406W} [26], UAS-tau^{P301L} [20], UAS-EGFP^{LUM} [20], and UAS-tau^{LUM} [20]
- Dissection microscope (Genesee Scientific)
- *Drosophila* CO_2 blowgun (Genesee Scientific)
- *Drosophila* CO_2 fly pad (Genesee Scientific)
- Paintbrush
- Fly food vials containing standard diet (Bloomington formulation) of cornmeal and yeast-based media

2.2 Procedure Steps

1. Maintain *Drosophila* stocks at 17°C; perform genetic crosses and experiments at 25°C with a 12 h light/dark cycle.
2. Genetic crosses:
   The *Drosophila* GAL4-UAS system is leveraged for pan-neuronal expression of human tau. When elav-Gal4 and UAS-driven transgenic flies are successfully mated, the elav promoter drives expression of the Gal4 transcription factor in progeny, which binds to UAS sequences and activates expression of the tau transgene in a tissue-specific manner.
   - Incubate elav-Gal4 female virgins with young males of each UAS-driven *Drosophila* stock. Transfer parents to a new food vial after 7 days to continue mating.
   - F1 progeny will begin eclosing approximately 10 days after setting up the cross. Genotypes of the resulting progeny are detailed in Table 1.
3. After eclosion, transfer progeny to fresh food vials and age them to 10 days, ensuring that they are transferred to new food every 3–4 days.

Table 1
***Drosophila* crosses**

Common name	Full genotype
Control	*elav-Gal4/+*
Tau^{WT}	*elav-Gal4/+; UAS-Tau^{WT}/+*
Tau^{R406W}	*elav-Gal4/+; UAS-Tau^{R406W}/+*
Tau^{P301L}	*elav-Gal4/+; UAS-Tau^{P301L}/+*
EGFP^{LUM}	*elav-Gal4/+; UAS-EGFP-NLuc, UAS-EGFP-CLuc/+*
Tau^{LUM}	*elav-Gal4/+; UAS-Tau^{P301L}-NLuc, UAS-Tau^{P301L}-CLuc/+*

2.3 Notes

- It is important to remove adults from mating vials after 7 days to avoid mixing with the eclosing progeny population.
- In this chapter, we have described protocols to study tau protein expression and multimerization using flies at day 10 of adulthood.
- Experimental sets mentioned in this chapter are referred to as:
  Non-LUM sets: Control, Tau^{WT}, Tau^{R406W}, and Tau^{P301L}
  LUM sets: $EGFP^{LUM}$ and Tau^{LUM}
- While we have provided a basic overview of preparing fly crosses, detailed information regarding *Drosophila* husbandry, virgin collection, balancer chromosomes, and genetics can be found in other excellent resources [27–29].

3 Protocol 2: SDS PAGE-Western Blot Using *Drosophila* Heads

3.1 Requirements

- *Drosophila* tau transgenic and control sets of interest

Buffers and Reagents

- 10× SDS (sodium dodecyl sulphate) PAGE (polyacrylamide gel electrophoresis) Running Buffer:
  - Glycine 144 g
  - Tris Base 30.3 g
  - SDS 10 g
  - Distilled water: Make up to 1 L
- 10× Transfer Buffer:
  - Glycine 144 g
  - Tris Base 30.3 g
  - Distilled water: Make up to 1 L
- 10× PBS (phosphate-buffered saline):
  - 10× PBS powder: one packet (Fisher)
  - Distilled water: Make up to 1 L
- 1× PBS-Tween (0.1%):
  - 10× PBS 100 mL
  - Tween20 1 mL
  - Distilled water: Make up to 1 L
- 2× Laemmli sample buffer (Bio-Rad) with 2-mercaptoethanol
- 4%–20% Mini-PROTEAN TGX stain-free precast protein gels, 15-well (Bio-Rad)
- EZ-Run prestained Rec protein ladder (Fisher)

Table 2
Antibodies for Western blots

Primary/ Secondary	Antibody (source)	Catalog no.	Host	Dilution
Primary	Tau (Dako)	A0024	Rabbit	1:2,000,000
Primary	EGFP (Invitrogen)	CAB4211	Rabbit	1:1,000
Primary	β-tubulin (Abcam)	ab179513	Rabbit	1:50,000
Secondary	Goat Anti-Rabbit Ig, Human ads-HRP (SouthernBiotech)	4010-05	Goat	1:20,000

- 0.2 μm nitrocellulose membrane (Bio-Rad)
- Methanol
- Ponceau S
- Nonfat milk powder
- Clarity Western ECL substrates (Bio-Rad):
  - Peroxide solution
  - Luminol/Enhancer solution
- Antibodies as detailed in Table 2

Equipment

- *Drosophila* CO_2 fly pad (Genesee Scientific)
- Dissection microscope (Genesee Scientific)
- Small glass or plastic tissue homogenizer pestle
- Tube heating block (Eppendorf)
- Gel running apparatus: Mini-PROTEAN Tetra Cell (Bio-Rad)
- Blot transfer apparatus: Mini Trans-Blot Cell (Bio-Rad)
- PowerPac power supply (Bio-Rad)
- Magnetic stirrer and stir bar
- Platform rocker
- ChemiDoc Imaging System (Bio-Rad)

Miscellaneous

- 1.5 mL microcentrifuge tubes
- Paintbrush
- Two sharp forceps
- Plastic petri dish
- 70% ethanol
- Gel releaser wedges (Bio-Rad)
- Filter paper

- Blunt-end filter forceps (Millipore)
- Medium-sized blot incubation boxes (LICOR)
- Ice pack cooling unit
- Large Pyrex dish

3.2 Preparation

- Anesthetize 10-day old flies on the *Drosophila* CO_2 fly pad.
- Transfer whole flies into labeled 1.5 mL microcentrifuge tubes; separate sexes into two different tubes.
- Freeze tubes at −80°C until ready for use.

3.3 Procedure Steps

Sample preparation

1. Before starting:
   - Gather sharp forceps, a large plastic petri dish and a tissue homogenizer pestle. Chill forceps and petri dish on dry ice.
   - Set a tube heating block to 95°C.
2. Retrieve the tubes with frozen flies and keep on dry ice.
3. Transfer the required number of frozen flies to the petri dish.
4. Using the forceps, remove the head of each frozen fly and transfer it to its individual microcentrifuge tube on dry ice. Six replicates per genotype (three males and three females) are typically analyzed, but this may differ based on your own experimental design.
5. Add 10 μL of 2× Laemmli sample buffer to each tube. Manually homogenize each fly head for 30 s to 1 min in the buffer on regular ice. Wash the pestle between samples.
6. Boil the samples for 10 min at 95°C on the heating block.
7. Spin the samples briefly on a tabletop microcentrifuge. If needed, samples can be stored at this step at −20°C.

SDS PAGE

8. Before starting:
   - Prepare 1 L 1× SDS PAGE running buffer from its 10× stock.
   - Rinse the mini-PROTEAN gel running apparatus with distilled water.
   - Retrieve a Bio-Rad 4%–20% mini-PROTEAN TGX stain-free precast protein gel from 4°C. Gently pull out its comb and remove the tape from the bottom edge.
   - Clamp two gels properly on either side of the electrode assembly of the gel running apparatus. Place this assembly in the gel tank and fill it completely with 1× running buffer, ensuring all the wells are submerged and devoid of bubbles.

In case of leaks, adjust the gels correctly so that they have correct contact with the electrode assembly. If only one gel is being used for the experiment, place a buffer dam on the other side of the assembly.

- Fill the gel tank with the remaining 1× running buffer. Rinse and straighten wells of the gel with running buffer.
- Retrieve Fisher EZ-Run prestained Rec protein ladder from −20°C and keep it on ice.

9. Load 2 μL of the protein ladder in the first well. Load the complete liquid lysates from each of the sample tubes onto the gel without disturbing their cuticle debris pellets. Record the order of samples loaded.
10. Run the SDS PAGE at 50 V for the first 5 min and 90–150 V for the next 45–60 min or until the running dye reaches the bottom of the gel.

Western blot transfer

11. Before starting:
    - Prepare 1 L 1× transfer buffer while the gel is running: Add 700 mL of distilled water to 100 mL of 10× transfer buffer stock and 200 mL methanol. Mix thoroughly, seal the container and let it chill at 4°C for at least an hour.
    - For each gel to be blotted, cut a 0.2 μm nitrocellulose membrane to its size and two filter papers slightly larger than its size. Handle the membrane carefully at all times.
    - Rinse the Bio-Rad mini trans-blot apparatus with distilled water.
12. Using a gel releaser wedge, carefully open the gel cassette. Cut the stacking gel edge and the bottom edge of the gel and discard these portions. Rinse the gel briefly with distilled water.
13. Soak the SDS PAGE gel, nitrocellulose membrane, filter papers, and sponges in chilled 1× transfer buffer for a few minutes.
14. Assemble the blot transfer sandwich within the transfer cassette such that the gel is oriented at the cathode and the membrane at the anode of the apparatus. Thus, the order of the transfer sandwich is: the black side of the transfer cassette → sponge → filter paper → SDS PAGE gel → nitrocellulose membrane → filter paper → sponge → the clear side of the transfer cassette. Ensure all the components are soaked in 1× transfer buffer and do not dry out. Use a glass or plastic pipette to gently roll out any bubbles while assembling

each layer of the transfer sandwich. Lock the transfer cassette firmly and slide it in the electrode assembly in the correct orientation. Up to two blot transfers can run in the same tank.

15. Place a small magnetic stir bar in the transfer tank. Place the electrode assembly containing the blot sandwiches in the tank along with an ice pack cooling unit. Fill the tank completely with 1× transfer buffer. Close the transfer tank with its lid securely. Set this apparatus on ice or in a cold room.
16. Place the transfer tank on a magnetic stirrer and set it to stir at medium-low speed.
17. Set the transfer to run at 90 V for 90 min.

Ponceau S staining of the membrane

18. Before starting:
    - Prepare 1 L of 1× PBS from its 10× stock. Pour it into a large Pyrex dish.
    - Prepare 50 mL of 2% milk blocking buffer fresh: Dissolve 1 g of milk powder in 1× PBS-Tween and make up the volume to 50 mL in a conical tube. Keep on ice for immediate use and store at 4°C overnight for next day use.
19. After the transfer is complete, open the transfer cassette. Using blunt-end forceps, remove the blot membrane and carefully trim the edges.
20. For antigen retrieval, gently submerge the membrane in 1× PBS, allowing it to settle at the bottom of the large Pyrex dish and close it with its lid. Microwave for 9 min.
21. After cooling, transfer the membrane to an opaque blot incubation box. Immerse the blot with minimal amount of Ponceau S stain and let in incubate with shaking for a few minutes.
22. Decant the Ponceau S stain back into its container as it is reusable. Rinse the blot several times with distilled water to remove excess stain.
23. Image the Ponceau S-stained membrane with a Bio-Rad ChemiDoc imaging system or equivalent. This helps visualize the protein bands, equal loading across the lanes and any bubbles in the blot.
24. Wash the membrane several times with 1× PBS-Tween on a platform rocker until the Ponceau S stain has disappeared.

Antibody incubations and detection

25. Retrieve primary antibodies from 4°C or −20°C and dilute them appropriately in freshly prepared blocking buffer. Working dilutions of the antibodies used in our experiment are provided in Table 2.

26. Incubate the membrane in diluted antibody overnight with gentle rocking at 4°C.
27. The next day, wash the blot with 1× PBS-Tween three times for 5 min each on the rocker.
28. Dilute appropriate HRP-conjugated secondary antibodies in cold blocking buffer and incubate the blot with it for 2 h at room temperature on the rocker.
29. Wash the blot with 1× PBS-Tween three times for 5 min each on the rocker and transfer to 1× PBS.
30. Transfer the membrane to a flat piece of parafilm. Mix 1 mL each of the two Bio-Rad ECL substrates (peroxide and luminol solutions) and pipette onto the membrane. Gently rock for 3 min.
31. Place the membrane between two pieces of plastic and image on your imaging system. Exposure time and conditions for optimal signal detection need to be determined based on the blot and imaging system. Also capture colorimetric images of the protein ladder on the blot. Merge the chemiluminescence and colorimetric images to generate a composite blot image.
32. Wash the membrane with PBS-Tween one to two times to remove ECL substrates. Store in 1× PBS in its incubation box at 4°C for short-term storage or in plastic wrap at 4°C for long-term storage.

3.4 Visualization and Analysis

- Western blot data from non-LUM-based models: As seen in Fig. 2A, all samples have equal loading as inferred by β-tubulin bands at ~55 kDa. Tau^{WT} and Tau^{R406W} *Drosophila* models are based on transgenic expression of the 0N4R human tau isoform and thus have a band at ~72 kDa. The Tau^{P301L} model expresses transgenic 2N4R human tau and thus runs at ~90 kDa. Controls have no human tau protein as expected. Tau protein levels are comparable across tau transgenic models.
- Western blot data from LUM-based models: As seen in Fig. 3A, all samples have equal loading as inferred by β-tubulin bands at ~55 kDa. $EGFP^{LUM}$ samples have EGFP-NLuc bands at 70 kDa and EGFP-CLuc bands at ~50 kDa. Tau^{LUM} samples have Tau-NLuc bands at ~120 kDa and Tau-CLuc bands at ~100 kDa.
- Quantitative blot analyses can be performed via the FIJI Image J software using the TIF files of the chemiluminescent images generated by the ChemiDoc system.

3.5 Notes

- The fly freezing steps need to be done quickly to avoid the flies from getting stressed or dying due to asphyxiation in the tubes.
- After removing the heads of the frozen flies in this protocol, their bodies can be discarded or saved for other experiments.

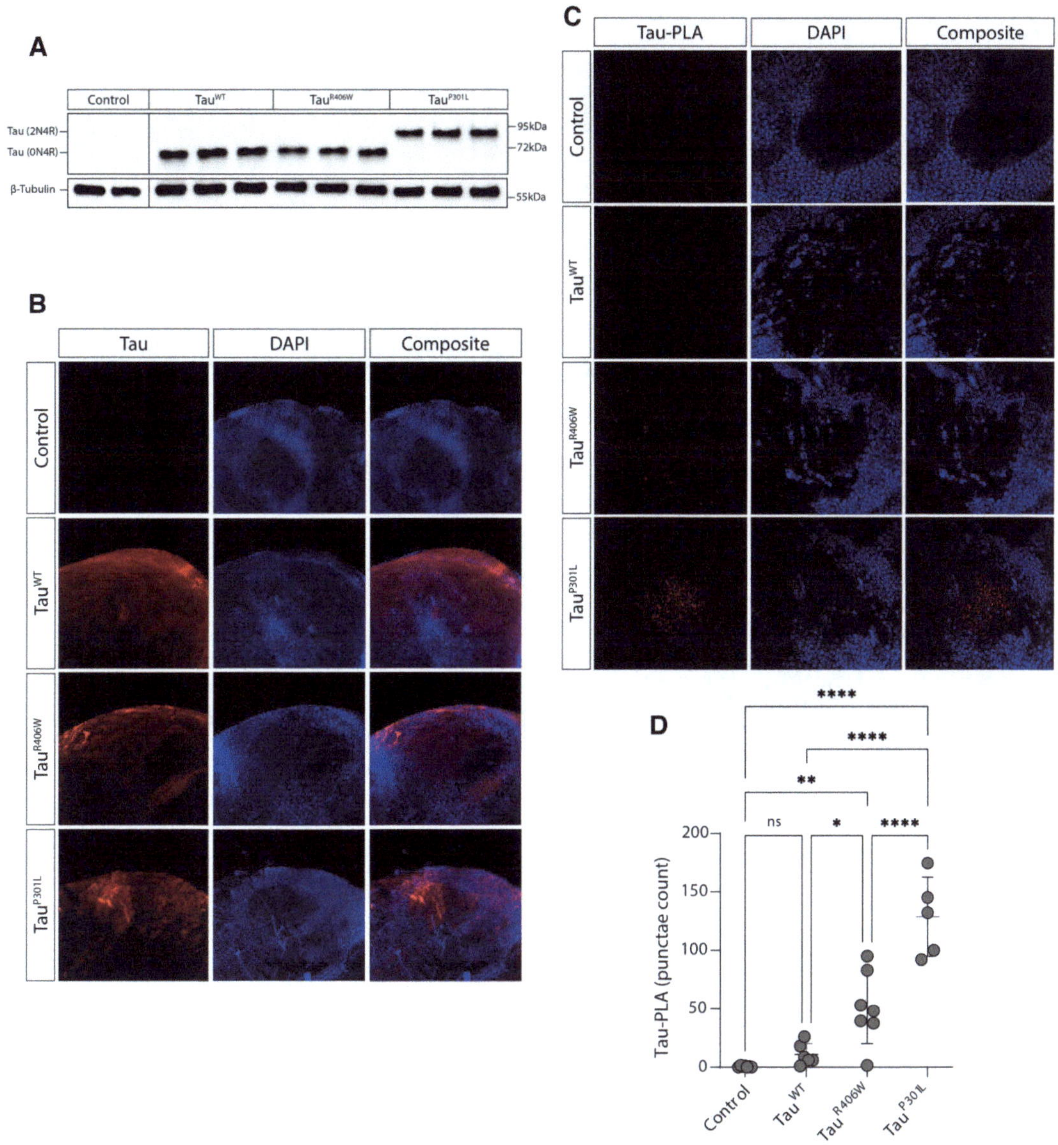

Fig. 2 Representative results from non-LUM *Drosophila* models. (**A**) SDS PAGE-Western blot detecting tau and β-tubulin protein expression from *Drosophila* head tissue lysates. (**B**) Immunofluorescence-based detection of tau protein in whole *Drosophila* brains: Imaged at 20× magnification and zoomed in to one hemisphere of the protocerebrum including the mushroom body. (**C**) In situ tau-PLA fluorescence and DAPI staining of whole *Drosophila* brains: Imaged at 63× magnification in the superior neuropils and mushroom body region. (**D**) Quantification and statistical analysis of the tau-PLA fluorescent puncta observed in (**C**). Data are represented as mean ± SD, one-way ANOVA with Sidak's multiple comparisons test, *****p*<0.0001, ****p*<0.001, ***p*<0.01, **p*<0.05, ns *p*>0.05

- Glass homogenizer pestles can be made in the lab by carefully melting the capillary tip of glass Pasteur pipettes to fit snugly in the tapered end of microcentrifuge tubes for efficient *Drosophila* tissue lysis.

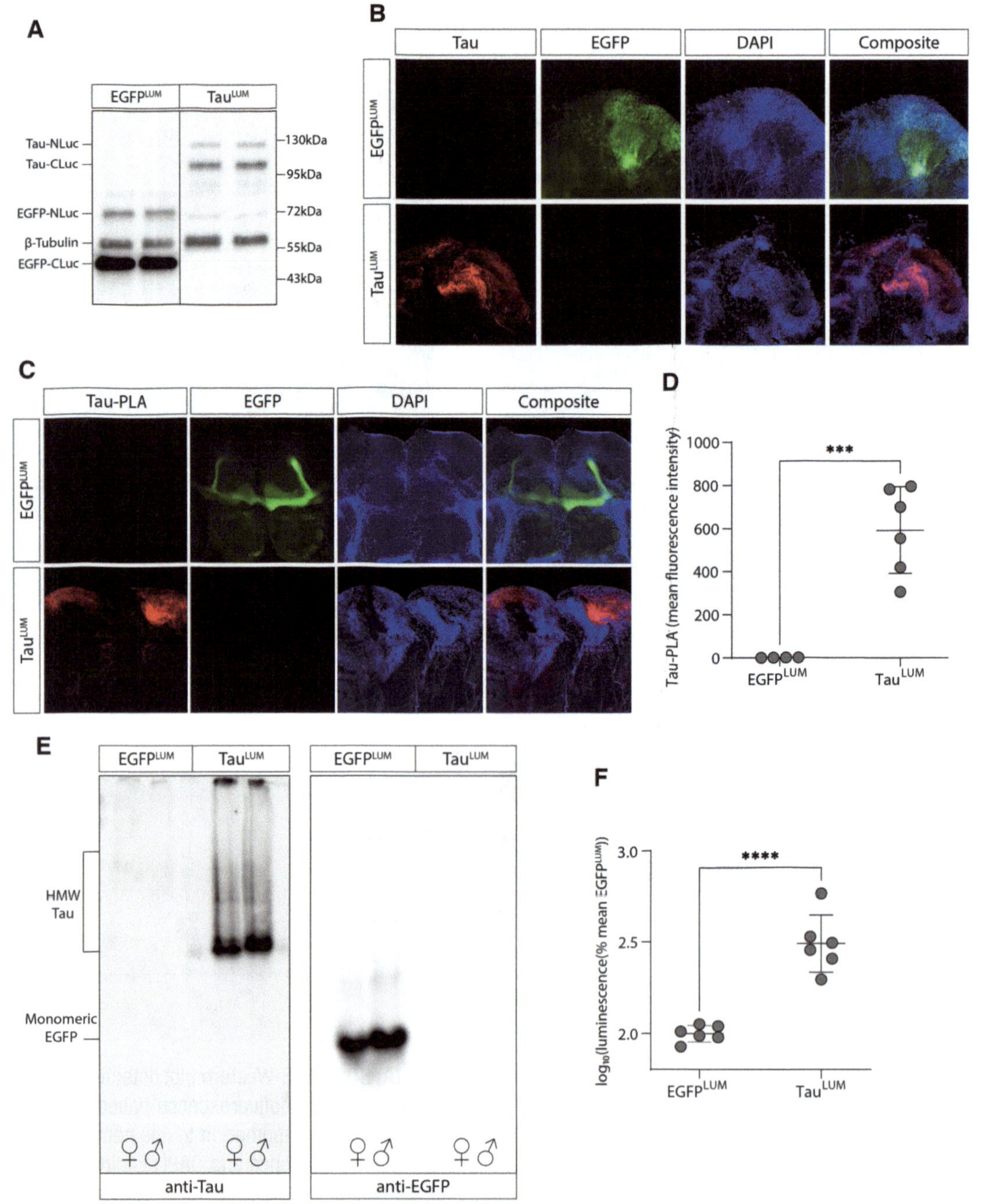

Fig. 3 Representative results from LUM *Drosophila* models. (**A**) SDS PAGE-Western blot detecting tau, EGFP, and β-tubulin protein expression from *Drosophila* head lysates. (**B**) Immunofluorescence-based detection of tau and EGFP proteins in whole *Drosophila* brains: Imaged at 20× magnification and zoomed in to one hemisphere of the protocerebrum including the mushroom body. (**C**) In situ tau-PLA fluorescence, EGFP and DAPI staining of whole *Drosophila* brains: Imaged at 20× magnification and zoomed into the central brain focused on the superior neuropils and mushroom body regions. (**D**) Quantification and statistical analysis of the tau-PLA fluorescent signal observed in (**C**). Data are represented as mean ± SD, unpaired *t*-test, ***$p<0.001$. (**E**) Native PAGE-Western blotting of tau and EGFP proteins in *Drosophila* head lysates. (**F**) In vivo luminescence assay measurements in EGFP^{LUM} and Tau^{LUM} flies. Data are represented as mean ± SD, unpaired *t*-test, ****$p<0.0001$

- After blot transfer, we include a modified antigen retrieval step in order to increase signal [30].

4 Protocol 3: *Drosophila* Whole Brain Dissection and Fixation

4.1 Requirements

- Fly vials containing *Drosophila* tau transgenic and control sets of interest

Reagents

- Ice cold 1× PBS
- Ice cold distilled water
- 70% ethanol
- 4% paraformaldehyde (PFA)

Equipment and Miscellaneous

- Dissection microscope (Genesee Scientific)
- *Drosophila* CO_2 blowgun (Genesee Scientific)
- Two sharp dissection forceps (Dumont)
- Whetstone
- Plastic transfer pipettes
- Small glass petri dish
- Microscopy glass slides (Fisherbrand Superfrost Plus)
- Hydrophobic barrier pen (ImmEdge)
- 9-well glass spot plate (Corning Pyrex)
- Covered ice bucket
- Tape
- Timer

4.2 Preparation

- Sharpen two dissection forceps to a fine point using the whetstone.
- Chill 1× PBS and distilled water on ice with their respective transfer pipettes.
- Fill a small petri dish with 70% ethanol.
- Paste a microscopy glass slide for dissection onto the bench and adjust to comfortably view and focus through the microscope eyepiece.
- On a new glass slide, i.e., the experiment slide, use a hydrophobic barrier pen to create a grid for the different sets/genotypes to be tested (a grid of up to 2×4 squares fits per slide). Let it

dry, then place a drop of 1× PBS in the square for the set currently being dissected. Label the slide appropriately. Firmly place this slide on ice in a covered ice bucket.

- Keep a few wet and dry Kimwipes handy for cleaning the forceps during dissections.
- Keep 4% PFA on ice.

4.3 Procedure Steps

Whole brain dissection

1. Anesthetize the flies using the CO_2 blowgun in the fly vials. Put one to two flies in the 70% ethanol petri dish for 1–2 min.
2. Hold a pair of forceps in each hand.
3. Place a drop of chilled 1× PBS on the dissection slide. Pick up one anesthetized fly and place it on its back in this drop and bring to focus under the microscope.
4. Hold it by its body with the forceps in the nondominant hand and by its proboscis with the forceps in the dominant hand.
5. Firmly pull the head upwards by its proboscis to separate it from the body, which is then discarded.
6. Pull the proboscis off of the head. Place one arm of each pair of forceps within the resulting groove and gently but firmly pull in opposite directions to expose the brain.
7. Gently remove cuticle, trachea, air sacs, and fibrous tissue surrounding the brain without damaging the brain.

Tissue fixation

8. In one well of the 9-well spot plate, pipette 10–20 μL of 4% PFA at the center. Gently transfer the freshly dissected brain with forceps to this spot, fully immersing it in 4% PFA. Incubate for 20 min at room temperature.
9. Remove the 4% PFA via pipetting and wash the fixed brain in a few drops of distilled water.
10. Place this brain in a drop of 1× PBS in the square for its respective set on the experiment slide grid, and maintain on ice in a covered ice bucket until all the brains for the experiment have been dissected and fixed in this manner.

4.4 Notes

- Ensure the flies are anesthetized and then euthanized in ethanol for a couple of minutes, else they may still be alive and moving during dissections.
- Once dissections have begun, brains should not be allowed to dry out during the entire experiment.
- Try to limit dissection time per brain to 3–5 min. This comes with practice. Using a multichannel timer and a 9-well spot

plate helps with having multiple brain samples being processed and fixed in tandem to minimize overall time required for these steps.

- The protocols cannot be held at any step. Once dissections begin, protocols should to be taken all the way through to overnight primary antibody incubation (for immunofluorescence) or probe incubation (for proximity ligation assays).
- Several resources provide detailed written and video protocols for performing fly brain dissections [31–33].
- As shown in elav-driven tau transgenic models in the following protocols, we find that the most consistent, and perhaps functionally relevant, tau expression and multimerization occurs primarily in the mushroom body and superior neuropils regions of the brain, which are the centers of learning, memory and higher behavioral function in *Drosophila* [34].

5 Protocol 4: Immunofluorescence (IF) of Whole *Drosophila* Brains

5.1 Requirements

- Experimental slide containing dissected and fixed brains from Protocol 3

Reagents

- 1× PBS
- 1× PBS-Triton (0.3%)
- BSA (bovine serum albumin)
- DAPI (4′,6-diamidino-2-phenylindole) Fluoromount G (SouthernBiotech)
- Antibodies as detailed in Table 3

Equipment and Miscellaneous

- Slide staining tray with black lid
- End-on-end rotator

Table 3
Antibodies for immunofluorescence

Primary/ Secondary	Antibody (source)	Catalog no.	Host	Dilution
Primary	MAPT/Tau (DSHB)	5A6-s	Mouse	1:150
Secondary	Goat anti-Mouse IgG (H+L), Highly Cross-Adsorbed, Alexa Fluor 555 (Invitrogen)	A42124	Goat	1:200

- 1.5 mL microcentrifuge tubes
- Microscopy glass slides (Fisherbrand Superfrost Plus)
- Microscopy glass coverslips, rectangles (Fisherbrand)
- Hydrophobic barrier pen (ImmEdge)

5.2 Procedure Steps

1. Dissect and fix fly brains as described in Protocol 3.
2. Carefully lift the experimental slides containing the sets of dissected brains floating in 1× PBS out of the ice bucket. Place on a clean bench and slowly remove the 1× PBS via pipetting from each square. Do not let the brain samples dry.

Permeabilization

3. Place a drop (approx. 20–50 μL) of 1× PBS-Triton to each square so as to completely immerse the brains.
4. Incubate for 15 min at room temperature.
5. Remove the reagent via pipetting from each square and wash the fly brains in place with a large drop of 1× PBS.

Blocking

6. Freshly prepare 2% BSA or 2% milk blocking buffer in 1× PBS-Triton and keep on ice. It can later be stored at 4°C for next day use.
7. Place the slide on a slide staining tray prefilled with water and remove the 1× PBS wash from the samples.
8. Completely cover the brains in each square with 20–50 μL of blocking solution. Place the lid on the tray and incubate for 30 min at room temperature.

Primary antibody incubation and washes

9. Prepare primary antibody diluted in the blocking solution. For this experiment, we use 2% BSA as the blocking solution for diluting the antibodies as per Table 3.
10. Replace the blocking solution on the samples with 20–50 μ L of the primary antibody solution and incubate as per your experimental conditions. For this experiment, we performed overnight Tau antibody incubation in a covered staining tray (prefilled with water) at room temperature.
11. The next day, carefully pipette each set of fly brains into separate microcentrifuge tubes and add 1 mL 1× PBS-Triton to each tube.
12. Set the tubes in an end-on-end tube rotator at low-medium speed. Repeat these washes twice for 5 min at room temperature. Allow the brains to settle at the bottom of the tubes before changing the washes.

Secondary antibody incubation and washes

13. During the previous washes, make a fresh grid slide and let it dry fully.
14. Carefully pipette each set of fly brains into their respective squares on this slide.
15. Wash the brains in place once with 1× PBS on the grid.
16. Remove the buffer from the samples and incubate with 20–50 μL of secondary antibody solution as per your experimental conditions. We perform this incubation in a covered staining tray (prefilled with water) at room temperature for 2 h.
17. Perform the washes in the end-on-end rotator as before with 1× PBS-Triton for 5 min, repeated twice.
18. Transfer brains to the grid and wash in place with 1× PBS, followed by a wash with distilled water. Ensure that the brains are spotted in the center of their square on the grid, and not near the edges.

Slide mounting

19. Carefully remove the liquid from the grid without disturbing the brains and let the surrounding liquid almost evaporate.
20. Add 50–60 μL DAPI Fluoromount G dropwise across the slide. Carefully lower the glass coverslip onto the slide, allowing the mounting media to spread evenly. Get rid of bubbles by gently tapping the coverslip.
21. Let the slide set overnight in the dark before viewing.

5.3 Visualization and Analysis

All images were captured using the confocal microscope Zeiss LSM 880 with the ZEN software at 10× and 20× magnification focused on the mushroom body and superior neuropils regions of the dissected brains. Samples were imaged in the red channel (561 nm) for tau immunofluorescence signal, the green channel (488 nm) for EGFP (only in the LUM sets) and blue channel (405 nm) for DAPI.

- Non-LUM Sets: As seen in Fig. 2B, Tau^{WT}, Tau^{R406W}, and Tau^{P301L} transgenic brains are similar in that they localize most intensely to regions of the brain that are dense with cell bodies. The three models vary in immunofluorescence-based tau patterns in the following ways:
  - In Tau^{WT} brains, tau protein appears very diffuse and does not localize to cell bodies.
  - In Tau^{R406W} and Tau^{P301L} brains, tau appears to be more strongly localized to cell bodies.
- LUM Sets: In Fig. 3B, as expected, $EGFP^{LUM}$ shows strong GFP signal but no tau, while Tau^{LUM} shows intense tau

immunofluorescence in the cell body-rich areas of the brain and in the mushroom bodies.

5.4 Notes

- To determine the optimal antibody conditions for immunofluorescence assays described here, it is recommended to refer to the set working dilutions for that antibody in formalin-fixed paraffin-embedded (FFPE) brain sections, and accordingly test its titrations in the whole fly brains. Our lab had standardized the working dilution of MAPT/Tau5A6 antibody as 1:500 in FFPE brain sections with overnight incubation at room temperature. Testing in dissected brains, we found ~3× of that dilution, i.e., 1:150 works optimally for the same incubation conditions.
- Changing the experimental grid slide after the overnight primary antibody incubation step is recommended because the blocking buffer used to dilute the antibody appears to leave a residue on the slide that contributes to background noise in the samples during imaging.
- The brains need to be handled carefully when transferring between the experimental slides and microcentrifuge tubes. They are susceptible to getting lost in transit because they tend to adhere to the interior of pipette tips and the inner walls or lids of microcentrifuge tubes. It is, therefore, best to use P200 tips and only perform washes in microcentrifuge tubes that contain nonionic detergent in the buffer, e.g., 1× PBS-Triton, as it helps to prevent the brain samples from adhering to plastic surfaces. Washes that do not contain detergents are performed on the slide itself. Using low-bind microcentrifuge tubes may also help circumvent this issue.
- This protocol was designed for maximizing efficient usage of biological material and reagents, while minimizing waste in *Drosophila* whole brain immunohistochemical assays. An entire experiment of up to eight sets each comprising of ~3–8 biological replicates of fly brains can be run on a single grid slide. This is because the surface area of these brains and the area they occupy on the slide are greatly reduced compared to FFPE brain sections. By judiciously optimizing the various incubation concentrations and conditions as well as washes in the protocol, one can easily acquire ample observations and data from complete fly brains, while consuming only a fraction of the reagents and requirements of conventional methods.

6 Protocol 5: Proximity Ligation Assay (PLA) of Whole *Drosophila* Brains

Using the resources and application notes available on the Sigma-Aldrich Duolink® Proximity Ligation Assay website, especially the PLA Probes and DuoLink PLA Fluorescence Protocols, the following PLA protocol was designed and optimized for studying tau–tau protein interactions in situ in whole *Drosophila* brains.

6.1 Requirements

- Experimental slide containing dissected and fixed brains from Protocol 3

Reagents and Kits

- MAPT/Tau 5A6-s (DSHB) antibody
- 1× PBS
- 1× PBS-Triton (0.3%)
- DuoLink PLA reagents and kits as detailed in Table 4

Table 4
DuoLink (Sigma-Aldrich) PLA reagents used in this assay

Name	Catalog no.	Components used in this protocol	Used in protocol step	Storage conditions
DuoLink In Situ Probemaker: • PLUS • MINUS	DUO92009 DUO92010	• DuoLink PLA oligonucleotides (lyophilized) • Conjugation buffer • Stop reagent • Storage solution	Tau antibody-oligonucleotide conjugation to generate Tau-PLA probes	−20°C
		• Blocking solution	Blocking	
		• PLA probe diluent	Tau-PLA probe incubation	
DuoLink In Situ Detection Reagents Red (Fluorescent)	DUO92008	• 5× Ligation buffer stock • Ligase (1 U/μL)	Ligation	−20°C
		• 5× Amplification buffer stock • Polymerase (10 U/μL)	Amplification	
DuoLink In Situ Wash Buffers, Fluorescence: • Wash Buffer A • Wash Buffer B	DUO82049 DUO82046 DUO82048	Wash Buffers A and B, prepared as follows: For each buffer, dissolved one powder packet in 1 L of high purity water	Washes	4°C (long-term); room temperature (short term)
DuoLink In Situ Mounting Medium with DAPI	DUO82040	Mounting medium with DAPI	Slide mounting	4°C

Equipment and Miscellaneous

- 0.5 mL 100 kDa MWCO centrifugal filters (Amicon Ultra)
- Slide staining tray with black lid
- Humidity chamber
- End-on-end rotator
- Syringe filters (Merck Millipore)
- 5 mL syringes
- Aluminum foil
- 1.5 mL microcentrifuge tubes
- Microscopy glass slides (Fisherbrand Superfrost Plus)
- Microscopy glass coverslips, rectangles (Fisherbrand)
- Nail polish

6.2 Generation of Tau-PLA Probes Using the DuoLink Probemaker Kits

Here, we describe the method by which Tau-PLA probes were generated using the DuoLink Probemaker kits (detailed in Table 4) with the MAPT/Tau 5A6 (DSHB) antibody. Concentrations and calculations will vary depending on your antibodies and reagents.

Concentrating the MAPT/Tau 5A6 antibody stock

- The antibody to be conjugated needs to have 1 mg/mL concentration. 20 μg (=20 μL) of antibody is required per conjugation with each of the lyophilized oligonucleotides (PLUS and MINUS).
- The antibody must be in an amine-, carrier-, and preservative-free buffer, ideally 1× PBS.

1. As MAPT/Tau 5A6 antibody is in supernatant form with low concentration (~60 μg/mL), a large volume (~700 μL) of this stock is required as starting material in this process, calculated for a total of ~42–45 μg antibody (as needed for making the two probes).
2. To purify IgG-type proteins with molecular weights of ~150 kDa, use Amicon Ultra 0.5 mL 100 kDa MWCO centrifugal filters as per manufacturer instructions. Antibody A/G purification columns can be used as an alternative.
3. Perform buffer exchange with filter-sterilized 1× PBS prior to collecting the concentrated retentate.
4. Determine the protein concentration of the purified Tau antibody. Typically, ~10% loss in final protein yield is expected from using the Amicon Ultra filter-type concentrators.

Tau antibody conjugation to the PLUS and MINUS PLA oligonucleotides

- Vortex all liquid reagents in the DuoLink Probemaker kits before use.

For each probe reaction (utilizing the components of their respective kits):

5. Add 20 μL of purified Tau antibody solution having ~1 mg/mL concentration in a microcentrifuge tube. Make up this volume with sterile 1× PBS if required.
6. Add 2 μL of Conjugation Buffer and gently mix by pipetting.
7. Transfer this antibody mix to the glass vial of lyophilized oligonucleotides provided in the kit and mix gently by pipetting.
8. Incubate at room temperature overnight.
9. Add 2 μL of Stop Reagent to the reaction and gently mix by pipetting.
10. Incubate at room temperature for 30 min.
11. Add 24 μL of Storage Solution to the vial and mix well.
12. The Tau-conjugated PLUS and MINUS probes are stored at 4°C for several months.

6.3 Procedure Steps

1. Dissect and fix fly brains as described in Protocol 3.
2. Carefully lift the experimental slides containing the sets of dissected brains floating in 1× PBS out of the ice bucket. Place on a clean bench and slowly remove the 1× PBS via pipetting from each square of the grid. Do not let the brain samples dry.
3. Retrieve the DuoLink Blocking Solution and the DuoLink PLA Probe Diluent from −20°C and let them thaw on ice.
4. Retrieve Wash Buffers A and B from 4°C and maintain at room temperature for next-day use.

Permeabilization

5. Place a drop (approx. 20–50 μL) of 1× PBS-Triton to each square to completely immerse the fly brains. Incubate for 15 min at room temperature.
6. Remove the reagent via pipetting from each square carefully so as to not disturb or draw the brains into the tip.
7. On the slide, wash the fly brains in place with a large drop of 1× PBS twice, followed by a wash with distilled water.

Blocking

8. Fill a slide staining tray with water and place this slide on it.

9. Remove the previous wash from the samples. Add a drop (20–50 μL) of Blocking Solution (which is provided in the DuoLink Probemaker kit) to each square on the grid, so as to completely immerse the brains contained in them.
10. Cover the tray and place in the incubator/humidity chamber at 37°C for 1 h.

Tau-PLA probe incubation and washes

11. Retrieve the PLUS and MINUS Tau-PLA probe vials from 4°C and lightly vortex them before use.
12. Dilute the probes in the PLA Probe Diluent (which is provided in the DuoLink Probemaker kit) as per your experimental conditions. We optimized the concentration for Tau-PLA probes to be:
    - 0.3 μL of each probe in 400 μL probe diluent for LUM fly brains.
    - 0.5μL of each probe in 400 μL probe diluent for non-LUM fly brains.

    Gently mix the diluted Tau-PLA probe solution and keep on ice.
13. Retrieve the slide staining tray from the humidity chamber. Replace the blocking solution with 20–50 μL of the diluted Tau-PLA probe solution to completely immerse the brain sets on the slide.
14. Cover the tray and incubate at room temperature overnight (~16 h).
15. Label microcentrifuge tubes for each sample set. Gently pipette the brains from the grid into their respective tubes. Add 1 mL of room temperature Wash Buffer A (which is prepared as described in Table 4).
16. Set the tubes in an end-on-end tube rotator at low-medium speed to wash for 5 min. Allow the brains to settle at the bottom of the tubes before changing the washes.
17. Gently remove the used buffer and replace with fresh Buffer A. Repeat the wash for a total of two 5 min washes in Buffer A.
18. During the washes, rinse the experimental slide with distilled water and let it dry completely.
19. Place the grid slide on the staining tray (prefilled with water). Carefully pipette each set of fly brains into their respective squares on the grid.

Ligation and washes

20. Retrieve 5× Ligation Buffer stock, which is provided in the DuoLink Detection Reagents Red Fluorescence kit (detailed in Table 4), from −20°C and let it thaw on ice. Gently vortex to resuspend any precipitation in the tube.

21. Dilute to 1× Ligation Buffer in high purity water and mix gently. Calculate 1× Ligation Buffer at 25–40 μL per grid square so as to completely cover the brains in each sample set.

22. Just before adding to samples, retrieve the Ligase (1 U/μL stock) from −20°C and add it at 1:40 dilution to the 1× Ligation Buffer. Mix by pipetting gently on ice.

23. Remove the used Buffer A from the samples and allow to almost dry (not completely).

24. On the grid slide, apply the 1× Ligation Mix to each sample set. Cover the tray and incubate at 37°C in the humidity chamber for 30 min.

25. Retrieve the slide staining tray from the humidity chamber and wash the sample sets with Buffer A as described in Steps 15–19.

Amplification and washes

[Note: The amplification buffer is light-sensitive. Protect all solutions containing buffer from light.]

26. Retrieve 5× Amplification Buffer stock, which is provided in the DuoLink Detection Reagents Red Fluorescence kit, from −20°C and let it thaw on ice. Gently vortex to resuspend any precipitation in the tube.

27. Dilute to 1× Amplification Buffer in high purity water and mix gently in a dark or foil-wrapped microcentrifuge tube. Prepare the same volume of 1× Amplification Buffer as previously calculated for the 1× Ligase Buffer.

28. Just before adding to samples, retrieve the Polymerase (10 U/μL stock) from −20°C and add it at 1:80 dilution to the 1× Amplification Buffer. Mix by pipetting gently on ice.

29. Remove the used Buffer A from the samples and allow them to almost dry (not completely).

30. On the grid slide, apply the 1× Amplification Mix to each sample set. Cover the tray and incubate at 37°C in the humidity chamber for 100 min (i.e., 1 h 40 min).

31. Retrieve the slide staining tray from the humidity chamber, and keep the samples protected from light.

32. Pipette the brain sets into their respective foil-wrapped microcentrifuge tubes. Add 1 mL of room temperature Wash Buffer B (which is prepared as described in Table 4) per tube.

33. Set the tubes in an end-on-end tube rotator at low-medium speed to wash for 10 min. Allow the brains to settle at the bottom of the tubes before changing the washes.
34. Gently remove used buffer and replace with fresh Buffer B. Repeat the wash once for a total of two 10 min washes in Buffer B.
35. During washes, rinse the experimental slide with distilled water and let it dry completely.
36. To transfer the brain sets back onto the experimental slide, first coat a pipette tip internally with minimal Buffer A since it contains nonionic detergent. This ensures the brains do not get stuck in the tip during the transfer. Spot the brains in the center of their square on the grid, and not near the edges.
37. On the slide, wash the brain sets briefly in place with 20–50 μL Buffer B, followed by a wash in 20–50 μL 0.01× Buffer B (i.e., Buffer B diluted 100× in high purity water).

Slide mounting

38. Carefully remove the liquid from the grid and let the samples almost dry out.
39. Then add 50–60 μL DuoLink PLA Mounting Media with DAPI dropwise across the slide. Carefully lower the glass coverslip onto the slide, allowing the mounting media to spread evenly. Remove bubbles by gently tapping the coverslip. Seal with nail polish and let it set in the dark overnight.
40. Store the slide in the dark at 4°C up to 7 days to retain maximum PLA signal for imaging. These slides can also be stored at −20°C for a few months.

6.4 Visualization and Analyses

Here, we describe the methods by which data were acquired and analyzed, separately for non-LUM and LUM experimental sets. All images were captured using the confocal microscope Zeiss LSM 880 with the ZEN software.

Non-LUM Sets

- **Visualization:** Samples were imaged in the red channel (561 nm) for Tau-PLA signal and blue channel (405 nm) for DAPI. As the Tau-PLA signals produced in non-LUM sets are discrete red fluorescent puncta, we utilized 63× magnification to capture the images clearly. Most of the signal localizes to cell body-dense regions of the brain and the optical lobes. Tau-PLA signals, if present, are most consistently observed in the mushroom bodies and superior neuropils of the non-LUM fly brains. Two separate images of the regions of interest were acquired per sample.

- **Analyses:**
  - **Background measurements**: The complete (full frame) images of only the control brain samples in the red channel (Tau-PLA) were measured for total fluorescence intensity in the FIJI Image J software [Analyze→Measure]. The maximum gray value "Max" obtained per image is averaged for replicates of each brain sample. Single values per control brain were averaged again to get the mean "Max" value for the controls, which is assigned as the lower threshold in the macro for experimental image quantification. The lower threshold was determined to be 70 in the example shown in Fig. 2C and D.
  - **Image analysis**: Given that the Tau-PLA signals in non-LUM sets form puncta, we performed particle analysis in the red channel. The following macro was used to analyze all images generated in this experiment:

    ```
    setAutoThreshold("Default dark");
    //run("Threshold...");
    setThreshold(70, 255, "raw");
    run("Analyze Particles...", "display exclude include summarize overlay add composite");
    close();
    ```

    The "Count" value obtained in the readings per image is averaged for the replicates of each brain sample and is thus the representative value for the individual samples used in further analysis.
  - **Statistical analysis:** The sample sizes (*n*) range is 5—7 for the sets in this experiment. The calculated "Count" values for individual brain samples in each set were entered as grouped data columns in the GraphPad Prism 10 software. Data were analyzed using one-way ANOVA with Sidak's multiple comparisons test for all pairwise comparisons.
- **Results and inference:** As expected, Tau-PLA signal is not detectable in controls. Tau-PLA is negligible in brains of Tau^{WT} flies. In contrast, Tau^{R406W} and Tau^{P301L} brains show significant Tau-PLA puncta, with the signal from Tau^{P301L} being considerably higher as shown in Fig. 2C and D. There is some variation in positive Tau-PLA signals that may be sex-dependent.

LUM Sets

- **Visualization:** Samples were imaged in the red channel (561 nm) for Tau-PLA, green channel (488 nm) for EGFP and blue channel (405 nm) for DAPI. As the Tau-PLA signals produced in the Tau^{LUM} fly brains are bright with discernible red fluorescence, images were captured at 10× and 20× magnifications, focusing on the central brain, either in its entirety (at

10×) or specifically the mushroom body and superior neuropil regions (at 20×) where the signals are most strongly and consistently observed. As in the case with non-LUM sets, Tau-PLA signal is also present in other cell body-dense regions of the brain and optical lobes in Tau^{LUM} flies. Replicate images per brain sample are not necessary as imaging at low magnifications captures the entire regions of interest.

- **Analyses:**
  - **Background measurements**: Images of EGFP^{LUM} control brain samples in all three channels were opened in the FIJI Image J software. For each image, the region of interest was outlined in the blue or green channel using the Freehand tool, copied [Ctrl+C], and imposed [Ctrl+Shift+E] onto the red channel image. The red channel signal from this selected region is measured [Analyze→Measure] and the maximum gray values "Max" were averaged across control brain samples. This mean "Max" value is assigned as the lower threshold in the macro for experimental image quantification. In the example in Fig. 3C and D, we analyzed 20× images, wherein the lower threshold was determined to be 21.
  - **Image analysis**: As the Tau-PLA signal in Tau^{LUM} brain samples is non-discrete and continuous, total fluorescence intensity was quantified in the regions of interest. Thus, for each image, the region to be analyzed was first outlined as described above, and then the following macro was executed for all images in the red channel:

    ```
    setAutoThreshold("Default dark");
    //run("Threshold...");
    setThreshold(21, 255, "raw");
    run("Measure");
    close;
    ```

    Of interest are the mean gray value "Mean" and area fraction "% Area" generated in the readings for each image. We multiplied "Mean" * "%Area" per image to obtain true Tau-PLA mean fluorescence intensity, which is thus the representative value for the individual sample used in further analysis.
  - **Statistical analysis:** The sample size (n) of the EGFP^{LUM} set was 4 and that of Tau^{LUM} was 6 in this experiment. The Tau-PLA mean fluorescence intensity values for individual brain samples in each set were entered as grouped data columns in the GraphPad Prism 10 software. Data were analyzed using an unpaired two-tailed t-test.
- **Results and inference:** As expected, Tau-PLA signal is absent in EGFP^{LUM} controls, but it is present in 10-day old Tau^{LUM}

fly brains. The Tau-PLA signal observed in the Tau^{LUM} model is not discrete puncta as is expected in the assay but appears as extensive coalescent fluorescence in the fly brain. This could be because the Tau^{LUM} transgenic construct expressed pan-neuronally in these flies comprise of two copies of Tau^{P301L}, which likely results in higher levels of tau protein interactions in this model.

6.5 Notes

- In this protocol, the antibody-PLA oligonucleotide conjugation step is essential because homomeric tau interactions are being investigated here. However, the standard PLA protocols for heteromeric interactions do not require this step and instead probes are added after primary antibody incubation.
- While we have used a supernatant-form antibody to generate the antibody-PLA probes, use of high purity antibodies for this process may be more ideal, as they contain less impurities and/or other protein species that may interfere with the conjugation process. For cases in which non-purified low concentration antibodies are required for conjugation, using antibody-specific purification and protein estimation methods provide the most accurate results.
- Applying the same primary antibody conditions used in the IF protocol, we performed antibody-PLA probe titrations to obtain optimal probe dilutions for PLA:
  - Based on (i) Tau5A6 antibody stock concentration and (ii) its working dilution for IF in whole fly brains, we determined the effective antibody concentrations applied in those experiments.
  - As we know that the antibody concentration in each Tau-PLA probe is ~0.4 μg/μL (since 20 μg Tau5A6 was used in ~50 μL conjugation reaction), we can calculate a probe dilution that is equivalent to the effective antibody concentration in IF experiments, and thus test the Tau-PLA probe titrations accordingly.

  Hence, we have optimized the total Tau-PLA probe dilutions in these PLA experiments at 1.5–3× of effective antibody concentration used in IF experiments for overnight room temperature incubations.
- As with the IF protocol, this PLA protocol in whole *Drosophila* brains has similarly been designed for efficient reagent usage while maximizing consistent results.
- Besides optimization of the Tau-PLA probe dilutions for adopting PLA in whole fly brains, following the standard DuoLink PLA fluorescence protocol worked well for our experiments. However, based on your experimental requirements, the timing,

temperatures or conditions of the incubations, enzymatic processes and washes may require alterations as they affect the output.

- In addition to the several useful materials on the Sigma-Aldrich DuoLink PLA website, other informative resources also helped in the development of our PLA protocol [9, 35].

7 Protocol 6: Native PAGE-Western Blot Using *Drosophila* Heads

The protocol for native PAGE-Western blots is similar to that of SDS PAGE-Western blots (Protocol 2), with some key differences that are highlighted below:

7.1 Requirements

- Most requirements are the same as those in Protocol 2. However, as the protein samples are run under nondenaturing and nonreducing conditions in native PAGE, followed by modified Western blot steps, Table 5 lists the changes made in this protocol:

Table 5
Requirements specific to the native PAGE-Western blot protocol

Component	In this protocol	In Protocol 2
Gel running buffer	10× Native PAGE Running Buffer stock (which is same as the regular Western 10× Transfer Buffer stock composition): • Glycine 144 g • Tris Base 30.3 g • Distilled water to 1 L	10× SDS PAGE Running Buffer
Sample buffer	2× Native Tris-Glycine Sample Buffer (Novex)	2× Laemmli Sample Buffer (Bio-Rad) with 2-mercaptoethanol
SDS	Added 0.05% w/v fresh to 10X Native PAGE Running Buffer and volumed up with distilled water to make 1× Native Transfer Buffer	Added to 10× SDS PAGE Running Buffer stock
Blot membrane	0.2 μm Immun-Blot PVDF (polyvinylidene fluoride) membrane (Bio-Rad)	0.2 μm Nitrocellulose membrane (Bio-Rad)
Methanol	Added to activate the PVDF membrane	Added fresh to make regular Western 1× Transfer Buffer
Blocking agent	BSA	Non-fat milk powder
Blot incubation box	Large-sized (LICOR)	Medium-sized (LICOR)

7.2 Preparation

- See Protocol 2.

7.3 Protocol Steps

Sample preparation

1. Gather sharp forceps, a large plastic petri dish and a tissue homogenizer pestle. Chill forceps and petri dish on dry ice.
2. Retrieve the tubes with frozen flies and keep on dry ice.
3. Transfer the required number of frozen flies to the petri dish.
4. Each lysate sample is prepared from homogenizing three fly heads together. Hence, using forceps, remove the heads of three frozen flies per sex per genotype and transfer them to individual microcentrifuge tubes on dry ice.
5. Add 12 μL of 2× native sample buffer to each tube. Manually homogenize the fly head samples for 30 s to 1 min in the buffer on regular ice. Wash the pestle between samples.
6. Centrifuge the samples at 5000 rpm for 5 min at 4°C to pellet their cuticle debris.

Native PAGE

7. Before starting:
   - Prepare 1 L 1× native PAGE running buffer from its 10× stock.
   - Follow the substeps at this stage as in Protocol 2, using 1× native PAGE running buffer instead of 1× SDS PAGE running buffer.
   - Move this gel running setup to a cold room.

8. Load the complete liquid lysate from each sample tube onto the gel, without disturbing their pellet. Record the order of samples loaded.
9. Optionally, load a regular protein ladder, not as a reference for the size of protein complexes but as visual indicator of the native PAGE run; use of a native protein ladder is preferable.
10. Run the native PAGE at 50 V for the first 10 min and then at 70–90 V for the next 5.5–6 h in the cold room.

Western blot transfer

11. Before starting:
    - Prepare 1 L 1× native transfer buffer while the gel is running: To 100 mL of 10× native PAGE running buffer stock, add 0.5 g SDS (0.05% w/v) and make up the volume to 1 L with distilled water. Mix thoroughly, seal the container and let it chill at 4°C for at least an hour.
    - For each gel to be blotted, cut a 0.2 μm PVDF membrane to the full gel size and two filter papers.
    - Rinse the Bio-Rad mini trans-blot apparatus thoroughly with distilled water.

12. Carefully open the gel cassette using a gel releaser wedge. Cut just above the bottom of the wells and the bottom edge of the gel, and discard these portions, but preserve the complete stacking and resolving regions of the gel. Rinse the gel briefly with distilled water.
13. Activate the PVDF membrane in methanol for 1–2 min on the rocker. Then rinse it well with distilled water.
14. Follow the rest of the steps pertaining to transfer sandwich preparation, assembly and loading as in Protocol 2, using 1× native transfer buffer instead of regular 1× transfer buffer.
15. In a cold room, place the transfer tank on a magnetic stirrer and set it to stir at medium-low speed.
16. Set the transfer to run at 24 V for 15–16 h overnight.

Ponceau S staining of the membrane

17. Before starting:
    - Prepare 1 L of 1× PBS from its 10× stock. Pour it into a large Pyrex dish.
    - Prepare 50 mL of 5% BSA blocking buffer fresh: Dissolve 2.5 g of BSA in 1× PBS-Tween and make up the volume to 50 mL in a conical tube. Keep on ice for immediate use and store at 4°C overnight for next day use.
18. Follow the steps for antigen retrieval and Ponceau S staining as in Protocol 2. Use large blot incubation boxes for this protocol.

Antibody incubations and detection

19. Block the PVDF membrane with freshly prepared 5% BSA blocking buffer at room temperature for 1 h.
20. Follow the remaining steps of Protocol 2 for antibody incubations, washes and signal detection.

7.4 Visualization and Analysis

- For native PAGE-Western blot data, each lane comprises a sample generated from three fly heads lysed together. All analyses are performed at day 10 of adulthood.
- For the LUM sets, all samples were equally loaded as inferred from the Ponceau S staining of these blots. As seen in Fig. 3E, when probed with the EGFP antibody, the $EGFP^{LUM}$ samples show strong EGFP bands in the monomeric range and very faint bands in the high molecular weight (HMW) range. When probed with the Tau antibody, the Tau^{LUM} samples show intense HMW tau smears, indicating significant tau multimerization in brains of these tau transgenic models.

7.5 Notes

- In this protocol, fly head lysates are not subject to denaturing (i.e., SDS treatment and sample boiling) or reducing (i.e., 2-mercaptoethanol) conditions during gel running so that the sample proteins maintain their native conformation. Care is taken to wash the gel running apparatus thoroughly with distilled water before use to remove any traces of SDS.
- Here, we modified a previously used lab protocol for native PAGE-Western blots [20], primarily at the blot transfer step, wherein in we removed methanol from the transfer buffer and included 0.05% SDS instead. This is because it has been reported that methanol provides little advantage while low concentrations of SDS (up to 0.25%) facilitate improved transfer of large proteins and complexes [36–38]. This is especially useful for blot transfers onto PVDF membranes.
- This protocol can be further optimized by performing blot transfers onto 0.45 μm instead of 0.2 μm PVDF membranes. Including a tau standard such as purified tau and/or preaggregated recombinant tau proteins would help estimate oligomeric tau species observed in this experiment [39]. Incorporating a native gel protein ladder would also be useful for this purpose.
- To determine the ideal overnight blot transfer conditions (time and voltage), use double membranes at a starting condition of 20–30 V for 16 h transfer. If proteins appear to have passed through the first membrane and onto the second membrane based on Ponceau S staining, then modify the blot transfer conditions accordingly.

8 Protocol 7: In Vivo Luminescence Assay in *Drosophila*

8.1 Requirements

- $EGFP^{LUM}$ and Tau^{LUM} *Drosophila* sets

Reagents

- Pierce D-Luciferin Monopotassium Salt, 1 g (Thermo Scientific #88294)
- 100 mM D-luciferin stock:
  - Completely dissolve 1 g D-luciferin salt in 31.4 mL autoclaved water in an aluminum foil-wrapped 50 mL conical tube, as this reagent is light-sensitive.
  - Aliquot this stock solution into aluminum foil-wrapped microcentrifuge tubes.
  - Store the 100 mM D-luciferin stock tubes at −20°C.
- Previously prepared fly food bottles

Equipment and Miscellaneous

- Glass beaker
- Thermometer
- 2 mL or 5 mL syringes
- Fresh empty fly food vials and cotton plugs
- Aluminum foil
- Dissection microscope (Genesee Scientific)
- *Drosophila* CO_2 blowgun (Genesee Scientific)
- *Drosophila* CO_2 fly pad (Genesee Scientific)
- Paintbrush
- White flat-bottom 96-well plate (Nunc MaxiSorp)
- Sharp and blunt-end forceps
- Clear plastic adhesive lid
- Timer
- Microplate reader (Promega GloMax Discover)

8.2 Procedure Steps

24 h prior to performing the luminescence assay

1. Prepare the fly food containing 15 mM D-luciferin as follows:
   - Either freshly prepared fly food or previously prepared fly food, i.e., food that has been bottled and stored at 4°C, can be used in this protocol. Collect 40–50 mL solid fly food in a glass beaker.
   - Heat this food in the microwave in 15s intervals until it just boils. Stir well and remove any chunks. Note: Significantly more food than required is collected at this step to avoid burning during heating.
   - Let the food cool to ~55°C.
   - Thaw 100 mM D-luciferin stocks on ice. Add 300 μL of this stock to empty new fly vials, calculated to provide a final concentration of 15 mM D-luciferin in fly food.
   - Using a 2 mL or 5 mL syringe, quickly dispense 2 mL food to vials, mix well, seal with cotton plug and fully wrap the vial with aluminum foil.
   - Keep vials at 4°C for at least 30 min to let cool and solidify.
2. Anesthetize and collect 9-day old $EGFP^{LUM}$ and Tau^{LUM} flies on a CO_2 fly pad. Transfer up to 15 flies per set to the luciferin-containing food vials and incubate at 25°C for 24 h.

On the day of the luminescence assay

3. Using the *Drosophila* CO_2 blowgun, anesthetize the now 10-day old flies and align them on the CO_2 fly pad in the same

orientation that they are to be placed in the white flat-bottom 96-well plate. Ensure that all the flies across sets are exposed to approximately the same duration of CO_2 treatment.

4. Cover the white flat-bottom 96-well plate with a clear plastic adhesive lid. Opening one column at a time, quickly place one fly per well by holding a wing with blunt-end forceps. Close each column as they are done. After loading all flies to the plate, poke two small holes over each well using sharp forceps to allow the flies to breathe. Make a note of the order of the samples.
5. Place this plate in a Promega GloMax Discover microplate reader or equivalent and allow the flies to recover from anesthesia. Make sure that the flies are alive by gently tapping on the plate. This may take up to 10 min. Censor any dead flies.
6. On the plate reader software, precisely select all the wells containing flies to be measured for luminescence and additional empty wells for blank readings. The luminescence measurement parameter is set to 10 s integration time per well.
7. 30 min after completely setting up the plate, initiate the luminescence measurements on the plate reader.
8. Once completed, save and export the data as Microsoft Excel files.

After the experiment

9. These flies can be reused for other experiments such as Western blotting, brain dissections, or behavioral assays. To retrieve flies from the plate, place the plate on ice for 10 min to anesthetize the flies and then collect them for further experimentation.

8.3 Visualization and Analysis

- Luminescence readings generated in this assay are in relative light units (RLU) and are also color-coded based on their levels for easy visualization.
- As we have previously reported that the luminescence data follows lognormal distribution, we perform $\log_{10}$-based calculations for formal analysis:
  - Find the mean of all the luminescence values of the control, i.e., $EGFP^{LUM}$ data set.
  - Perform normalization by dividing the individual luminescence values of the $EGFP^{LUM}$ and Tau^{LUM} sets with the mean $EGFP^{LUM}$ value. Multiply these values with 100 to calculate the luminescence (% of $EGFP^{LUM}$) value for each sample.
  - Perform $\log_{10}$ transformation on the luminescence (% of $EGFP^{LUM}$) values.

- Graph the resultant individual values as grouped data sets in the GraphPad Prism software and perform unpaired two-tailed t-test.
- While the EGFP^{LUM} flies have luminescence levels similar to that of the blank wells of the plate, Tau^{LUM} flies exhibit high luminescence levels indicative of significant tau–tau protein interactions. This is also confirmed in the log normalized data in Fig. 3F. This assay shows robust statistical significance even with a sample size of $n = 6$ flies per set.

8.4 Notes

- We have followed the protocol previously established by our group [20]. The assay described here measures luminescence at 30 min after the flies are transferred onto the white 96-well plate. This is an optimal endpoint because by this time, the flies have typically stabilized from any stress caused by the anesthesia and/or transfer process that may affect luminescence readings. As the flies are placed in empty plate wells without luciferin-treated food in this assay, prolonging the measurements to a later time-point could also adversely affect the luminescence readings.
- In case flies used in this assay are required for further experiments, organize the plate in such a way such that there are empty columns between the sets tested. This avoids any mix-ups during fly collection from the assay plate.

9 Future Directions

While the methods we have presented provide a sound foundation for studying various paradigms of tauopathy in *Drosophila*, other techniques can be implemented to supplement the approaches that we have described. Sarkosyl insolubility assays can be combined with native PAGE to investigate tau protein aggregation [40, 41]. Tau-PLA experiments can be combined with immunofluorescence [35], Western blotting [42], or be multiplexed with high throughput proteomic techniques to identify other protein interactors [43]. The Tau^{LUM} fly model is especially valuable for an extensive range of uses: (i) Luminescence signals from dissected *Drosophila* brains can be directly visualized and quantified using luminescence microscopy, thereby corroborating in vivo measurements [44], (ii) Functions as a versatile platform for genetic, pharmacological and behavioral screening to identify genetic modifiers of tau aggregation and small molecules that suppress tau multimerization, (iii) Amenable to straightforward modifications to study other tau mutations or proteotoxic proteins in neuronal and non-neuronal tissues, and (iv) Housing in 96-well plates containing luciferin-enriched food, facilitating luciferase measurements over the course of *Drosophila* aging.

A misfolding-prone luciferase variant has recently been developed in *Drosophila* as a sensor of subcellular proteostasis, including testing the effect of pathogenic tau [45], further highlighting the significance of luciferase reporters in *Drosophila* and their value as a screening platform. Overall, *Drosophila melanogaster* continues to offer an excellent framework to develop biological innovations that enhance our understanding of neurodegenerative and proteotoxic disorders.

Conflict of Interest

The authors have no conflicts of interest to declare that are relevant to the content of this chapter.

References

1. Cowan CM, Sealey MA, Quraishe S et al. (2011) Modelling tauopathies in *Drosophila*: insights from the fruit fly. Int J Alzheimer's Dis 2011:598157
2. Gistelinck M, Lambert JC, Callaerts P et al. (2012) *Drosophila* models of tauopathies: what have we learned?. Int J Alzheimer's Dis 2012:970980
3. Jeon Y, Lee JH, Choi B et al. (2020) Genetic dissection of Alzheimer's disease using *Drosophila* models. Int J Mol Sci 21:884
4. Giong H-K, Subramanian M, Yu K et al. (2021) Non-rodent genetic animal models for studying tauopathy: review of *Drosophila*, zebrafish, and *C. elegans* models. Int J Mol Sci 22:8465
5. Jiang Y, MacNeil LT (2023) Simple model systems reveal conserved mechanisms of Alzheimer's disease and related tauopathies. Mol Neurodegener 18:1–23
6. Fredriksson S, Gullberg M, Jarvius J et al. (2002) Protein detection using proximity-dependent DNA ligation assays. Nat Biotechnol 20:473–477
7. Söderberg O, Gullberg M, Jarvius M et al. (2006) Direct observation of individual endogenous protein complexes in situ by proximity ligation. Nat Methods 3:995–1000
8. Greenwood C, Ruff D, Kirvell S et al. (2015) Proximity assays for sensitive quantification of proteins. Biomol Detect Quantif 4:10–16
9. Wang S, Yoo SH, Kim HY, et al. (2015) Detection of in situ protein-protein complexes at the *Drosophila* larval neuromuscular junction using proximity ligation assay. J Vis Exp 20(95), e52139.
10. Mannix KM, Starble RM, Kaufman RS, et al. (2019) Proximity labeling reveals novel interactomes in live *Drosophila* tissue. Development 146(14), dev176644.
11. Wang M, Han X, Liu C et al. (2021) Intracellular trafficking of Notch orchestrates temporal dynamics of Notch activity in the fly brain. Nat Commun 12:1–15
12. Sharp KA, Khoury MJ, Wirtz-Peitz F, et al. (2021) Evidence for a nuclear role for *Drosophila* Dlg as a regulator of the NURF complex. Mol Biol Cell 32(21), ar23.
13. Ashley J, Carrillo RA (2025) Using the proximity ligation assay to visualize colocalization of proteins at the *Drosophila* larval neuromuscular junction. Cold Spring Harb Protoc 2025:1–6
14. Gu GJ, Wu D, Lund H et al. (2013) Elevated MARK2-dependent phosphorylation of tau in Alzheimer's disease. J Alzheimer's Dis 33:699–713
15. Gu GJ, Lund H, Wu D et al. (2013) Role of individual MARK isoforms in phosphorylation of tau at Ser 262 in Alzheimer's disease. NeuroMolecular Med 15:458–469
16. Bengoa-Vergniory N, Velentza-Almpani E, Silva AM et al. (2021) Tau-proximity ligation assay reveals extensive previously undetected pathology prior to neurofibrillary tangles in preclinical Alzheimer's disease. Acta Neuropathol Commun 9:1–20
17. Puangmalai N, Sengupta U, Bhatt N et al. (2022) Lysine 63-linked ubiquitination of tau oligomers contributes to the pathogenesis of Alzheimer's disease. J Biol Chem 298:101766
18. Romero-Fernandez W, Carvajal-Tapia C, Prusky A et al. (2023) Detection, visualization and quantification of protein complexes in human Alzheimer's disease brains using proximity ligation assay. Sci Rep 13:1–13
19. Zhang ZY, Harischandra DS, Wang R et al. (2023) TRIM11 protects against tauopathies and is downregulated in Alzheimer's disease. Science 381:eadd6696

20. Levy SA, Zuniga G, Gonzalez EM et al. (2022) Tau^{LUM}, an in vivo *Drosophila* sensor of tau multimerization, identifies neuroprotective interventions in tauopathy. Cell Rep Methods 2:100292
21. Sanders DW, Kaufman SK, DeVos SL et al. (2014) Distinct tau prion strains propagate in cells and mice and define different tauopathies. Neuron 82:1271–1288
22. Chandupatla RR, Flatley A, Feederle R et al. (2020) Novel antibody against low-n oligomers of tau protein promotes clearance of tau in cells via lysosomes. Alzheimer's Dement Transl Res Clin Interv 6:e12097
23. Lo CH (2021) Recent advances in cellular biosensor technology to investigate tau oligomerization. Bioeng Transl Med 6:e10231
24. Lopez DM, Maltby CJ, Warming H et al. (2023) A luminescence-based reporter to study tau secretion reveals overlapping mechanisms for the release of healthy and pathological tau. Front Neurosci 17:1196007
25. Cecon E, Oishi A, Luka M, et al. (2023) Novel repertoire of tau biosensors to monitor pathological tau transformation and seeding activity in living cells. Elife 12, e78360.
26. Wittmann CW, Wszolek MF, Shulman JM et al. (2001) Tauopathy in *Drosophila*: neurodegeneration without neurofibrillary tangles. Science (80-) 293:711–714
27. Greenspan, RJ, (2004). Fly pushing: the theory and practice of *Drosophila* genetics, 2nd edn. Cold Spring Harbor Laboratory Press, NY
28. Roote J, Prokop A (2013) How to design a genetic mating scheme: a basic training package for *Drosophila* Genetics. G3 3:353–358
29. Flyer-Adams, JG, Barbagallo, B, Griffith, LC, (2020) FlyBuilder: A Multimodal Dry Lab Curriculum Teaches Mendelian Genetics through the Lens of *Drosophila* Balancer Chromosomes. Advances in Biology Laboratory Education.
30. Patiño-García D, Rocha-Pérez N, Moreno RD et al (2019) Antigen retrieval by citrate solution improves western blot signal. MethodsX 6:464–468
31. Wu JS, Luo L (2006) A protocol for dissecting *Drosophila melanogaster* brains for live imaging or immunostaining. Nat Protoc 1:2110–2115
32. Williamson WR, Hiesinger PR (2010) Preparation of developing and adult *Drosophila* brains and retinae for live imaging. J Vis Exp 37, e1936.
33. Kelly SM, Elchert A, Kahl M (2017) Dissection and immunofluorescent staining of mushroom body and photoreceptor neurons in adult *Drosophila* melanogaster brains. J Vis Exp 129, e56174.
34. Robinson JW, Bechard AT, Frame AK et al. (2025) The *Drosophila* adult brain: short overview of structure, function, and resources Graphical review paper. Curr Res Insect Sci 7:100113
35. Hegazy M, Cohen-Barak E, Koetsier JL et al. (2020) Proximity ligation assay for detecting protein-protein interactions and protein modifications in cells and tissues in situ. Curr Protoc Cell Biol 89:e115
36. Kurien BT, Hal Scofield R (2015) Other notable protein blotting methods: a brief review. Methods in molecular biology. Springer, New York, 487
37. Mishra M, Tiwari S, Gomes AV (2017) Protein purification and analysis: next generation Western blotting techniques. Expert Rev Proteomics 14:1037
38. Heda GD, Shrestha L, Thapa S et al. (2020) Optimization of western blotting for the detection of proteins of different molecular weight. Biotechniques 68:318–324
39. Wegmann S, Maury EA, Kirk MJ et al. (2015) Removing endogenous tau does not prevent tau propagation yet reduces its neurotoxicity. EMBO J 34:3028–3041
40. Mohideen SS, Yamasaki Y, Omata Y et al. (2015) Nontoxic singlet oxygen generator as a therapeutic candidate for treating tauopathies. Sci Rep 5:1–14
41. Chi H, Sun L, Shiu RH et al. (2020) Cleavage of human tau at Asp421 inhibits hyperphosphorylated tau induced pathology in a *Drosophila* model. Sci Rep 10:1–15
42. Liu Y, Gu J, Hagner-McWhirter Å et al. (2011) Western blotting via proximity ligation for high performance protein analysis. Mol Cell Proteomics 10:O111.011031
43. Tracy TE, Madero-Pérez J, Swaney DL et al. (2022) Tau interactome maps synaptic and mitochondrial processes associated with neurodegeneration. Cell 185:712–728.e14
44. Schubert FK, Helfrich-Förster C, Rieger D (2020) Single-cell resolution long-term luciferase imaging in cultivated *Drosophila* brains. MicroPubl Biol.
45. Curley M, Rai M, Chuang CL, et al. (2024) Transgenic sensors reveal compartment-specific effects of aggregation-prone proteins on subcellular proteostasis during aging. Cell Rep Methods 4(10), 100875.

Check for updates

Chapter 15

Testing Small Molecules for Activity Against Neurological Phenotypes in a Zebrafish 4-Repeat Tauopathy Model

Oluwaseyi Pearce, Qing Bai, Hugh S. Hammer and Edward A. Burton

Abstract

Progressive supranuclear palsy (PSP) is a neurodegenerative disease characterized clinically by multiple motor, bulbar, oculomotor, and cognitive deficits, and pathologically by neuronal loss, neuroinflammation, and the accumulation of 4-repeat Tau in the central nervous system (CNS). The pathophysiology downstream of 4R-Tau accumulation is incompletely understood and there are no effective treatments for the disease. Zebrafish models offer unique opportunities to carry out unbiased small molecule screens in vivo. There has been substantial interest in exploiting this approach to investigate PSP and other tauopathies, given the potential to gain new insights into mechanisms and targets, identify small molecules that might provide a basis for the development of effective drugs, or repurpose existing drugs as treatments. We recently developed a transgenic zebrafish model expressing human WT 0N/4R-Tau conditionally in CNS neurons, leading to the rapid onset of multiple PSP-like neurological, biochemical, and pathological phenotypes amenable to unbiased screening approaches. Here, we provide a detailed protocol for carrying out a small molecule screen in this novel model, using hypokinesia and neurodegeneration phenotypes as robust assay endpoints that are unbiased to preconceptions about molecular mechanisms.

Keywords Tau, Tauopathy, Progressive supranuclear palsy, Central nervous system, Zebrafish, GAL4-UAS, Small molecule, Chemical screen, Movement assay, Acridine orange

1 Introduction

1.1 Progressive Supranuclear Palsy

Progressive supranuclear palsy (PSP) is a neurodegenerative disease that affects motor function, balance, speech, swallowing, ocular movements, and cognition [1]. Pathologically, PSP is characterized by neuronal loss, gliosis, and formation of neurofibrillary tangles (NFTs) in the basal ganglia and brainstem [2]. NFTs in PSP are composed predominantly of 4-repeat (4R) Tau isoforms that are misfolded, hyperphosphorylated, and aggregated, and correspondingly, PSP is classified as a primary 4R tauopathy [2–4]. Although the cause of PSP is unknown, genetic evidence links 4R-Tau to pathophysiology. PSP is disproportionately more common amongst people who are homozygous for the MAPT H1 haplotype that is

Jessica S. Fortin (ed.), *Translational Research Methods for Tauopathies*, Neuromethods, vol. 233,
https://doi.org/10.1007/978-1-0716-5234-3_15,

associated with a small increase in expression of 4R-Tau [5]. Even after correction for the H1 haplotype, genome-wide association studies (GWAS) identified risk-conferring single-nucleotide polymorphisms (SNPs) near the MAPT locus in PSP [6]. In addition, a rare Mendelian PSP phenocopy is caused by mutations in the MAPT gene that encodes Tau, including missense mutations that selectively alter the amino acid sequence of 4R-Tau, and splice site mutations that increase expression of WT 4R-Tau isoforms [7]. Finally, MAPT gene duplications, presumably leading to Tau over-expression, were reported recently in two cases of sporadic neurological disease with unusual and rapid clinical presentations but typical PSP pathology [8]. Together, these data suggest that 4R-Tau plays an important role in pathophysiology. However, exactly how 4R-Tau accumulation causes disease is unclear, and since the common genetic risk factors are alone inadequate to cause PSP, it is likely that other genetic and environmental factors are important [9].

1.2 Zebrafish as a Model for Studying Human Disease

Zebrafish are increasingly used as a model for human neurological diseases, because they combine vertebrate genetics and brain organization with an experimentally accessible system that presents unique opportunities for analysis. Following fertilization, single-cell zebrafish embryos develop rapidly and externally from their mother. By 3 days post fertilization (dpf), the central nervous system (CNS) is recognizably divided into forebrain, midbrain, and hindbrain (reviewed in [10]). By 5 dpf, larval zebrafish hunt prey (paramecia) using visual guidance and exhibit a repertoire of sensory stimulus-evoked and spontaneous motor behaviors that can be measured readily using automated machine vision approaches [11, 12]. Genetic modification of zebrafish is relatively efficient using CRISPR/Cas9-mediated mutagenesis or transposon-mediated transgenesis, permitting gene knockout or over-expression studies. Transgenesis further allows cell types of interest to be labeled by genetically encoded reporters for high-resolution imaging in vivo, exploiting the external development and optical transparency of zebrafish larvae (reviewed in [13]).

The applicability of zebrafish models to human disease is an important consideration, given the evolutionary distance between zebrafish and humans (last common ancestor approximately 450 million years ago, compared with 112 million years between mice and humans) [14]. Many aspects of brain structure, function, neural circuitry, and neurotransmitter systems with relevance to disease are conserved between humans and zebrafish (reviewed in [13, 15–19]). The zebrafish CNS also contains cell types involved in human neurodegenerative diseases, such as dopaminergic neurons (which have a central role in motor function and exhibit cell-intrinsic pacemaker activity in zebrafish, similar to their mammalian counterparts [20, 21]), astrocytes [22], microglia [23, 24], and oligodendrocytes (reviewed in [25]). Zebrafish have at least one ortholog for approximately 70% of human genes, including many involved in neurodegenerative diseases such as LRRK2 [26], Huntingtin [27], DJ1 [28], Parkin [29], PINK1 [30], Synucleins [31, 32]. In addition, a whole genome duplication event that occurred during the evolution of teleost fish led to the appearance of paired zebrafish paralogues of some human genes. Examples include γ-Synuclein [31, 32], TDP43 [33, 34], and Tau [35]. The two zebrafish Tau paralogues, *mapta* and *maptb* are expressed in the developing CNS. Both mRNA transcripts are thought to undergo alternative splicing, resulting in multiple Tau isoforms, similar to humans [35]. The functions of these genes and the relative abundance of different isoforms during development have not yet been resolved. The premise supporting the validity of zebrafish as a model for human neurological disease was recently discussed in detail elsewhere [13].

The ability to breed and analyze large numbers of zebrafish was initially used to isolate mutants with essential roles in development (reviewed in [36]). More recently, zebrafish have been employed in small molecule screens to identify compounds that alter phenotypes of interest (reviewed in [37]). Groups of larval zebrafish can develop normally in as little as 2 mL E3 embryo buffer (reviewed in [38]). Since many small molecules enter zebrafish passively from the medium [39], larvae can be exposed by dissolving compounds in the water, potentially allowing a large number of different molecular structures to be screened in vivo against phenotypes of interest. Not surprisingly, there has been significant interest in the possibility of using this chemical biology approach in zebrafish models for drug repurposing and drug discovery. This is particularly attractive for neurological disease, since neurological function can be evaluated as early as 5 dpf using automated assays that are unbiased to molecular and neurological mechanisms of action. Screening and imaging applications in zebrafish neurological disease models were recently discussed in detail elsewhere [38].

1.3 Transgenic Zebrafish Expressing Human Tau

A zebrafish 4R-tauopathy model with a measurable phenotype during larval development would unlock the possibility of screening annotated small molecule collections to provide insights into pathophysiological mechanisms or to accelerate drug discovery. This is of considerable interest, since the molecular mechanisms underlying PSP are incompletely understood, and historically, a large proportion of first-in-class drugs were discovered by phenotypic screening [40]. Several transgenic zebrafish models expressing human Tau in the brain have been reported previously, including transient expression of a Tau-GFP fusion protein [41], constitutive expression of wildtype (WT) 0N/4R [42] or mutant (P301L) human Tau [43, 44], and conditional expression of the human FTDP17 Tau mutant (P301L) [45] or a variant (A152T) that may be associated with PSP [46].

Recently, we developed a conditional model in which human WT 0N/4R-Tau (the Tau isoform that accumulates most prominently in PSP [3, 4]) is over-expressed in neurons throughout the CNS

[47]. These zebrafish showed impaired survival, neurodegeneration, microgliosis, synapse loss, and multiple biochemical and histological abnormalities consistent with human tauopathy, in addition to PSP-like clinical phenotypes including hypokinesia and abnormal ocular movements. Importantly, the neurological abnormalities at 5 dpf were sufficiently severe that it was possible to use automated assays of motor function as an unbiased endpoint for a small molecule rescue screen. This allowed us to interrogate a library of inhibitors of epigenetic readers, writer and erasers against hypokinesia, using automated machine vision approaches. This work unexpectedly identified a bromodomain inhibitor that rescued hypokinesia by preventing microglial synaptic elimination [47]. Since this new model and innovative approach are potentially of broad interest and applicability, the purpose of this chapter is to provide a detailed protocol for breeding and selecting transgenic zebrafish expressing human 0N/4R-Tau, exposing larval zebrafish to small molecules, and then completing in vivo assays for motor function or cell death to detect rescue of disease-relevant phenotypes.

1.4 Genetics of the Conditional Zebrafish PSP Model

The transgenic PSP model employs a bipartite expression system based on Gal4-UAS genetics [48, 49]. The model depends on two transgene alleles that can be propagated separately in healthy adult zebrafish (Fig. 1):

- The **driver line**, Tg(*elavl3:Gal4-VP16*), expresses a fusion protein containing the DNA-binding domain from the yeast transcription factor Gal4 and the transactivation domain from the HSV1 virion protein VP16, under the control of the *elavl3* pan-neuronal promoter [50] (this line was a kind gift from Dr. Josh Bonkowsky, University of Utah). Tg(*elavl3:Gal4-VP16*) zebrafish are usually maintained crossed with Tg(*UAS:egfp*) zebrafish, enabling non-invasive identification of driver zebrafish by fluorescence microscopy for GFP.
- The **responder line**, Tg(*UAS:Hsa.MAPT-p2A-nls-mCherry*), encodes human WT 0N/4R-Tau, and nls-mCherry under a minimal promoter coupled to 5 copies of the UAS yeast enhancer (the cognate DNA binding sequence for Gal4). Since these zebrafish do not express any transgene until the UAS enhancer is activated by Gal4, there is no visible reporter gene. Consequently, this line is maintained by fin clip PCR genotyping.

Importantly, neither parental line expresses human Tau, meaning that the alleles can be maintained in the absence of Tau-induced phenotypes that impair viability or breeding potential. When the driver and responder are crossed, Gal4-VP16 (expressed in neurons from the driver allele) transactivates the responder allele in double transgenic *Tg(elavl3:Gal4-VP16); Tg(UAS:Hsa.MAPT-p2A-nls-mCherry)*

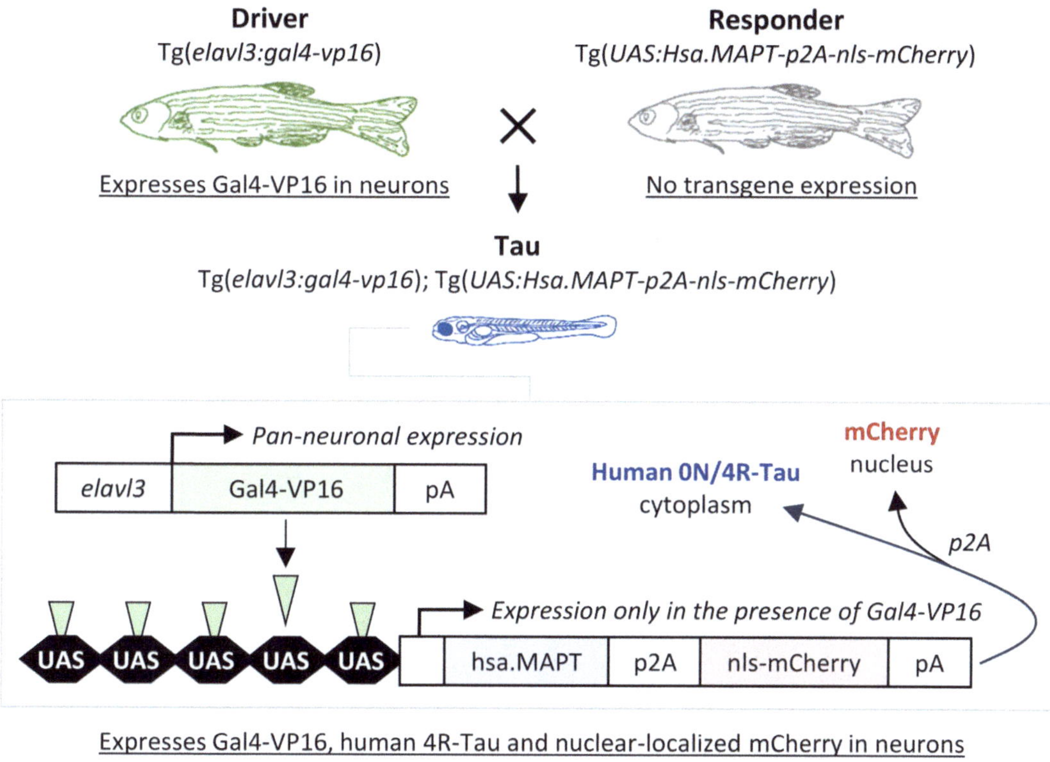

Fig. 1 Generation of Tau zebrafish. The driver line, Tg(*elavl3:gal4-vp16*), expresses a fusion protein containing the DNA-binding domain of Gal4 and the transactivation domain of VP16, under control of the pan-neuronal *elavl3* promoter. The responder line, Tg(*UAS:Hsa.MAPT-p2A-nls-mCherry*), encodes WT human 0N/4R-Tau and nls-mCherry under transcriptional control of the UAS enhancer that contains the cognate DNA binding sequence for Gal4. In the absence of Gal4, the responder allele does not express any transgene. About 25% of the offspring from a [driver x responder] cross are double transgenic Tg(*elavl3:gal4-vp16*); Tg(*UAS:Hsa.MAPT-p2A-nls-mCherry*) Tau zebrafish, in which Gal4-VP16 from the driver cassette transactivates the responder cassette to express human 0N/4R-Tau and mCherry in neurons. Although both transgenes are encoded by the same mRNA, the 2A peptide causes their separation into two discrete proteins during translation. A signal peptide at the N-terminus of mCherry results in nuclear import, so the two proteins are directed to different subcellular compartments. For convenience, the driver line is usually maintained crossed with a Tg(*UAS:egfp*) allele, which allows driver zebrafish to be identified by fluorescence microscopy. The GFP transgene does not influence phenotypes in the Tau model, but its presence increases the number of different genotypes resulting from the [driver x responder] cross (see Table 1)

progeny (referred to as "Tau" zebrafish for brevity). These transgenic Tau zebrafish express human 4R-Tau in most neurons and show robust and rapid onset PSP-like phenotypes suitable for screening. Tau zebrafish also express mCherry from the same bicistronic mRNA as 4R-Tau, enabling their identification by fluorescence microscopy [47]. Importantly, the p2A peptide separating 4R-Tau and mCherry ensures their expression as separate proteins, preventing artifacts from fusion proteins with altered biological properties. The siblings

Table 1
Genotypes resulting from a [Tg(*elavl3:gal4-vp16*); Tg(*UAS:egfp*)] X Tg(*UAS:Hsa.MAPT-p2a-nls-mCherry*) cross

Genotype	Number of transgenes	Transgene expression	Fluorescent marker	Phenotype
WT	0	None	None	None
Tg(*elavl3:gal4-vp16*)	1	Gal4-VP16	None	None
Tg(*UAS:egfp*)	1	None	None	None
Tg(*UAS:Hsa.MAPT-p2a-nls-mCherry*)	1	None	None	None
Tg(*elavl3:gal4-vp16*); Tg(*UAS:egfp*)	2	Gal4-VP16 GFP	Green fluorescence in neurons	None
Tg(*UAS:Hsa.MAPT-p2a-nls-mCherry*); Tg(*UAS:egfp*)	2	None	None	None
Tg(*elavl3:gal4-vp16*); Tg(*UAS:Hsa.MAPT-p2a-nls-mCherry*)	2	Gal4-VP16 4R-hTau mCherry	Red fluorescence in neurons	Neurodegeneration
Tg(*elavl3:gal4-vp16*); Tg(*UAS:Hsa.MAPT-p2a-nls-mCherry*); Tg(*UAS:egfp*)	3	Gal4-VP16 4R-hTau mCherry GFP	Green and red fluorescence in neurons	Neurodegeneration

of Tau zebrafish that do not express mCherry (or Tau) are referred to as "Sib" zebrafish and consist of a 1:1:1 proportion of driver, responder, and WT zebrafish. Sib zebrafish have no phenotype and are used as controls for screening applications.

Maintaining the driver line crossed with a GFP responder line for convenience means that additional genotypes are generated by each cross (Table 1). The presence (or absence) of the GFP transgene does not seem to affect the motor phenotype of Tau zebrafish, or normal swimming function in Sibs, and so it might be possible to include GFP-expressing larvae in downstream experiments (which is advantageous as it increases the proportion of each cross with genotypes suitable for analysis). However, it is recommended to confirm that the GFP transgene does not alter measured outcomes in any downstream assays before including GFP^{+} zebrafish in these analyses.

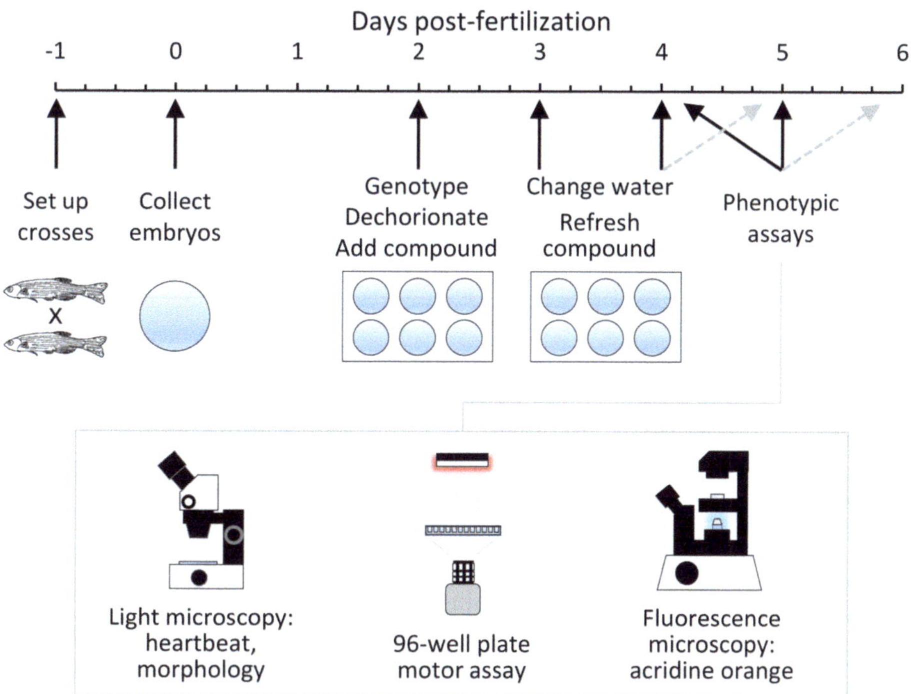

Fig. 2 Overview of the experimental protocol. After setting up crosses the day before spawning, zebrafish embryos are collected following fertilization on day 0 and raised in E3 buffer in 100 mm × 15 mm petri dishes. By 2 dpf, major organogenesis is complete, and there is strong transgene expression. At this timepoint, larvae are genotyped by microscopy for mCherry fluorescence, dechorionated, and then transferred to 6-well plates, where they are exposed to small molecules or carrier alone. Water changes, including the addition of fresh compound, are completed at 3 dpf and 4 dpf. Phenotypic assays include: light microcopy to detect morphological defects or loss of viability resulting from small molecule exposure (5 dpf); 96-well plate automated motor function assays, to detect rescue of hypokinesia (5 dpf); and acridine orange labeling, to detect rescue of cell death (4 dpf). Optionally, small molecule exposure can be extended (grey dashed arrows) and movement assays completed at 6 dpf; motor phenotypes are more severe at this time point and provide a stringent test of rescue

1.5 Overview of Protocol

Small molecules can be tested for activity in Tau zebrafish in vivo over an experimental time course lasting one week (Fig. 2). During this time, Tau zebrafish are generated, identified, exposed to small molecules and then tested for changes in phenotypic outcomes. This protocol is correspondingly divided into four sections:

- *Section 1: Maintaining lines and breeding transgenic zebrafish*—how to maintain both the driver and responder lines and how to identify transgenic Tau zebrafish and controls for assays.
- *Section 2: Small molecule exposure*—how to use a small molecule library for screening in Tau zebrafish, including empiric determination of the maximum tolerated concentration (MTC), and exposure over the time course during which pathogenesis occurs in the 4R-Tau model.
- *Section 3: Rescue of motor function*—how to run and analyze motor function assays to detect rescue of neurological phenotypes in Tau zebrafish exposed to small molecules.
- *Section 4: Rescue of neurodegeneration*—how to quantify neuronal death in live Tau zebrafish using acridine orange labeling coupled with intravital microscopy and rapid manual counting.

A "**Notes**" section is included to highlight potential problems that may be encountered, as well as ways to identify and address them.

2 Materials

Guidelines for handling these materials are listed in the corresponding "**Notes**" sections.

2.1 Equipment

- Basic lab equipment: pipettors; precision balance, spatulas, vortex mixer, heat blocks or baths, centrifuges, refrigerators, and freezers.
- Zebrafish breeding tanks. Several types are available, most of which include an inner tank with a perforated bottom that allows fertilized embryos to pass into an external tank from where they can be recovered. A divider to separate males and females until breeding is also useful, as reviewed in [51]. Our facility uses 1.7 L sloping breeding tanks (Tecniplast) (*see* **Notes 1 and 2**).
- 28.5°C incubator with internal white light illumination on 14 hour:10 hour light-dark cycle, illuminance approximately 200 lx at the plane of the zebrafish.
- Dumont #5 forceps × 2 (Cat # 11254-20; Fine Science Tools; Foster City, CA).
- Eyelash with handle (Cat # 113; Ted Pella, Inc.; Redding, CA).
- System for measuring zebrafish motility by tracking the location of the zebrafish centroid marker in each frame of a video

stream or recording, for a single zebrafish in each well of a 96-well plate. We developed an inexpensive custom apparatus and open-source MATLAB apps, *LSRtrack* and *LSRanalyze*, for this purpose [12, 20, 52], but a number of excellent proprietary solutions are available with similar functionality (see Section: Measurement of zebrafish movement below). The system must allow the visual motor response (VMR) to be elicited, which generally entails infrared (IR) videography allowing separate control of ambient illumination with continuous video capture in darkness or light (*see* **Note 8**).

2.2 Consumables and Plasticware

- Pipette tips, 10 mL serological pipettes, glass Pasteur pipettes, 1.5 mL microcentrifuge tubes with locking caps, 15 mL centrifuge tubes, polystyrene weighing dishes, 100 mm × 15 mm polystyrene petri dishes, 6- and 48-well plates, aluminum foil.
- 96-well plates to house zebrafish during motor assay. The plates should be compatible with the system being used to quantify larval zebrafish movement. Plates with black well surrounds and an optically transparent bottom (Cat # CLS3720; Corning Incorporated; Corning, NY) are required for our open-source MATLAB-based system [12].
- 35 mm MatTek dishes with glass coverslip bottoms (Cat # P35G-1.5-14-C; MatTek Corporation; Ashland, MA).

2.3 Chemicals and Solutions

- E3 zebrafish embryo buffer: 5 mM NaCl, 0.17 mM KCl, 0.33 mM $CaCl_2$, 0.33 mM $MgSO_4$ (Sigma, St. Louis, MO).
- E3 + 0.1% methylene blue.
- Small molecule collection or library (*see* **Note 6**).
- Small molecule solvent, for example, Dimethyl sulfoxide; DMSO (Cat # 472310-100 ML; Sigma-Aldrich; St. Louis, MO). *See* **Note 5**.
- Acridine Orange hemi(zinc chloride) salt (Cat # A6014; Sigma; St. Louis, MO).
- Low melting-point agarose (LMA) (Cat # A9414-100G; Sigma; St. Louis, MO).
- 0.4% buffered tricaine in E3 (Cat # A5040; Sigma-Aldrich; St. Louis, MO).

2.4 Microscopes

- Stereo zoom microscope equipped for widefield transmitted/reflected white light illumination (used for dechorionation and positioning zebrafish in agarose for microscopy) and epifluorescence imaging (used for non-invasive identification of transgenic zebrafish). The working distance should be sufficient to allow manipulation of zebrafish under visualization. We use a Leica S6E for white light illumination and an Olympus MVX10 for epifluorescence.

- Inverted epifluorescence microscopes with bandpass filters for visualizing acridine orange labeling. The working distance should be sufficient to visualize the zebrafish spinal cord in a laterally positioned anesthetized zebrafish larva through a cover slip. For manual counting of acridine orange-labelled spinal cord cells, we use an Olympus CKX41 inverted microscope with a 20×/NA 0.75 dry objective and 10× eyepieces. For capturing illustrative images of acridine orange-labelled spinal cord cells in the figures, we used a Nikon Ti2E inverted microscope and Nikon AXR resonance scanning confocal system with a 20×/ NA 1.15 long working distance water immersion objective.

2.5 Zebrafish

- WT zebrafish (AB).
- Tg(*elavl3:Gal4-VP16*$_{413\text{-}470}$)^{zc87} driver line [50]. This line was crossed with Tg(*UAS:egfp*) zebrafish [21] to obtain Tg(*elavl3:Gal4-VP16*); Tg(*UAS:egfp*), allowing non-invasive identification of driver zebrafish by fluorescence microscopy for GFP (Fig. 1) (*see* **Notes 2 and 3**).
- Tg(*UAS:Hsa.MAPT-p2A-nls-mCherry*)^{Pt423} responder zebrafish (Fig. 1) [53] (*see* **Notes 2 and 3**).
- Access to an aquatics facility to provide clean circulating system water at the correct temperature and composition for raising and housing adult zebrafish.

3 Methods

3.1 Ethical Considerations

Zebrafish are vertebrates; consequently, experiments involving zebrafish require ethical review and approval. Regulations vary according to location, and it is the investigator's responsibility to ensure work is carried out in accordance with all applicable laws and guidelines. All experiments shown here were completed at the University of Pittsburgh with approval from the University of Pittsburgh Institutional Animal Care and Use Committee (IACUC), in accordance with the NIH Guide for the Care and Use of Laboratory Animals.

3.2 Maintaining Adult Zebrafish

There are multiple ways to house and maintain adult zebrafish that are beyond the scope of this chapter. At the University of Pittsburgh, zebrafish are housed in a recirculating aquaculture system designed and built by Aquaneering LLC. The volume of the system is approximately 50,000 L. Municipal water from the City of Pittsburgh is filtered by a 5-μm sediment filter, multiple granular activated carbon filters, and reverse osmosis filtration prior to storage. This base water is modified by the addition of sea salt (Instant Ocean Reef Crystals) to a target conductivity of 1200 μS and the addition of sodium bicarbonate to a target pH of 7.5. Water temperature is maintained at 28 ± 2°C with submersible electric heaters (Aqualogic Inc). All water quality parameters are maintained with optimal parameters for zebrafish [54]. The recirculating aquaculture system features propeller washed bead filters (Aquaculture Systems Technologies) for mechanical filtration; large, fluidized sand biological filters (Aquaneering LLC); degassing columns (Aquaneering LLC); secondary mechanical filters (Pentair Aquatic Ecosystems); and UV disinfection (Aquaneering LLC). Approximately 10% of the water is exchanged and replaced each week when filters are backwashed. This system operates on a 14 h light/10 h dark cycle with a stocking density maximum of 10 zebrafish/L. Adult zebrafish are housed in single-sex tanks to facilitate breeding when transferred to breeding tanks.

3.3 Breeding Larval Zebrafish

Small-molecule *toxicity* experiments are completed in WT zebrafish. Small molecule *phenotypic rescue* experiments require Tau and Sib zebrafish obtained from a Tg(*elavl3:Gal4-VP16*) x Tg(*UAS:Hsa.MAPT-p2A-nls-mCherry*) cross.

1. On the afternoon before spawning, assemble a 1.7 L slope breeding tank (Tecniplast) and fill with system water. The breeding tank consists of an external tank, an internal tank with a gravel-like opaque bottom through which fertilized embryos fall after spawning, a divider, and a lid. The gravel-like-opaque bottom of the internal tank prevents the ingestion of ova by adult zebrafish but allows fertilization by adult males (*see* **Note 1**).
2. Transfer zebrafish to a breeding tank, ensuring males and females are separated by the divider, and leave overnight (*see* **Notes 1 and 2**).
3. Spawning is stimulated the following morning by the initiation of the aquarium light cycle, which occurs at 08:00 in our system. Immediately after the lights in the aquarium turn on, remove the divider from the breeding tank so the male and female zebrafish can access each other for spawning (*see* **Note 1**).
4. To collect embryos, transfer the internal tank with adult zebrafish into a new external tank containing fresh system water. Pour the contents of the original external tank through a tea strainer to recover the embryos, then gently flush embryos from the strainer into 100 mm × 15 mm Petri dishes (no more than 60 embryos per Petri dish) using a wash bottle filled with system water.
5. Replace system water with 60 mL E3 buffer containing 0.1% methylene blue to inhibit mold growth then transfer the petri dishes to a 28.5°C light cycle incubator. This incubator should operate on a 14 h light/10 h. dark cycle (ours has white light illumination at 200 lx and color temperature of ≈3700 K).

6. Adult zebrafish may breed again over the next few hours, and additional embryos can be harvested as appropriate. After embryo collection is completed, return adult male and female zebrafish to their single-sex housing tanks (*see* **Note 1**).

3.4 Embryo Growth and Selection

1. Ensure E3 buffer is changed at least daily.
2. At 1 dpf, remove any dead or malformed embryos to prevent bacteria and mold growth in petri dishes, then replace buffer with E3 that does not contain methylene blue.
3. At 2 dpf, identify Tau zebrafish by fluorescence microscopy. Transfer petri dishes containing 2 dpf embryos to a stereo zoom fluorescence microscope. Using the mCherry channel (excitation $\lambda = 587$ nm, emission $\lambda = 610$ nm [55]), identify Tau zebrafish by red fluorescence in the brain and spinal cord (Fig. 3), and move them to a separate dish containing fresh E3 by gently pulling them into a Pasteur pipette with a wide flame-polished aperture (*see* **Note 4**).
4. Both mCherry$^+$ (Tau) and mCherry$^-$ (Sib) zebrafish can be rescreened using the GFP channel to remove larvae that are GFP$^+$, if this will interfere with downstream assays (such as labeling with acridine orange).

3.5 Small Molecule Exposure

It is advisable to dechorionate embryos for small molecule exposures to avoid possible issues with chorionic penetration.

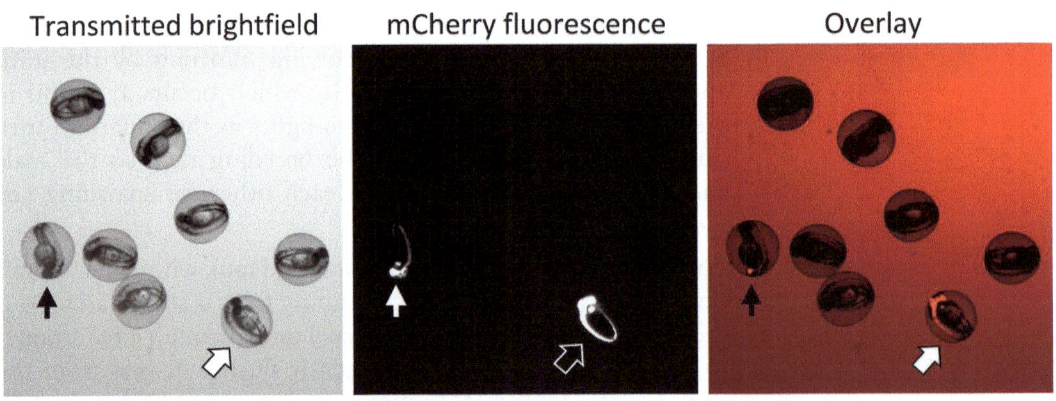

Fig. 3 Identification of Tau and Sib zebrafish. Crosses between adult driver and responder zebrafish generate embryos that are approximately 25% Tau zebrafish and 75% siblings. The siblings comprise a 1:1:1 mixture of WT, driver, and responder zebrafish (see Table 1). Tau zebrafish are identified readily by bright red fluorescence in the brain and spinal cord (large arrow) under illumination for the mCherry reporter (peak excitation $\lambda = 587$ nm, peak emission $\lambda = 610$ nm). This allows selection of Tau zebrafish for downstream experiments using a stereo zoom microscope. Occasional Tau zebrafish show decreased mCherry expression attributable to variegation (small arrow). Since the phenotypes in this model are expression-level dependent, embryos showing obvious variegation should not be used for screening experiments, as they may show apparent rescue even in the absence of treatment

1. For Tau zebrafish, switch from epifluorescence to white light reflected/transmitted illumination after genotyping. For WT zebrafish go straight to Step 2.
2. Using two pairs of Dumont #5 forceps, pinch each embryo's chorion at opposite ends and gently pull the forceps in opposite directions, tearing the chorion and releasing the embryo into the medium.
3. Refresh medium to remove chorion debris.

Small Molecule Exposure: Determining Maximum Tolerated Concentration (Fig. 4)

1. Dissolve small molecules at 50 mM (1000× stock solution), using the appropriate solvent, usually DMSO. Ensure complete dissolution by vortexing.
2. In a separate tube, dilute 50 mM stock solution by 1000-fold in E3 buffer to prepare 1 mL of 50 μM working solution. Prepare 500 μL each of 1:1 serial dilutions of working solution in E3 (i.e., 25 μM, 12.5 μM, 6.3 μM, etc.) (*see* **Note 5**).
3. Transfer three dechorionated WT embryos to each well of a 48-well plate at 2 dpf.
4. Replace E3 buffer in each well with 500 μL small molecule solution at each serial dilution. A control well should contain E3 buffer with the highest DMSO concentration in the dilution series (*see* **Note 5**). Place the 48-well plate in the 28.5°C incubator.
5. After removing sufficient working stock for a further two to three buffer changes, aliquot the remaining 50 mM stock solution and store at −80°C.
6. Refresh media in the 48-well plate daily between 3 and 5 dpf with fresh E3 + small molecule at the same concentration, as described above (Step #2).
7. At 5–6 dpf inspect larvae using a stereo zoom microscope for viability (movement, heartbeat) and morphological abnormalities (body axis deficits, size abnormalities, pericardial edema, craniofacial malformations, neural tube defects, etc.).
8. The MTC is defined as the highest concentration at which no overt signs of toxicity are seen in any of the larvae. It may be necessary to iteratively test progressively lower concentrations to arrive at this value, see workflow in Fig. 4.

Small Molecule Exposure: Testing for Activity Against Neurological Phenotypes

1. Dilute small molecule stock solution to the previously determined MTC, using E3 buffer to a volume of 3 mL.
2. Transfer dechorionated Tau embryos and Sibs at 2 dpf to wells of a 6-well plate in fresh E3 buffer (12 larvae/well).

   Each well should be assigned to one of the following groups:

   - Tau zebrafish exposed to small molecules (test groups, 12 zebrafish/well in each group, maximum six groups, each exposed to a different small molecule—see below)

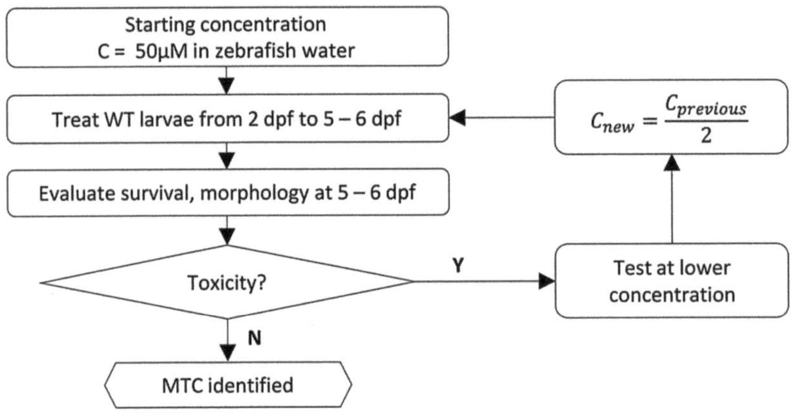

Fig. 4 Determination of maximum tolerated concentration (MTC) of a small molecule. WT zebrafish larvae are exposed to each small molecule at an initial concentration of 50 μM from 2 dpf to 5 dpf (or 6 dpf if phenotypic endpoints will be measured subsequently at 6 dpf instead of 5 dpf). If evidence of morphological abnormalities (e.g., body axis or growth deficits, craniofacial abnormalities, pericardial edema) or loss of heartbeat is seen, a new batch of larvae is exposed at half of the previous concentration. MTC is determined iteratively for each compound as the first dilution that does not result in death or any morphological abnormalities

- Tau zebrafish not exposed to small molecules (essential control, 12 zebrafish/well, one group).
- Sibs without mCherry expression (essential control, 12 zebrafish/well, one group) (*see* **Note 7**).

3. Replace E3 buffer in each well with 3 mL E3 buffer containing small molecule at MTC. For untreated Tau and Sib controls replace E3 buffer with 3 mL fresh E3 buffer containing the same DMSO concentration present in the small molecule solution. Refresh [E3] or [E3 + small molecule] in each well every 24 h as above (*see* **Note 6**).
4. At the desired timepoint for phenotypic assays, replace medium in all wells with fresh E3 without compound then begin phenotypic testing.

3.6 Phenotypic Test 1: Evaluation of Motor Function in an Automated 96-Well Plate Assay

Measurement of Zebrafish Movement

Multiple options are available for evaluating zebrafish motor activity in 96-well plates. These include commercially available turnkey solutions, such as the popular *Zebrabox* (ViewPoint, Montreal, Canada) or *Daniovision* (Noldus, Leesburg, VA) systems, and custom-built apparatus and software [12]. A full discussion of the relative advantages of each type of system is outside the scope of this chapter. Briefly, commercial systems are usually capable of acquiring reliable data immediately after installation but are prohibitively expensive for many labs. Custom solutions can be built with significantly less financial outlay, offer flexibility concerning experimental configuration, and can take advantage of several freely available zebrafish tracking applications. However, custom systems can involve hidden expenses such as time spent sourcing parts and assembling equipment, and they may require significant troubleshooting [12].

Our recent publication, in which we screened a small molecule library against the motor phenotype of a human 4R-Tau-expressing zebrafish, used an inexpensive custom system that we reported previously [12, 52]. The detailed instructions provided in [12] include a step-by-step protocol for building this system and using it to record and track videos of zebrafish swimming in multi-well plates, along with troubleshooting tips that will not be repeated here. It is important to note, however, that any system could be used to detect and quantify the hypokinetic phenotype of Tau zebrafish in a small molecule screen, provided it satisfies the following minimum specifications:

- Measurement of x-, y-plane zebrafish displacement in 96-well plates viewed from above or below.
- Recordings are carried out using infrared (IR) illumination so that ambient visible light illumination can be controlled separately, allowing the VMR to be elicited and analyzed.
- Tracking software allows the determination of (x, y) location for each zebrafish in each frame of the video. z-coordinate (depth location) tracking is not necessary.
- Tracking is reliable with minimal tracking errors (such as lost zebrafish or mis-assigned object errors).
- Signal/noise ratio is sufficient that the majority of the reported displacement is accounted for by zebrafish movement and not by stochastic pixel noise. This is determined empirically by comparing measured movement in swimming zebrafish larvae with identical larvae that are anesthetized and not moving. This value is usually >200 in the simple system we reported previously [12].
- Data output includes (x, y, t) data for every zebrafish throughout the video recording or live stream.

Experimental Design

The essential elements of the experimental design are as follows (Fig. 5):

a. Every assay includes internal controls:
    - At least 12 Sib zebrafish that have not been exposed to small molecules and that have no phenotype.

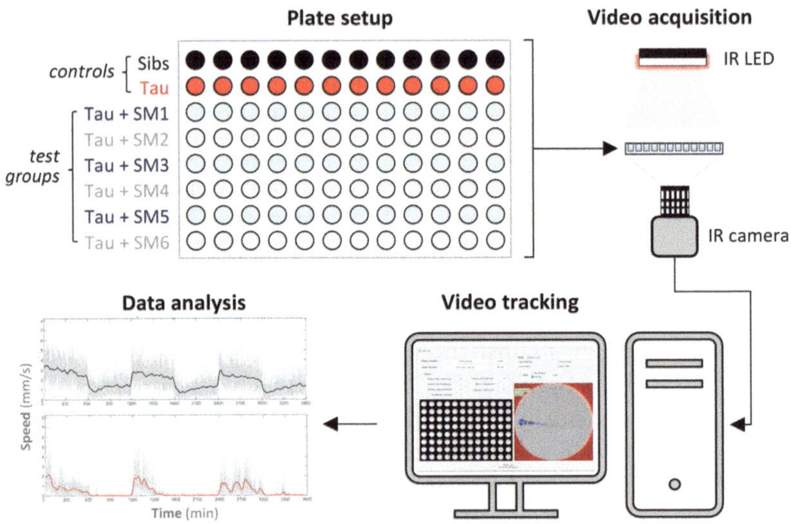

Fig. 5 Overview of motor function rescue assay. Tau larvae that were exposed to each small compound at MTC from 2 to 5 dpf, are transferred to a 96-well plate at 5 dpf (one larva per well), along with groups of untreated Sibling and Tau controls. The minimum replicate size that yields acceptable screening metrics in this assay is 12 zebrafish per group, shown here by each experimental group occupying one row of the 96-well plate. Assay metrics can be improved by using larger replicate sizes, although this reduces compound throughput. It is critical that the untreated Tau and Sibling controls are run on every plate, because rescue activity outputs are reported relative to the assay window defined by these controls. Larval motor function is recorded using a video camera under infrared (IR) illumination during cyclic white-light ambient illumination, then tracked, quantified, and analyzed offline using open-source software [12, 52]. Alternatively, commercially available zebrafish neurobehavioral systems can be used to measure motor function, as long as data are output in a format that can be analyzed as detailed in the text

- At least 12 Tau zebrafish that have not been exposed to small molecules and have an unaltered phenotype as reported in the initial characterization of the model [47].
- Each test group contains at least 12 Tau zebrafish exposed to a small molecule at MTC as above (*see* **Note 7**).

b. All groups are evaluated by video tracking through at least three cycles of the VMR elicited by exposure to 10 min dark followed by 10 min white light at 200 lx (*see* **Note 8**).

c. The mean VMR light phase swimming speed is calculated in 30-s bins for each larva across the 20-min averaged stimulus cycle.

Controls are **essential** for two reasons:

- Since the quantitative phenotypes defined by mean VMR light phase swimming speed are variable between assays, it is essential to compare every test group with controls from the same zebrafish clutch that are recorded and analyzed in the same assay.

- The controls allow calculation of quality control (QC) metrics to evaluate whether the assay gave interpretable data.

Hypokinesia can be quantified in Tau zebrafish at 5 dpf [12, 47]. The movement phenotype is more severe at 6 dpf, but running a movement assay at 6 dpf involves an extra day of small molecule exposure and results in a lower throughput.

Running a 96-Well Plate VMR Assay

1. Ensure white light illumination is calibrated at the plane of the 96-well plate to 200 lx using a light meter.
2. Program the experiment controller to capture zebrafish movement continuously during at least three sequential cycles of ambient illumination [10 min light-off–10 min light-on]. Optionally, a prolonged light phase can be captured at the end of the recording (as shown in Fig. 5b of [47]).
3. Pipette 200 μL E3 buffer into each well of the 96-well plate then gently transfer larvae to their assigned wells (1 larva/well; Fig. 5).
4. Place zebrafish into the apparatus at 28.5°C and allow them to habituate for at least 10 min (*see* **Note 8**).
5. Zebrafish motion during the assay is captured using an IR-sensitive video camera with an IR-pass filter to ensure changes in white light illumination during the assay do not affect the video stream or tracking algorithm (Fig. 5). If using *LSRtrack*, follow the published detailed protocol [12] for video capture, tracking and data analysis; alternatively, follow the manufacturer's instructions for using a proprietary system. In our published work [47], we recorded movement at a low frame rate of 4 frames/s. This makes it easier to distinguish movement from noise and ensures video file sizes are manageable while capturing most activity. However, any recording or video stream that satisfies the criteria listed above (Section: Measurement of zebrafish movement) should be suitable, and many proprietary solutions capture video at higher frame rates or offer live tracking.
6. Record (or analyze a live video stream of) zebrafish swimming during alternating light-dark cycles as above.
7. Once the experiment is completed, remove the 96-well plate from the 28.5°C incubator, euthanize larvae following all applicable laws, local regulations, and approved protocols, then rinse the 96-well plate thoroughly using first tap water followed by distilled water to remove any chemical residue and allow the plate to be reused as necessary.
8. If the system completes zebrafish tracking offline, the recording of the experiment should now be analyzed (Fig. 5). Otherwise, save the data from the live tracking stream.

Analysis of Movement Data

1. At completion of Section: Running a 96-well plate VMR assay, a set of (x, y, t) coordinates of the centroid position of each zebrafish larva in each frame of the video during the experiment should be available for downstream analysis. These data can be used to calculate larval displacement D at each frame transition between frame $f-1$ and f:

$$D(f) = \sqrt{(x_{(f)} - x_{(f-1)})^2 + (y_{(f)} - y_{(f-1)})^2}$$

Instantaneous scalar speed S for frame f can then be calculated from framewise displacement as:

$$S(f) = D(f) \times framerate\ (Hz)$$

The mean speed for each zebrafish is then averaged in each 30-s time bin across the stimulus cycle and then the values from each zebrafish are averaged over the three stimulus cycles. This provides a single value for each zebrafish in each time bin, allowing the mean and SE to be calculated for each group in each time bin.

2. The expected results for untreated Tau and sibling controls in this analysis are shown in Fig. 6. Cyclic ambient illumination causes striking changes in WT zebrafish motor behavior [56]. At a light-dark transition, zebrafish show an initial large amplitude turning movement called an "O"-bend followed by elevated motility that slowly declines over time in ongoing darkness. At a dark-light transition, zebrafish show an abrupt decline in motility that slowly recovers to a steady-state value in ongoing stable ambient illumination. This response is severely attenuated in Tau zebrafish, such that there is wide separation between Tau and Sib larvae in these assays. The stereotypical and repetitive nature of the response allows for averaging responses over multiple stimulus cycles, reducing variability and improving statistical power for detection of phenotypic rescue, in samples of zebrafish as small as 12.

3. Assay QC is essential to ensure that the assay yields interpretable results (Fig. 7).

   a. In each time bin across the response, Assay Window (AW) is how much faster Sib moved than Tau larvae (Fig. 7A, B). Phenotype Ratio (PR) is the assay window divided by the sibling swimming speed (Fig. 7C):

$$Assay\ window = Sib - Tau$$

$$Phenotype\ ratio = \frac{[Sib - Tau]}{Sib}$$

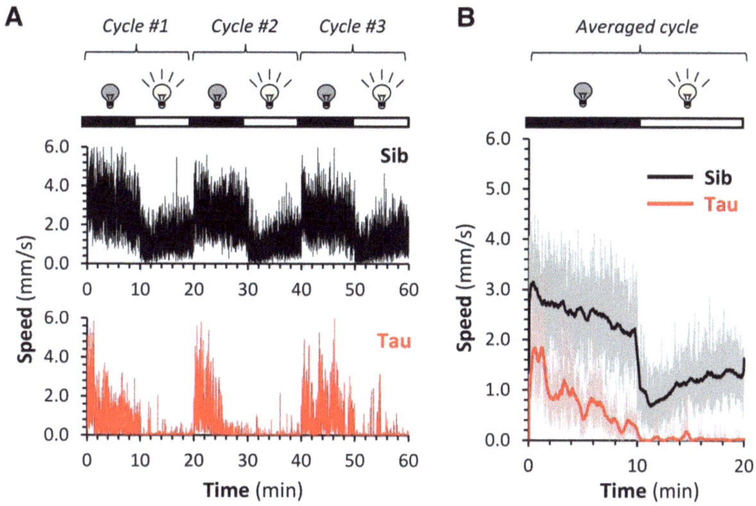

Fig. 6 Visual motor responses in sibling and Tau zebrafish. (**A**) The mean group displacement of 12 Sib (black; above) and 12 Tau (red; below) 5 dpf zebrafish larvae at each video frame transition is scaled to show mean instantaneous swimming speed in mm/s (y-axis). The x-axis shows time, as the larvae are exposed to cyclic ambient illumination (10 min dark—10 min light; indicated above the graphs) for a total of 60 min. Zebrafish larval movement increases in darkness and decreases under bright light in both groups. However, Tau zebrafish are clearly hypokinetic compared to siblings in all phases of the response. (**B**) To reduce variability and facilitate quantitative analysis, responses are averaged over the three stimulus cycles shown in panel (**A**). The average traces for each group are shown in light grey and smoothed trace (showing the mean of a rolling 5-s window) is shown as a bold line. This analysis clearly illustrates the severe neurological deficit in Tau zebrafish, particularly during the light phase of the VMR

AW should be >0.3 mm/s (and ideally >0.8 mm/s). PR is ideally close to 1. These values indicate a sufficiently robust phenotype in Tau zebrafish to allow meaningful comparison of each test group to the controls.

b. More complex metrics for assay QC are described in our published work [47]. A QC score was assigned to each time bin depending on AW and the *p* value resulting from a t-test of Sib vs. Tau (a perfect QC score of 1 was assigned to any bin in which AW > 0.3 mm/s AND uncorrected $p < 0.01$; an unusable QC score of 0 was assigned to any time bin in which AW < 0.3 mm/s OR $p > 0.05$; intermediate *p* values were assigned a QC value that was weighted by *t*) (Fig. 7D). If the mean QC score for all time bins was <0.95, the assay was rejected. These stringent criteria ensured robust and replicable screening outputs by removing assays that did not show sufficiently robust phenotypes in Tau zebrafish to enable quantitative comparison to controls, or in which the controls gave uninterpretable results.

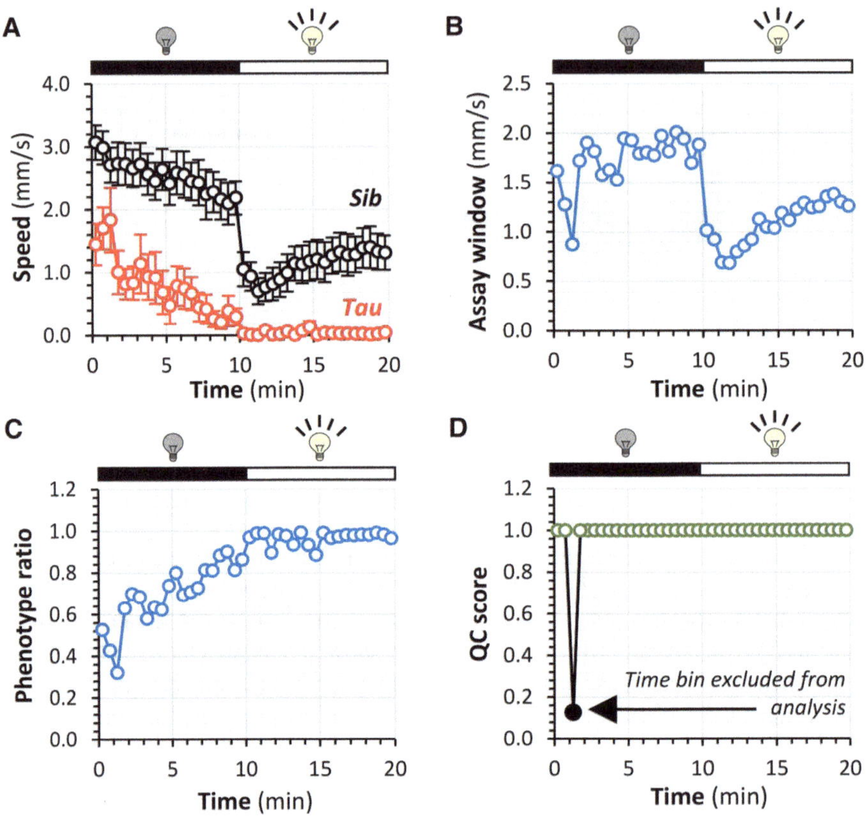

Fig. 7 96-well plate motor assay quality control (QC). Stringent QC measures, based on untreated Sib and Tau controls, ensure that each assay performs within acceptable limits that allow calculation of interpretable rescue scores for Tau zebrafish exposed to small molecules. (**A**) Data from Fig. 6B are shown with the smoothed trace replaced by group mean ± SE speed in each contiguous 30-s time bin over the duration of the VMR cycle. (**B**) The assay window (AW) is the difference in mean speed between sibling and Tau zebrafish in each time bin (mm/s). (**C**) The phenotype ratio (PR) is the assay window divided by the mean speed of sibling controls in the same time bin (dimensionless ratio). (**D**) A QC score, based on AW and statistical criteria (as detailed in the text) determines which time bins are acceptable and can be included in analysis. In the example shown, the third time bin is excluded because Tau and sibling controls are not adequately separated, so any rescue score would be unreliable

This was a small number of assays in the reported screen, but the phenotype of Tau zebrafish in this assay is dependent on careful maintenance of lines, including continual selection for strong expression (*see* **Note 3**).

4. If the assay passed QC, a rescue score can then be calculated for each test compound in each time bin (Fig. 8):

$$\%\ Rescue = 100 \times \frac{[Test - Tau]}{[Sib - Tau]}$$

where Sib, Tau and Test indicate the mean VMR light phase swimming speed of larvae in the respective group. In this scheme, *rescue*=0% indicates that a test group of small molecule-exposed Tau zebrafish showed mean swimming speed undistinguishable from untreated Tau zebrafish in the same assay, suggesting the molecule is inactive. Conversely, *rescue*=100% indicates that the test group showed mean swimming speed undistinguishable from sibling controls in the same assay, indicating that the phenotype was fully rescued, and the molecule is active. Examples of data from small molecules that were inactive or active in this model are shown in Fig. 8.

5. The bin-by-bin approach to QC in Step #3 also allows the rescue score to be weighted by a *t*-test comparing [Tau] versus [Tau+compound] in each time bin, and to eliminate any time bins with QC=0 from final rescue calculations.

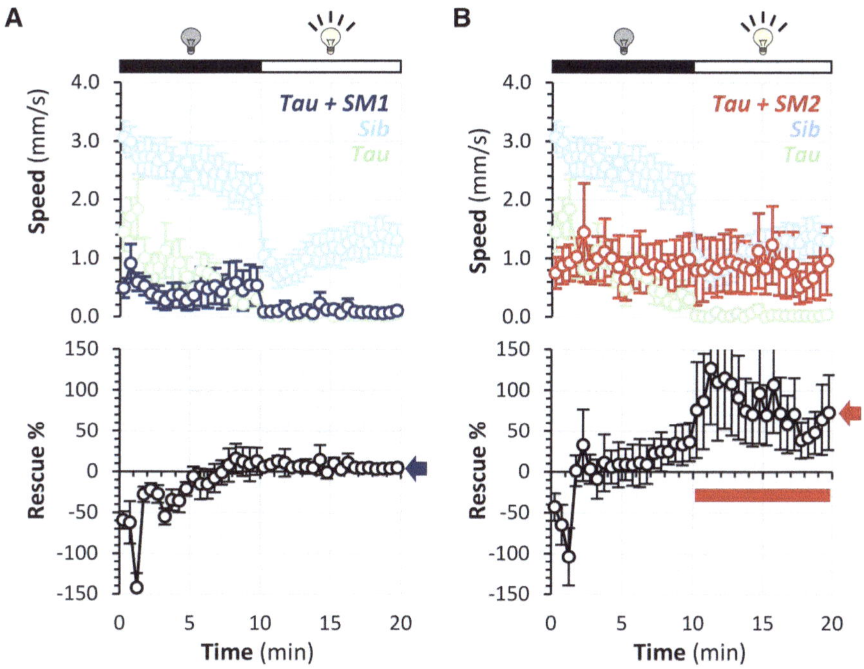

Fig. 8 Example small molecule rescue assays. In the upper graphs, mean ± SE swimming speed of untreated sibling (light blue) and Tau (light green) zebrafish are shown (data from Fig. 7A) with the responses of Tau zebrafish exposed to different small molecules superimposed over these controls for comparison. In the lower graphs, rescue % is shown in each time bin across the stimulus cycle, calculated as: $Rescue\ \% = 100 \times \frac{[Test - Tau]}{[Sib - Tau]}$. Consequently, 100% rescue describes a situation in which treated Tau zebrafish are indistinguishable from siblings whereas 0% rescue indicates that treated Tau zebrafish are indistinguishable from untreated Tau zebrafish. (**A**) Small molecule #1 (SM1; dark blue) does not rescue hypokinesia in Tau zebrafish and is inactive in this assay (blue arrow). (**B**) Small molecule #2 (SM2; dark red) is strongly active during the VMR light phase (red arrow)

Accepting removal of 1–2 of the time bins from calculations and presuming that rescue within any time bin was 0 unless statistically significant, together ensure that final mean rescue scores for each compound are not influenced by time bins in which controls did not behave as expected.

6. The mean rescue score for all time bins across the VMR light phase can then be calculated for each test compound. Since this is a ratio relative to the controls, rescue scores can be compared across assays. This is useful as the screen is completed six compounds at a time, so a 150-compound library requires a minimum of 25×96-well plate motor assays to screen.
7. These simple calculations can be automated using MATLAB or similar. The calculations are straightforward but depend on formatting the data readouts from whichever tracking system is used to provide inputs to the algorithm.
8. We used a cut-off of [library mean rescue + 3 SD] to define "hits" in our initial screen [47], although other definitions can be used.
9. It is critical to verify "hits" in post-screen assays using repurchased or resynthesized compounds, if the hits were identified from a screening library. The following additional verification steps are recommended:
   a. Concentration-response assays in Tau larvae to define the relationship between activity and compound concentration.
   b. Testing compound in WT larvae at the same concentrations to test whether the compound stimulates zebrafish motility non-specifically.
   c. Completing mCherry fluorescence microscopy in Tau zebrafish treated with compound at the same concentration to test whether the compound inhibits transgene expression.
   d. Additional phenotypic assays such as acridine orange labeling and optokinetic reflex assays.
10. Compounds passing these verification steps can then be further evaluated for mechanisms of action and molecular targets.

3.7 Phenotypic Test 2: Evaluation of Cell Death by Acridine Orange Labeling

Intravital labeling and counting of apoptotic cells in the spinal cord using acridine orange (Fig. 9) provides a complementary phenotypic assay to evaluate the activity of small molecules in Tau zebrafish. This approach does not require apparatus to measure zebrafish motor function as above (Sect. 3.6), but instead requires a microscope with fluorescence imaging capability.

We recently found that motor function and cell death do not necessarily report the same outcomes in Tau zebrafish—the top hit in our initial screen rescued motor function by preventing synapse loss, but did not prevent spinal cord cell death [47], suggesting that different hits may require different post-motor screen verification experiments. We have not yet completed a screen for compounds that prevent cell death, but this should be possible given the large increase (tenfold) in acridine orange-labeled cells in Tau zebrafish at 4–5 dpf compared with controls [47].

Preparation of Acridine Orange and 1.5% LMA

1. Dissolve 5 mg acridine orange salt in 1 mL Milli-Q H_2O by briefly vortexing. Owing to the light-sensitive nature of acridine orange, preparation of the solution should be done in dim room lighting, and the tube containing the solution should be covered in foil. Acridine orange solution should be stored at 4°C in darkness (*see* **Note 9**).
2. Add 15 mg LMA powder to 1 mL E3 buffer in a 1.5 mL microfuge tube and vortex briefly to disperse.
3. Place microfuge tube in a 95°C bath or heat block until the agarose has dissolved and the resulting gel melted.
4. Transfer the molten LMA to a 45°C bath and allow it to equilibrate to this temperature before use.

Exposure of Larval Zebrafish to Acridine Orange Solution

1. Prepare acridine orange working solution by adding 1 μL acridine orange stock (5 mg/mL) to 1 mL E3 buffer to yield a final concentration of 5 μg/mL (*see* **Note 9**).
2. Replace E3 buffer from Step #4 in Section: Small molecule exposure: testing for activity against neurological phenotypes with acridine orange master-mix (2 mL/well) and incubate in the dark for 30 min at room temperature.
3. After incubation, wash larvae three times in fresh E3 at room temperature to remove unbound acridine orange (*see* **Note 9**).

Quantification of Acridine Orange-Labeled Spinal Cord Cells

1. Anaesthetize larvae by adding 50 μL 0.4% buffered tricaine to 2 mL E3 buffer.
2. Under a stereo zoom microscope with transmitted/reflected white light illumination, position larvae so that the lateral surface of the trunk is in contact with the glass coverslip of a MatTek dish (1 larva/coverslip) (*see* **Note 9**).
3. Gently cover the larva in melted 1.5% LMA at ≤45°C. Use the pig eyelash to adjust position, keeping the lateral surface of the larva in contact with the glass coverslip. Allow agarose to solidify into a gel, thereby immobilizing the anesthetized zebrafish for microscopy (*see* **Note 9**).
4. Transfer embedded larvae to an inverted epifluorescence microscope then image using appropriate fluorescence filters for

acridine orange (excitation λ = 502 nm, emission λ = 526 nm). The excitation and emission spectra of both acridine orange and mCherry are shown in Fig. 10 to assist in selection of the correct filter sets (*see* **Note 9**).

5. Using a 20× objective, locate the spinal cord. This can be done by using either: (i) anatomical landmarks—the cord is a highly cellular structure running along the length of the body axis, deep to the axial muscles, and dorsal to the notochord; or (ii) mCherry fluorescence—the nuclei of spinal cord neurons should be brightly fluorescent in Tau zebrafish allowing easy identification of the cord (Fig. 9A, B) (*see* **Note 9**).
6. When bound to DNA in apoptotic cells, acridine orange fluoresces brightly. Acridine orange-labeled degenerating cells are seen clearly as bright yellow/green spots within the spinal cord (Fig. 9C). Manually count labeled cells by focusing through the plane of the spinal cord in contiguous microscopic fields moving rostrally from its caudal extremity to the boundary with the hindbrain. This can be done quickly and allows quantification of acridine orange-labeled cells over the whole spinal cord (*see* **Note 9**).
7. [Optional] Instead of Steps 5 and 6, the spinal cord can be imaged by confocal microscopy of the acridine orange and mCherry channels (Fig. 9 and as shown in Fig. 2a of [47]. The

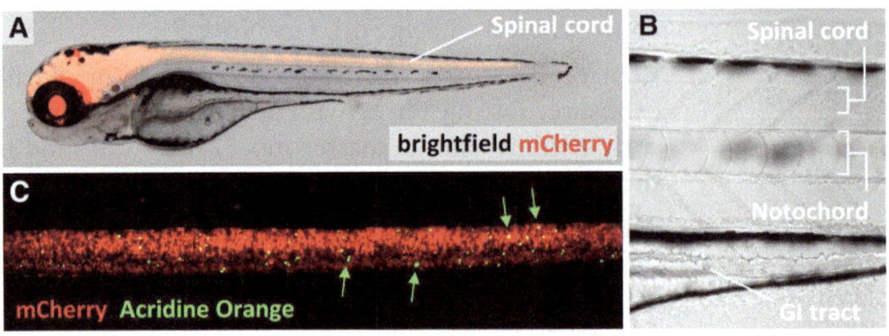

Fig. 9 Acridine orange cell death assay. As a complementary endpoint to the neurological function assays shown above, labeling dead or dying cells with acridine orange is a rapid way to determine whether small molecules alter neurodegeneration in Tau zebrafish. The example shown in the images was incubated in acridine orange at 4 dpf, then anesthetized and embedded as described in the text. (**A**) Low magnification lateral image showing a Tau zebrafish. The brightfield (greyscale) and mCherry epifluorescence (red) images are superimposed to show the expression pattern of the transgene in the CNS and to illustrate the location of the spinal cord. (**B**) Higher magnification lateral brightfield image of the larval zebrafish in panel (**A**). The notochord is readily identified running rostro-caudal along the center of the body axis. The spinal cord is seen as a highly cellular cylindrical column just dorsal to the notochord. (**C**) Confocal Z-plane maximum intensity projection of the spinal cord showing acridine orange (peak excitation λ = 502 nm, peak emission λ = 526 nm, when bound to nuclear dsDNA; pseudo-colored green), and mCherry (red) labeled neurons, in a Tau zebrafish. Green arrows show examples of acridine orange-labeled cells. Abundant dead and dying cells are clearly visible along the spinal cord in the Tau model at this time point

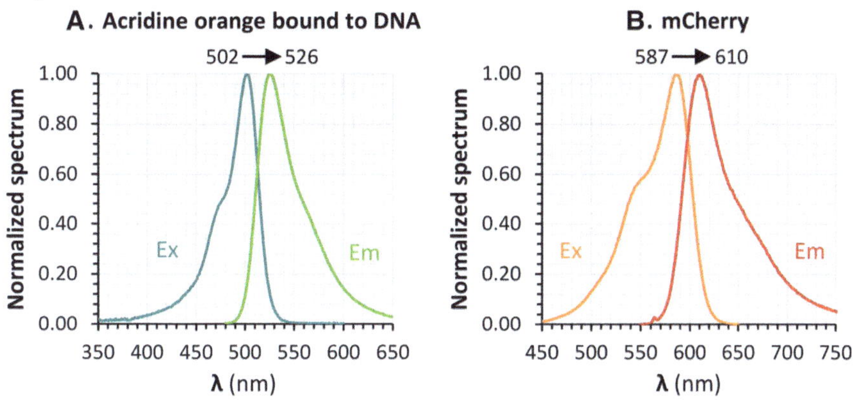

Fig. 10 Excitation and emission spectra for acridine orange and mCherry. Normalized excitation and emission spectra are shown for: (**A**) acridine orange bound to DNA (the source of the fluorescent acridine orange signal in dying and dead neurons), and (**B**) mCherry. These data are provided to assist in the selection of filters and light sources for imaging experiments involving acridine orange-labeled Tau zebrafish

outline of the cord revealed by mCherry labeling is then used to define a binary mask, and an automated algorithm counts labeled cells within the mask. Data acquisition in this method is slower but the approach yields a set of images as experimental records and employs automated algorithms with less potential for bias. Note that Sib zebrafish do not express mCherry, so if this method is used, it is necessary to employ a different control. A Tg(*UAS:p2A-nls-mCherry*) responder line that does not express Tau can be used to label the spinal cord in this case, as shown in Fig. 2a of [47] (*see* **Note 9**).

8. The sample size should be at least 12 zebrafish per group, and it is advisable to run controls, including untreated Tau and Sib, for comparison.

4 Notes

1. Breeding tank environmental enrichment, such as red marbles or artificial plastic leaves, may facilitate breeding. Overcrowding can impact breeding negatively; do not exceed a density of 10 zebrafish per 1.7 L of water (an equal number of both sexes is ideal). A significant change in water temperature during overnight housing in a breeding tank can negatively impact breeding; this is sometimes seen adjacent to air conditioning vents, which should be avoided. Refreshing the system water in the breeding tank at the start of the aquarium light cycle can facilitate spawning. An adequate recovery time after each spawning (1–2 weeks) also seems to improve yield.

2. Both the driver and responder alleles can be passed to progeny by parental zebrafish of either sex; it does not seem to make a significant difference whether the male or the female parent carries either the driver or responder allele and Tau zebrafish with robust expression and phenotypes can be generated either way (Fig. 1).
3. Similar to other UAS responder alleles, epigenetic variegation can occur in Tau zebrafish. This manifests as decreased or regionally absent mCherry expression in individual larvae. Zebrafish within the breeding colony should be selected for producing offspring with minimal variegation. It is advisable to set up crosses to yield more larvae than the minimum necessary for the experiment, as this allows selection for Tau zebrafish with uniformly strong CNS mCherry expression that will produce robust and replicable phenotypes (Fig. 3).
4. Once Tau zebrafish have been identified and selected, it is important to assign zebrafish to experimental and control groups **randomly** to avoid introducing bias.
5. The maximum DMSO concentration that larval zebrafish can tolerate is 1% [57]. This should not limit library analysis if 50 mM small molecule stocks are prepared, because dilution in E3 to 50 μM results in a DMSO concentration of 0.1%. However, it is occasionally necessary to increase the DMSO concentration above 0.1% if the small molecule precipitates when diluted in E3, in which case it is critical to keep the concentration below 1% and to treat controls with the same concentration of DMSO without small molecule to ensure any observed effects are not attributable to DMSO.
6. It is occasionally necessary to treat zebrafish in darkness if a small molecule is light-sensitive. In this case, it is essential to also keep controls in darkness, as this may alter the development of light-evoked responses that could complicate interpretation of VMR data.
7. Rigorous evaluation of the VMR light phase motor phenotype used in the published screen [47] showed that a sample size of 12 zebrafish in each group was the minimum that consistently gave a positive Z'. If fewer than six compounds are tested on a 96-well plate, the assay performance can be improved by including more than 12 zebrafish in each group. It is not recommended to use fewer than 12 zebrafish as this adversely affects the reliability of the assay for detecting phenotypic rescue.
8. Including several cycles of alternating ambient illumination [5 min darkness–5 min light] during acclimatization to the behavioral chamber may improve the replicability of the responses over all three subsequent stimulus cycles.

9. Since acridine orange is light-sensitive, preparation of solution should be completed in dim room lighting and the tubes containing acridine orange stock and working solution should be covered in foil. To prevent photobleaching of acridine orange label, zebrafish should remain protected from light before embedding and exposed to minimal light while embedding and during imaging.

5 Conflict of Interest

The authors have no conflicts of interest to declare that are relevant to the content of this chapter.

References

1. Steele JC, Richardson JC, Olszewski J (1964) Progressive supranuclear palsy: a heterogeneous degeneration involving the brain stem, basal ganglia and cerebellum with vertical gaze and pseudobulbar palsy, nuchal dystonia and dementia. Arch Neurol 10:333–359. https://doi.org/10.1001/ARCHNEUR.1964.00460160003001
2. Dickson DW, Rademakers R, Hutton ML (2007) Progressive supranuclear palsy: pathology and genetics. Brain Pathol 17:74. https://doi.org/10.1111/J.1750-3639.2007.00054.X
3. Arai T, Ikeda K, Akiyama H et al (2001) Distinct isoforms of tau aggregated in neurons and glial cells in brains of patients with Pick's disease, corticobasal degeneration and progressive supranuclear palsy. Acta Neuropathol 101:167–173. https://doi.org/10.1007/s004010000283
4. Sergeant N, Wattez A, Delacourte A (1999) Neurofibrillary degeneration in progressive supranuclear palsy and corticobasal degeneration: tau pathologies with exclusively "Exon 10" isoforms. J Neurochem 72:1243–1249. https://doi.org/10.1046/j.1471-4159.1999.0721243.x
5. Caffrey TM, Joachim C, Paracchini S et al (2006) Haplotype-specific expression of exon 10 at the human MAPT locus. Hum Mol Genet 15:3529–3537. https://doi.org/10.1093/hmg/ddl429
6. Höglinger GU, Melhem NM, Dickson DW et al (2011) Identification of common variants influencing risk of the tauopathy progressive supranuclear palsy. Nat Genet 43:699. https://doi.org/10.1038/NG.859
7. Hutton M, Lendon CL, Rizzu P et al (1998) Association of missense and 5-splice-site mutations in tau with the inherited dementia FTDP-17. Nature 393(6686):702–705. https://doi.org/10.1038/31508
8. Chen Z, Chen JA, Shatunov A et al (2019) Genome-wide survey of copy number variants finds MAPT duplications in progressive supranuclear palsy. Mov Disord 34:1049–1059. https://doi.org/10.1002/mds.27702
9. Litvan I, Lees PSJ, Cunningham CR et al (2016) Environmental and occupational risk factors for progressive supranuclear palsy: Case-control study. Mov Disord 31:644–652. https://doi.org/10.1002/mds.26512
10. D'Amora M, Giordani S (2018) The utility of zebrafish as a model for screening developmental neurotoxicity. Front Neurosci 12:976. https://doi.org/10.3389/fnins.2018.00976
11. Semmelhack JL, Donovan JC, Thiele TR et al (2014) A dedicated visual pathway for prey detection in larval zebrafish. Elife 3:e04878. https://doi.org/10.7554/eLife.04878
12. Zhou Y, Cattley RT, Cario CL et al (2014) Quantification of larval zebrafish motor function in multi-well plates using open-source MATLAB® applications. Nat Protoc 9:1533. https://doi.org/10.1038/NPROT.2014.094
13. Burgess HA, Burton EA (2023) A critical review of zebrafish neurological disease models−1. The premise: neuroanatomical, cellular and genetic homology and experimental tractability. Oxf Open Neurosci 2:kvac018. https://doi.org/10.1093/OONS/KVAC018
14. Kumar S, Hedges SB (1998) A molecular timescale for vertebrate evolution. Nature 392(6679):917–920. https://doi.org/10.1038/31927
15. Fulwiler C, Gilbert W (1991) Zebrafish embryology and neural development. Curr

Opin Cell Biol 3:988–991. https://doi.org/10.1016/0955-0674(91)90118-I
16. Blader P, Strähle U (2000) Zebrafish developmental genetics and central nervous system development. Hum Mol Genet 9:945–951. https://doi.org/10.1093/HMG/9.6.945
17. Panula P, Sallinen V, Sundvik M et al (2006) Modulatory neurotransmitter systems and behavior: towards zebrafish models of neurodegenerative diseases. Zebrafish 3:235–247. https://doi.org/10.1089/ZEB.2006.3.235
18. Schmidt R, Strähle U, Scholpp S (2013) Neurogenesis in zebrafish – from embryo to adult. Neural Dev 8:3. https://doi.org/10.1186/1749-8104-8-3
19. Wasel O, Freeman JL (2020) Chemical and genetic zebrafish models to define mechanisms of and treatments for dopaminergic neurodegeneration. Int J Mol Sci 21:1–14. https://doi.org/10.3390/IJMS21175981
20. Farrell TC, Cario CL, Milanese C et al (2011) Evaluation of spontaneous propulsive movement as a screening tool to detect rescue of Parkinsonism phenotypes in zebrafish models. Neurobiol Dis 44:9–18. https://doi.org/10.1016/j.nbd.2011.05.016
21. Ilin VA, Bai Q, Watson AM et al (2021) Mechanism of pacemaker activity in zebrafish DC2/4 dopaminergic neurons. J Neurosci 41:4141–4157. https://doi.org/10.1523/JNEUROSCI.2124-20.2021
22. Chen J, Poskanzer KE, Freeman MR, Monk KR (2020) Live-imaging of astrocyte morphogenesis and function in zebrafish neural circuits. Nat Neurosci 23:1297–1306. https://doi.org/10.1038/s41593-020-0703-x
23. Silva NJ, Dorman LC, Vainchtein ID et al (2021) In situ and transcriptomic identification of microglia in synapse-rich regions of the developing zebrafish brain. Nat Commun 12(1):5916. https://doi.org/10.1038/s41467-021-26206-x
24. Herbomel P, Thisse B, Thisse C (2001) Zebrafish early macrophages colonize cephalic mesenchyme and developing brain, retina, and epidermis through a M-CSF receptor-dependent invasive process. Dev Biol 238:274–288. https://doi.org/10.1006/dbio.2001.0393
25. Czopka T (2016) Insights into mechanisms of central nervous system myelination using zebrafish. GLIA 64:333–349. https://doi.org/10.1002/glia.22897
26. Sheng D, Qu D, Kwok KHH, et al. (2010) Deletion of the WD40 domain of LRRK2 in zebrafish causes parkinsonism-like loss of neurons and locomotive defect. PLoS Genet 6(4):e1000914. https://doi.org/10.1371/journal.pgen.1000914
27. Karlovich CA, John RM, Ramirez L et al (1998) Characterization of the Huntington's disease (HD) gene homolog in the zebrafish Danio rerio. Gene 217(1-2):117–125. https://doi.org/10.1016/s0378-1119(98)00342-4
28. Bai Q, Mullett SJ, Garver JA et al (2006) Zebrafish DJ-1 is evolutionarily conserved and expressed in dopaminergic neurons. Brain Res 1113:33–44. https://doi.org/10.1016/j.brainres.2006.07.057
29. Flinn L, Mortiboys H, Volkmann K et al (2009) Complex i deficiency and dopaminergic neuronal cell loss in parkin-deficient zebrafish (Danio rerio). Brain 132:1613–1623. https://doi.org/10.1093/brain/awp108
30. Flinn LJ, Keatinge M, Bretaud S et al (2013) TigarB causes mitochondrial dysfunction and neuronal loss in PINK1 deficiency. Ann Neurol 74:837–847. https://doi.org/10.1002/ana.23999
31. Milanese C, Sager JJ, Bais Q et al (2012) Hypokinesia and reduced dopamine levels in zebrafish lacking β- and γ1-synucleins. J Biol Chem 287:2971–2983. https://doi.org/10.1074/jbc.M111.308312
32. Sun Z, Gitler AD (2008) Discovery and characterization of three novel synuclein genes in zebrafish. Dev Dyn 237:2490–2495. https://doi.org/10.1002/dvdy.21569
33. Schmid B, Hruscha A, Hogl S et al (2013) Loss of ALS-associated TDP-43 in zebrafish causes muscle degeneration, vascular dysfunction, and reduced motor neuron axon outgrowth. Proc Natl Acad Sci U S A 110:4986–4991. https://doi.org/10.1073/pnas.1218311110
34. Shankaran SS, Capell A, Hruscha AT et al (2008) Missense mutations in the progranulin gene linked to frontotemporal lobar degeneration with ubiquitin-immunoreactive inclusions reduce progranulin production and secretion. J Biol Chem 283:1744–1753. https://doi.org/10.1074/jbc.M705115200
35. Chen M, Martins RN, Lardelli M (2009) Complex splicing and neural expression of duplicated tau genes in zebrafish embryos. J Alzheimers Dis 18:305–317. https://doi.org/10.3233/JAD-2009-1145
36. Fuentes R, Letelier J, Tajer B et al (2018) Fishing forward and reverse: advances in zebrafish phenomics. Mech Dev 154:296–308
37. Williams CH, Hong CC (2016) Zebrafish small molecule screens: taking the phenotypic plunge. Comput Struct Biotechnol J 14:350–356
38. Burton EA, Burgess HA (2023) A critical review of zebrafish neurological disease models–2. Application: functional and neuroanatomical phenotyping strategies and chemical screens. Oxf Open Neurosci 2:kvac019. https://doi.org/10.1093/oons/kvac019

39. Long K, Kostman SJ, Fernandez C et al (2019) Do zebrafish obey Lipinski rules?. ACS Med Chem Lett 10:1002–1006. https://doi.org/10.1021/acsmedchemlett.9b00063
40. Swinney DC, Anthony J (2011) How were new medicines discovered?. Nat Rev Drug Discov 10:507–519. https://doi.org/10.1038/nrd3480
41. Tomasiewicz HG, Flaherty DB, Soria JP, Wood JG (2002) Transgenic zebrafish model of neurodegeneration. J Neurosci Res 70:734–745. https://doi.org/10.1002/JNR.10451
42. Bai Q, Garver JA, Hukriede NA, Burton EA (2007) Generation of a transgenic zebrafish model of Tauopathy using a novel promoter element derived from the zebrafish eno2 gene. Nucleic Acids Res 35:6501. https://doi.org/10.1093/NAR/GKM608
43. Cosacak MI, Bhattarai P, Bocova L et al (2017) Human TAUP301L overexpression results in TAU hyperphosphorylation without neurofibrillary tangles in adult zebrafish brain. Sci Rep 7(1):12959. https://doi.org/10.1038/S41598-017-13311-5
44. Giong HK, Hyeon SJ, Lee JG et al (2024) Tau accumulation is cleared by the induced expression of VCP via autophagy. Acta Neuropathol 148(1):46. https://doi.org/10.1007/s00401-024-02804-z
45. Paquet D, Bhat R, Sydow A et al (2009) A zebrafish model of tauopathy allows *in vivo* imaging of neuronal cell death and drug evaluation. J Clin Invest 119:1382–1395. https://doi.org/10.1172/JCI37537
46. Lopez A, Lee SE, Wojta K et al (2017) A152T tau allele causes neurodegeneration that can be ameliorated in a zebrafish model by autophagy induction. Brain 140:1128–1146. https://doi.org/10.1093/BRAIN/AWX005
47. Bai Q, Shao E, Ma D et al (2024) A human Tau expressing zebrafish model of progressive supranuclear palsy identifies Brd4 as a regulator of microglial synaptic elimination. Nat Commun 15(1):8195. https://doi.org/10.1038/s41467-024-52173-0
48. Asakawa K, Kawakami K (2009) The Tol2-mediated Gal4-UAS method for gene and enhancer trapping in zebrafish. Methods 49:275–281. https://doi.org/10.1016/j.ymeth.2009.01.004
49. Köster RW, Fraser SE (2001) Tracing transgene expression in living zebrafish embryos. Dev Biol 233:329–346. https://doi.org/10.1006/dbio.2001.0242
50. Stevenson TJ, Trinh T, Kogelschatz C et al (2012) Hypoxia disruption of vertebrate CNS pathfinding through EphrinB2 is rescued by magnesium. PLoS Genet 8(4):e1002638. https://doi.org/10.1371/journal.pgen.1002638
51. Nasiadka A, Clark MD (2012) Zebrafish breeding in the laboratory environment. ILAR J 53:161–168. https://doi.org/10.1093/ILAR.53.2.161
52. Cario CL, Farrell TC, Milanese C, Burton EA (2011) Automated measurement of zebrafish larval movement. J Physiol 589:3703–3708. https://doi.org/10.1113/jphysiol.2011.207308
53. Van Laar VS, Chen J, Zharikov AD et al (2020) α-Synuclein amplifies cytoplasmic peroxide flux and oxidative stress provoked by mitochondrial inhibitors in CNS dopaminergic neurons *in vivo*. Redox Biol 37:101695. https://doi.org/10.1016/J.REDOX.2020.101695
54. Hammer HS (2019) Water quality for zebrafish culture. In: Cartner S, Eisen JS, Farmer SF (eds) The zebrafish in biomedical research. Biology, husbandry, diseases, and research applications, 1st edn. Academic, San Diego, CA, USA, pp 321–335
55. Shaner NC, Campbell RE, Steinbach PA et al (2004) Improved monomeric red, orange and yellow fluorescent proteins derived from Discosoma sp. red fluorescent protein. Nat Biotechnol 22:1567–1572. https://doi.org/10.1038/nbt1037
56. Burton CE, Zhou Y, Bai Q, Burton EA (2017) Spectral properties of the zebrafish visual motor response. Neurosci Lett 646:62. https://doi.org/10.1016/J.NEULET.2017.03.002
57. Hoyberghs J, Bars C, Ayuso M et al (2021) DMSO concentrations up to 1% are safe to be used in the zebrafish embryo developmental toxicity assay. Front Toxicol 3:804033. https://doi.org/10.3389/ftox.2021.804033

Chapter 16

Tau Transgenic Mice

Natalie G. Horgan, Emma N. Poston, Jessica S. Fortin and David G. Morgan

Abstract

Tauopathies encompass a class of almost 30 neurodegenerative diseases characterized by the accumulation of tau protein aggregates within neurons. These aggregates are associated with abnormal behavior, aphasia, cognitive decline, and movement disorders. Tau is produced from the alternative splicing of the microtubule-associated protein tau (MAPT) gene, and its aggregation leads to the formation of neurofibrillary tangles, a marker of tauopathies. Given the association between tau accumulation and neuropsychiatric symptoms, researchers have developed genetically modified mouse models with tauopathy to better understand these diseases and investigate possible treatments. This chapter presents a protocol for using the PS19 mouse model, which expresses mutant MAPT, to determine the efficacy of brain-penetrant compounds in reducing tau aggregation and formation.

Keywords Aggregation, Compounds, Microtubule-associated protein tau (MAPT), PS19 mouse model, Tauopathy

1 Introduction

Tauopathies are a large group of neurodegenerative diseases that are distinguished by the abnormal aggregation of the microtubule-associated protein tau (MAPT) within neurons [1–4]. These diseases include Alzheimer's disease (AD), frontotemporal lobar degeneration with tau inclusions (FTLD-tau), progressive supranuclear palsy (PSP), chronic traumatic encephalopathy (CTE), Pick's disease, and Parkinsonism linked to chromosome 17 (FTDP-17). In tauopathies, tau undergoes hyperphosphorylation (p-tau), making it more susceptible to aggregation and leading to the formation of intraneural lesions known as neurofibrillary tangles (NFTs), which are associated with the disruption of neuronal function [5]. This process follows an anatomical progression that is correlated with cognitive decline in Alzheimer's disease [6, 7]. Furthermore, tau aggregation actively drives neurodegeneration through mechanisms argued to involve toxic gain-of-function mutations, synaptic dysfunction, and trans-synaptic propagation [8–10].

Jessica S. Fortin (ed.), *Translational Research Methods for Tauopathies*, Neuromethods, vol. 233,
https://doi.org/10.1007/978-1-0716-5234-3_16,

To better understand the underlying mechanisms of tauopathy and develop targeted therapeutics, researchers have used genetically modified mouse models due to their structural similarities to the human brain. In particular, transgenic mice that express mutant tau proteins, such as the PS19 model, have aided in exploring tau's role in neurodegeneration [11]. Previous research has proven that human tau can interact with mouse tau to exhibit phenotypic traits like those in human neurodegenerative diseases [12]. Misfolded and phosphorylated tau is thought to propagate through a prion-like seeding mechanism [13–17]. This process has been demonstrated through intracerebral inoculation of misfolded tau fibrils in animal models, which shows tau's ability to spread through neural networks [18–23].

This chapter outlines a comprehensive protocol to investigate the efficacy and safety of brain-penetrant compounds in reducing p-tau-induced tauopathy, glial activation, and neurodegeneration in vivo, using the PS19 P301S transgenic mouse model. The PS19 mouse model is characterized by the overexpression of the human 1N4R tau isoform with the P301S mutation, under the control of the murine protein promoter [11]. This model exhibits a quick onset of AD-like symptoms. By 6–9 months, neuroinflammation, neuron death, and widespread tangle development are evident [24], and in some instances, mice develop muscle weakness and paralysis after 10 months of age.

The experimental protocol follows three stages. In the first stage, six different compounds will be administered to separate groups of 16 mice each using brain-implanted minipumps to determine their ability to reduce preformed tangles. The second stage will evaluate three different compounds at both low and high dosages in separate groups of twenty mice each using chow to test the compound's ability to prevent tauopathy without inducing sickness behavior. The third stage will study the lead compound in a group of twenty mice to assess its potential to reduce tauopathy associated with amyloidosis.

Efficacy will be assessed through neurochemical and immunohistological analyses, measuring tau pathology, neuroinflammation, and neuronal integrity. Western blot and enzyme-linked immunosorbent assay (ELISA) will quantify tau aggregation and autophagy markers, while quantitative reverse transcription polymerase chain reaction (qRT-PCR) will verify transgene expression. One-way ANOVA and Tukey's range test will be used for statistical analysis, with $p < 0.05$ and beta power of 0.8. We hypothesize that brain-permeant compoundswill reduce NFT formation and glial activation, thereby preserving neuronal integrity and function compared to untreated mice. Further research is needed to identify key molecular drivers of tau pathology, optimize disease models, and refine tau-targeted therapeutics. Advancing our understanding of tau biology will be essential for developing effective treatments for tauopathies.

2 Materials

2.1 Animals

- PS19 P301S transgenic mice.
- Non-transgenic littermates (for control groups).
- Age at study initiation:
  - Stage I: 8 months old.
  - Stage II: 4 months old.
  - Stage III: 12 months old (amyloid-enhanced tauopathy model).
- Number of mice (male and female, balanced per group):
  - Stage I: 16 mice per group × number of groups.
    - One vehicle group.
    - Six treatment groups.
    - One nontransgenic group.
  - Stage II: 20 mice per group × number of groups
    - One vehicle group.
    - Six treatment groups (three compounds with two groups each—low and high dose).
    - One non-transgenic group.
  - Stage III: 20 mice per group × number of groups
    - One vehicle group.
    - One treatment group.
    - One nontransgenic group.

2.2 Compounds

- Six brain-penetrant tau-targeting test compounds.
  - Formulated for chow/ minipump incorporation.
- Vehicle (placebo) for control chow groups.

2.3 Animal Husbandry and Housing Supplies

- Standard mouse chow + customized compound chow formulation.
- Housing cages, bedding, enrichment.
- Water bottles or an automatic watering system.
- Cage labels for blinding and identification.

2.4 Equipment

- Dissection tools (scissors, forceps, scalpels, etc.).
- Thermal heating pad (to maintain body temp during euthanasia).

- Microtome or cryostat (for brain sectioning).
- Digital scanning microscope.
- Centrifuge.
- Homogenizer.
- qPCR machine (e.g., ABI QuantStudio).
- Western blot system (gel electrophoresis, transfer, imaging).
- ELISA plate reader.
- Stereology software/system (e.g., Stereo Investigator).

2.5 Reagents for Histology and Immunohistochemistry

- Paraformaldehyde or formalin (for perfusion and fixation).
- Cryoprotectant (e.g., sucrose).
- Nissl stain.
- Gallyas silver stain kit.
- Antibodies:
  - Total tau (e.g., Tau-5).
  - TOC1 (tau oligomers).
  - Alz50 or MC1 (misfolded tau).
  - AT8 (pSer202/Thr205).
  - PHF1 (pSer396/404).
  - CP13 (pSer202).
  - AT180 (pThr231).
  - GFAP (astrocyte marker).
  - Iba-1 (microglia marker).
  - CD45, CD68, MHCII (inflammation and immune markers).
- Secondary antibodies (HRP- or fluorescence-conjugated).
- Blocking buffer (e.g., 5% BSA, normal serum).
- Mounting medium (with or without DAPI).
- DAB or fluorescence substrate kits.

2.6 Reagents for Biochemistry

- Lysis buffer (for soluble/insoluble protein extraction).
- Protease/phosphatase inhibitor cocktail.
- Antibodies for Western blot:
- Total tau, p-tau (same as above), TOC1, p62, and LC3.
- ELISA kits (e.g., TOC1 ELISA for tau oligomers).
- RNA extraction kit (e.g., Qiagen RNeasy).
- Reverse transcription reagents (RT kit).

- qPCR reagents (SYBR Green or TaqMan-based).
- Primers for human tau transgene and endogenous mouse tau.

2.7 Consumables

- Pipette tips (filter and standard).
- Microcentrifuge tubes.
- 96-well plates (for ELISA and qPCR).
- Western blot membranes (PVDF or nitrocellulose).
- Electrophoresis gels.
- Transfer buffers, SDS-PAGE reagents.
- Coverslips, glass slides.
- Cryomolds, embedding cassettes.
- Syringes and needles (for perfusion).
- Gloves, lab coats, and other PPE.

2.8 Euthanasia and Perfusion Supplies

- Anesthetic (e.g., isoflurane or ketamine/xylazine).
- Perfusion buffer (PBS, followed by fixative).
- Heating pad.
- Perfusion pump (optional but helpful).

3 Methods

The primary goal of this study is to identify which compound most effectively reduces tau aggregate formation in PS19 mice and in a model of amyloid-enhanced tauopathy. To accomplish this, we will use a three-stage funnel approach (Fig. 1). Stage I will assess six compounds that are administered directly to the central nervous system (CNS) over a 30-day period via minipumps. Stage II will study three compounds over a 5-month period via chow administration. Stage III will test the most effective compound for 5 months in the amyloid-enhanced tauopathy model via chow administration.

For stage I, before starting the treatment at 8 months, all mice will be evaluated for neuropsychological behavioral impairments using open field, elevated plus maze, nesting behavior, and marble burying. Mice that exhibit behavioral deficits that are associated with tauopathy and neurochemical abnormalities will be randomly assigned to a treatment or a vehicle control group. Additionally, sickness behavior will be monitored throughout the study using four standardized measures: (1) 2-min in-cage observations on days 1, 4, 7, 14, and 20 to assess lethargy, ptosis, and piloerection; (2) 2-min observations during open field activity; (3) food intake over a 24-h period; and (4) changes in body weight. For all stages,

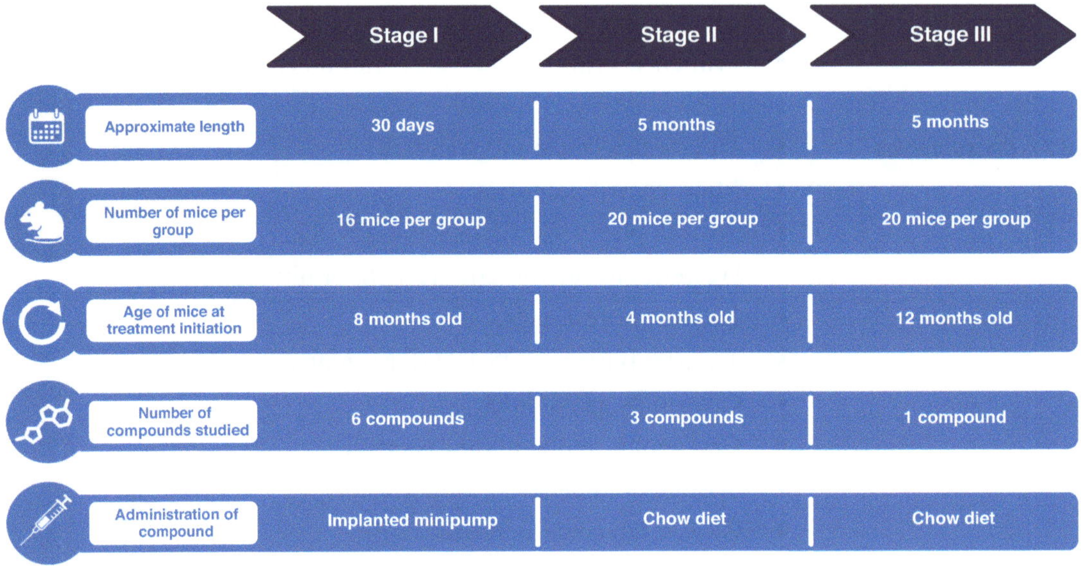

Fig. 1 Three-stage funnel of experimental design

the same neuropsychological behaviors will be tested at 9 months and cognitive function will be assessed by the Y maze and radial arm water maze [25].

After the designated study period, mice will be humanely euthanized and perfused transcardially with saline while their body temperature is maintained with a heating pad (avoids increased p-tau due to anesthesia-induced reduction in body temperature [26]. The left hemisphere will be used for immunohistochemistry to assess markers for tau pathology, glial activation, and neurodegeneration, while the right hemisphere will undergo dissection to isolate the hippocampus and anterior cortex for neurochemical analysis. More specifically, tau pathology will be measured histologically using markers for total tau (HT7), early-phase phosphorylated tau aggregates (AT8), late-phase aggregates (pTau396), and neurofibrillary tangles (Gallyas silver staining). In addition, ELISA will be used to quantify soluble and insoluble total tau, pTau199, pTau396, and oligomeric tau (TOC1 assay). Glial activation will be determined by markers for astrocytes (GFAP), total microglia (Iba-1), antigen-presenting microglia (MHC-2), and phagocytic microglia (CD68). Finally, neurodegeneration will be evaluated by estimation of the hippocampal volume via histology, detection of synaptophysin and PSD95 via ELISA, and measurement of the hippocampal weight. In stage III, amyloid load will also be measured using ELISA to determine the impact of the most effective compound in the amyloid-enhanced tauopathy model.

3.1 30-Day Administration of Six Compounds via Minipumps

The first stage will test six different compounds to examine their efficacy and potential sex differences. Each compound will be administered to a group of 16 PS19 mice (eight males and eight females) via implanted minipumps that deliver the compound directly to their brains over a 30-day period. A non-transgenic group will be included to serve as normal baseline values for behavioral measures, and to determine the magnitude of the transgene effects on neurochemical and histological measurements. After the study period, the mice will be re-tested for the neuropsychological behaviors, then humanely euthanized, and their brain tissues will be collected to assess tau pathology, glial activation, and neurodegeneration by t-tests with $p < 0.05$ and a beta power of 0.8. The expected outcome is for at least one agent to reduce tauopathy in the tissues. The three most efficacious agents will move onto stage II, in which they will undergo sub-chronic testing to determine their ability to reduce tau aggregates while avoiding sickness behavior.

Stage I Experimental Protocol

1. Select six brain-penetrating compounds to evaluate.
2. Obtain seven groups of 16 eight-month-old PS19 mice each and one group of non-transgenic mice.
3. Pre-screen the mice for behavioral impairments using:
   a. Open field test.
   b. Elevated plus maze.
   c. Nesting behavior.
   d. Marble burying test.
4. Randomly assign the mice with behavioral impairments to either treatment or vehicle groups (Compound A, Compound B, Compound C, Compound D, Compound E, Compound F, and the control group), ensuring a balance of sex and litter.
5. Surgically implant an Alzet infusion cannula into the left lateral ventricle of each mouse, affixed to the skull.
6. Connect the cannula via polyethylene tubing to a subcutaneously placed minipump in the flank.
7. Begin direct CNS administration of each compound via minipump.
   a. Calculate the dosage for each agent based on:
      i. Brain volume: 500 μL.
      ii. Cerebrospinal fluid turnover: 10 times/day.
      iii. Pump delivery rate: 0.11 μL/h.
      iv. Maximal tolerated dose from toxicity studies.
8. Monitor the mice for sickness behavior throughout the experiment:
   a. Conduct 2-min in-cage observations on days 1, 4, 7, 14, and 20 to assess lethargy, ptosis, and piloerection.
   b. Perform 2-min open field activity observations weekly.

   c. Monitor 24-h food intake.
   d. Record daily body weight changes.
9. On day 21 of treatment, reassess the mice for behavioral impairments using the same methods used during the pre-screen.
10. After 30 days of treatment, humanely euthanize the mice and collect their tissues for histological and neurochemical measurements.
   a. Tau pathology:
      i. Histological assessment: HT7 (total tau), AT8 (pTau early phase aggregates), pTau 396 (later phase aggregates), and Gallyas silver staining (tangles).
      ii. ELISA analysis: total tau, pTau199, pTau396, and oligomeric tau (TOC1 assay).
   b. Glial activation:
      i. Measure using GFAP (astrocytes), Iba-1 (total microglia), MHC-2 (antigen-presenting microglia), and CD68 (phagocytic microglia).
   c. Neurodegeneration:
      i. Histological estimation of hippocampal volume.
      ii. ELISA measurement of synaptophysin and PSD96.
      iii. Weigh the hippocampus.
11. Perform statistical analysis of the histological and neurochemical endpoints using $p < 0.05$. Based on prior work with these measures, the sample sizes should provide a β power of 0.8.
   a. Use a t-test to compare all markers of tauopathy in each treatment group to the vehicle control group.
   b. Use ANOVA followed by Tukey's means testing to compare all non-tau markers in each treatment group to the vehicle and non-transgenic groups.

3.2 5-Month Chronic Administration of Three Compounds via Chow

Stage II involves the evaluation of the three most successful compounds from stage I to determine their effects after chronic administration over a 5-month period. Each group of twenty mice, divided equally between male and female, will receive one compound via chow. The experiment will start when the mice reach 4 months of age, before the mice begin developing tau aggregates, and will continue until 9 months of age. At 8.5 months of age, the mice will be tested for neuropsychological impairments, cognitive function (Y-maze alternation), and spatial navigation learning (radial arm water maze). After the study period, the mice will be humanely euthanized, and their brain tissues will be collected to measure tauopathy, glial activation, and neurodegeneration by one-way ANOVA and Tukey's range test with $p < 0.05$ and a beta power of 0.8. The expected outcome is for at least one agent to prevent tauopathy and reduce behavioral effects, as evidenced by reduced ELISA and histology markers, lower glial activation markers, and less neurodegeneration. The most effective compound will progress to stage III for its ability to reduce tauopathy, which is elevated by the presence of amyloidosis (occurs in Alzheimer's disease).

Another component of this stage will assess the efficacy and safety of two different doses for each compound—one at the max tolerated dose (MTD) and another at one-third of the MTD (Fig. 2). To do so, the PS19 mice will be randomized three different groups: the vehicle, low dose, and high dose of the compound. A fourth group of non-transgenic littermates of PS19 mice will be present to serve as untreated control group for baseline behavioral performance. The vehicle group will have a control diet for the entire timeline. The low dose group will have the control diet for 7 days (to measure baseline food consumption), then will enter the low dose diet starting on day 8 until the remainder of the timeline. The high dose group will have the control diet for 7 days, then will start the low dose diet from day 8 to day 14, then will start the high dose diet on day 15 and on. Starting on day 21, the mice will be offered a daily medicated bacon-flavored wafer (unless the drug's half-life is greater than 12 h) that contains the same drug content as their standard chow. When the mice consume the wafer, their nightly chow intake will be reduced by an amount equal to the wafer's content to ensure consistent drug exposure. Throughout the experiment, mice will be monitored for their food intake and sickness behavior.

Stage II Experimental Protocol

1. Select three efficacious brain-penetrating compounds based on results from stage I.
2. Obtain seven groups of 20 four-month-old PS19 mice.
3. Randomly assign the mice to the treatment groups (Compound A MTD, Compound A 1/3 MTD, Compound B MTD, Compound B 1/3 MTD, Compound C MTD, Compound C 1/3 MTD, and the control group), ensuring a balance of sex and litter.
   a. Include a group of non-transgenic littermates as baseline behavior controls.
4. Begin all treatment groups on the control chow diet for 7 days.
   a. The control group will continue this diet for the entire duration of the study.
5. Monitor mice for sickness behavior throughout the experiment:
   a. Conduct 2-min in-cage observations weekly.
   b. Monitor 24-h food intake.
   c. Record daily body weight changes.

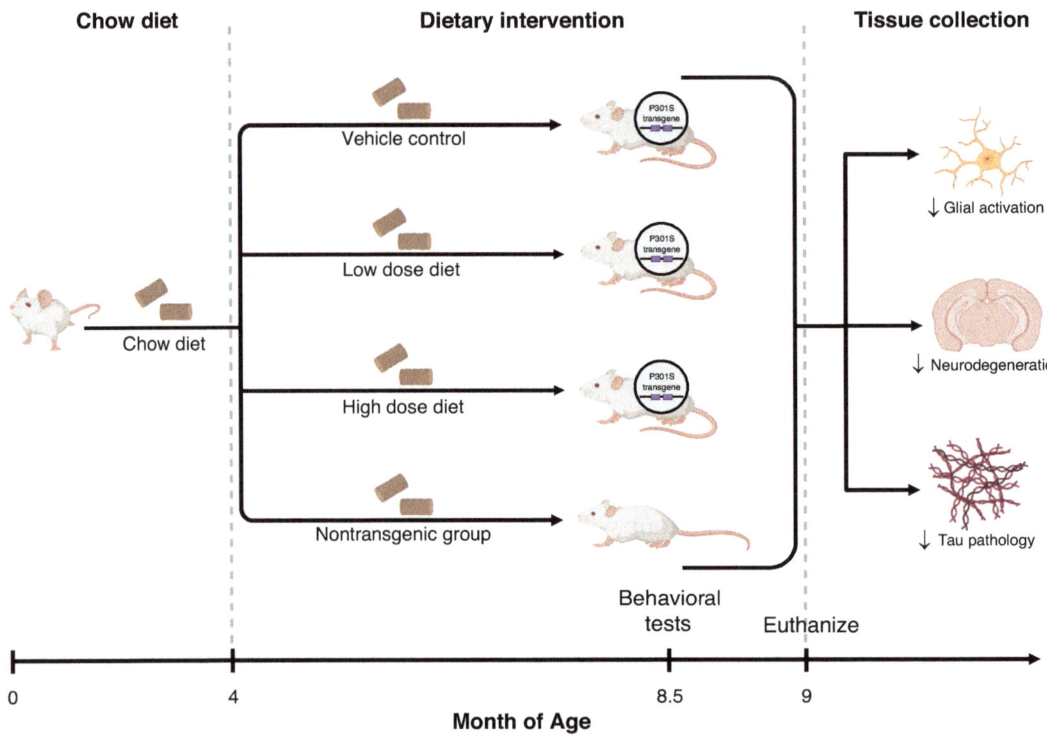

Fig. 2 Schematics of the experimental setup for studies in stage II

6. On day 8 of treatment, switch the low and high dose treatment groups to the low dose chow diet.
   a. The low-dose group (1/3 MTD) will continue this diet for the remainder of the study.
7. On day 15 of treatment, switch the high-dose treatment group to the high-dose chow diet.
   a. The high dose group (MTD) will continue this diet for the remainder of the study.
8. On day 21 of treatment, offer all mice a wafer each morning to maintain drug exposure throughout the 24-h period.
9. Once the mice reach age 8.5 months of age, or 4.5 months of treatment duration, conduct an extended battery of behavioral tests.
   a. Behavioral tests from stage I: open field, elevated plus maze, nesting behavior, marble burying
   b. Additional tests: Y-maze, radial arm water maze
10. Once the mice reach 9 months of age (5 months of treatment duration), humanely euthanize the mice and collect their tissues for histological and neurochemical measurements.

a. Tau pathology:

i. Histological assessment: HT7 (total tau), AT8 (pTau early phase aggregates), pTau 396 (later phase aggregates), and Gallyas silver staining (tangles).

ii. ELISA analysis: total tau, pTau199, pTau396, and oligomeric tau (TOC1 assay).

b. Glial activation:

i. Measure using GFAP (astrocytes), Iba-1 (total microglia), MHC-2 (antigen-presenting microglia), and CD68 (phagocytic microglia)

c. Neurodegeneration:

i. Histological estimation of hippocampal volume.

ii. ELISA measurement of synaptophysin and PSD96.

iii. Weigh the hippocampus.

11. Perform statistical analysis of the histological and neurochemical endpoints using $p < 0.05$. Based on prior work with these measures, the sample sizes should provide β power of 0.8.

a. Use a t-test to compare all markers of tauopathy in each treatment group to the vehicle control group.

b. Use ANOVA followed by Tukey's means testing to compare all non-tau markers in each treatment group to the vehicle and non-transgenic groups.

3.3 5-Month Administration of Lead Compound via Chow in the Amyloid-Enhanced Tauopathy Model

Stage III will assess the most promising compound identified in stages I and II for its ability to therapeutically treat an amyloid-enhanced tauopathy model. This model uses a model of amyloidosis that expresses both mutant amyloid precursor protein and presenilin-1 transgenes, and after amyloidosis is mature, it induces tau pathology using a systemic adeno-associated virus vector expressing P301L human tau at 8 months. By 12 months, these mice have both amyloid and tau pathology, with the amount of tau pathology elevated fivefold in the amyloid mice versus non-transgenic mice injected with the same AAV-tau vector [27]. Prior to the start of drug treatment at 12 months, tissue samples will be collected from one group of amyloid-enhanced tauopathy mice to establish baseline amyloid and tau values to ascertain if the drug removes pre-existing pathology (below baseline) or simply blocks further development of pathology at the end of the study (below vehicle treated group). A group of untreated non-transgenic littermates will serve as baseline controls for behavior. This compound will be tested for 5 months via chow consumption and titrated in the same fashion as described in stage II in 12-month-old mice that exhibit amyloidosis and tauopathy, with sickness behavior monitored. By 16 months of age, the mice will undergo testing for behavioral impairments. At 17 months, the mice will be humanely euthanized with their brain tissues collected to assess amyloid load, tau pathology, glial activation, and neurodegeneration by one-way ANOVA and Tukey's range test with $p < 0.05$ and a beta power of 0.8. The expected outcome of this particular study is that the treated mice will show reduced tauopathy, gliosis, and neurodegeneration compared to the vehicle chow group. If tau pathology in the treated mice is reduced compared to the baseline tau values, it will indicate that the compound is able to remove pre-existing tau pathology. If tau pathology in the treated mice is equal to the baseline tau values from the non-transgenic littermates, it will indicate that the compound is able to prevent further accumulation of tau pathology.

Stage III Experimental Protocol

1. Select the most promising compound based on the results from stages I and II.
2. Obtain three groups of 20 twelve-month-old amyloid-enhanced tau mice.
3. Collect tissue from one group of untreated amyloid-enhanced tau mice at 12 months for baseline tauopathy values.
4. Randomly assign the remaining mice to the treatment groups (Compound X and the vehicle control group), ensuring a balance of sex and litter.
   a. Include a group of non-transgenic littermates as baseline behavior controls.
5. Begin all treatment groups on the control chow diet for seven days.
   a. The control group will continue this diet for the entire duration of the study.
6. Monitor mice for sickness behavior throughout the experiment:
   a. Conduct 2-min in-cage observations weekly.
   b. Monitor 24-h food intake.
   c. Record daily body weight changes.
7. On day 8 of treatment, switch the compound treatment group to the low-dose chow diet.
8. On day 15 of treatment, switch the compound treatment group to the high-dose chow diet.
   a. The compound treatment group will continue this diet for the remainder of the study.
9. On day 21 of treatment, offer all mice a wafer to maintain drug exposure over 24 h.

10. Once the mice reach age 16 months of age, or 4 months of treatment duration, conduct an extended battery of behavioral tests.
    a. Behavioral tests from stage II: open field, elevated plus maze, nesting behavior, marble burying, radial arm water maze, and Y-maze.
11. Once the mice reach 17 months of age after 5 months of treatment duration, humanely euthanize the mice and maintain their body temperature by using a heating pad.
12. Collect the tissues of the mice for histological and neurochemical measurements.
    a. Amyloid load
        i. Histological assessment: Congo red (compacted plaques) and Aβ immunohistochemistry (compacted and diffuse deposits).
        ii. ELISA analysis: soluble, insoluble, and oligomeric Aβ42 and Aβ40.
    b. Tau pathology:
        i. Histological assessment: HT7 (total tau), AT8 (pTau early phase aggregates), pTau 396 (later phase aggregates), and Gallyas silver staining (tangles).
        ii. ELISA analysis: total tau, pTau199, pTau396, and oligomeric tau (TOC1 assay).
    c. Glial activation:
        i. Measure using GFAP (astrocytes), Iba-1 (total microglia), MHC-2 (antigen-presenting microglia), and CD68 (phagocytic microglia).
    d. Neurodegeneration:
        i. Histological estimation of hippocampal volume.
        ii. ELISA measurement of synaptophysin and PSD96.
        iii. Weigh the hippocampus.
13. Perform statistical analysis of the histological and neurochemical endpoints using $p < 0.05$ and a beta power of 0.8.
    a. Perform a one-way ANOVA to compare the baseline, treatment, and vehicle groups.
    b. Compare the means of the treatment groups using Tukey's range test.

4 Recommendations

For these animal studies, it is recommended to have both a transgenic control group and a non-transgenic group to assess treatment efficacy and baseline behavioral endpoints, respectively. These groups are necessary to determine the best treatment compound and its potential benefits against neurodegenerative diseases.

5 Troubleshooting

There are several limitations to this study. One limitation is the potential instability of the drugs in stage I at 37°C over 30 days, as well as the toxicity associated with direct administration to the CNS. If necessary, the study will be terminated early if more than 50% of the treated mice exhibit sickness behavior. In stages II and III, a limitation is that the mice may refuse to consume the drug-treated chow due to its bitter taste. One potential solution is to add 0.1% saccharin to mask the bitterness or to deliver all the drug via the nutritionally complete wafer, provided the mice will consume it in that form. A limitation specific to stage III is the advanced age of the mice. The experiment will begin once the mice reach 12 months of age, at which point they will already exhibit tauopathy and amyloidosis, making them potentially more prone to death. Although we have not previously observed this, we would increase sample sizes in a second cohort if this occurs.

Conflict of Interest

The authors have no conflicts of interest to declare that are relevant to the content of this chapter.

References

1. Hernandez F, Avila J (2007) Tauopathies. Cell Mol Life Sci 64(17):2219–2233. https://doi.org/10.1007/s00018-007-7220-x
2. Kulbe JR, Hall ED (2017) Chronic traumatic encephalopathy-integration of canonical traumatic brain injury secondary injury mechanisms with tau pathology. Prog Neurobiol 158:15–44. https://doi.org/10.1016/j.pneurobio.2017.08.003
3. Mohamed AZ, Cumming P, Gotz J et al (2019) Tauopathy in veterans with long-term posttraumatic stress disorder and traumatic brain injury. Eur J Nucl Med Mol Imaging 46(5):1139–1151. https://doi.org/10.1007/s00259-018-4241-7
4. Kovacs GG (2015) Invited review: neuropathology of tauopathies: principles and practice. Neuropathol Appl Neurobiol 41(1):3–23. https://doi.org/10.1111/nan.12208
5. Grundke-Iqbal I, Iqbal K, Tung YC et al (1986) Abnormal phosphorylation of the microtubule-associated protein tau (tau) in Alzheimer cytoskeletal pathology. Proc Natl Acad Sci U S A 83(13):4913–4917. https://doi.org/10.1073/pnas.83.13.4913
6. Arriagada PV, Growdon JH, Hedley-Whyte ET et al (1992) Neurofibrillary tangles but not senile plaques parallel duration and severity of Alzheimer's disease. Neurology 42(3 Pt 1):631–639. https://doi.org/10.1212/wnl.42.3.631

7. Nelson PT, Abner EL, Schmitt FA et al (2010) Modeling the association between 43 different clinical and pathological variables and the severity of cognitive impairment in a large autopsy cohort of elderly persons. Brain Pathol 20(1):66–79. https://doi.org/10.1111/j.1750-3639.2008.00244.x
8. Gendron TF, Petrucelli L (2009) The role of tau in neurodegeneration. Mol Neurodegener 4:13. https://doi.org/10.1186/1750-1326-4-13
9. Wu M, Zhang M, Yin X et al (2021) The role of pathological tau in synaptic dysfunction in Alzheimer's diseases. Transl Neurodegener 10(1):45. https://doi.org/10.1186/s40035-021-00270-1
10. Lamontagne-Kam D, Ulfat AK, Hervé V et al (2023) Implication of tau propagation on neurodegeneration in Alzheimer's disease. Front Neurosci 17:1219299. https://doi.org/10.3389/fnins.2023.1219299
11. Yoshiyama Y, Higuchi M, Zhang B et al (2007) Synapse loss and microglial activation precede tangles in a P301S tauopathy mouse model [published correction appears in Neuron 54(2):343-4]. Neuron 53(3):337–351. https://doi.org/10.1016/j.neuron.2007.01.010
12. Moore KBE, Hung TJ, Fortin JS (2023) Hyperphosphorylated tau (p-tau) and drug discovery in the context of Alzheimer's disease and related tauopathies. Drug Discov Today 28(3):103487. https://doi.org/10.1016/j.drudis.2023.103487
13. Clavaguera F, Bolmont T, Crowther RA et al (2009) Transmission and spreading of tauopathy in transgenic mouse brain. Nat Cell Biol 11(7):909–913. https://doi.org/10.1038/ncb1901
14. Frost B, Jacks RL, Diamond MI (2009) Propagation of tau misfolding from the outside to the inside of a cell. J Biol Chem 284(19):12845–12852. https://doi.org/10.1074/jbc.M808759200
15. Lasagna-Reeves CA, Castillo-Carranza DL, Sengupta U et al (2011) Tau oligomers impair memory and induce synaptic and mitochondrial dysfunction in wild-type mice. Mol Neurodegener 6:39. https://doi.org/10.1186/1750-1326-6-39
16. Wu JW, Hussaini SA, Bastille IM et al. (2016) Neuronal activity enhances tau propagation and tau pathology *in vivo*. Nat Neurosci 19(8):1085–1092. https://doi.org/10.1038/nn.4328
17. Wegmann S, Maury EA, Kirk MJ et al. (2015) Removing endogenous tau does not prevent tau propagation yet reduces its neurotoxicity. EMBO J 34(24):3028–3041. https://doi.org/10.15252/embj.201592748
18. Smolek T, Jadhav S, Brezovakova V et al. (2019) First-in-rat study of human Alzheimer's disease tau propagation. Mol Neurobiol 56(1):621–631. https://doi.org/10.1007/s12035-018-1102-0
19. Narasimhan S, Changolkar L, Riddle DM et al (2020) Human tau pathology transmits glial tau aggregates in the absence of neuronal tau. J Exp Med 217(2):e20190783. https://doi.org/10.1084/jem.20190783
20. He Z, McBride JD, Xu H et al (2020) Transmission of tauopathy strains is independent of their isoform composition. Nat Commun 11(1):7. https://doi.org/10.1038/s41467-019-13787-x
21. He Z, Guo JL, McBride JD et al (2018) Amyloid-beta plaques enhance Alzheimer's brain tau-seeded pathologies by facilitating neuritic plaque tau aggregation. Nat Med 24(1):29–38. https://doi.org/10.1038/nm.4443
22. Guo JL, Lee VM (2014) Cell-to-cell transmission of pathogenic proteins in neurodegenerative diseases. Nat Med 20(2):130–138. https://doi.org/10.1038/nm.3457
23. Buee L, Bussiere T, Buee-Scherrer V et al (2000) Tau protein isoforms, phosphorylation and role in neurodegenerative disorders. Brain Res Rev 33(1):95–130. https://doi.org/10.1016/s0165-0173(00)00019-9
24. Maruyama M, Shimada H, Suhara T et al (2013) Imaging of tau pathology in a tauopathy mouse model and in Alzheimer patients compared to normal controls. Neuron 79(6):1094–1108. https://doi.org/10.1016/j.neuron.2013.07.037
25. Alamed J, Wilcock DM, Diamond DM et al (2006) Two-day radial-arm water maze learning and memory task; robust resolution of amyloid-related memory deficits in transgenic mice. Nat Protoc 1(4):1671–1679. https://doi.org/10.1038/nprot.2006.275
26. Planel E, Richter KE, Nolan CE et al (2007) Anesthesia leads to tau hyperphosphorylation through inhibition of phosphatase activity by hypothermia. J Neurosci 27(12):3090–3097. https://doi.org/10.1523/JNEUROSCI.4854-06.2007
27. Finneran DJ, Desjarlais T, Henry A et al (2024) Induction of tauopathy in a mouse model of amyloidosis using intravenous administration of adeno-associated virus vectors expressing human P301L tau. Alzheimers Dement (N Y) 10(2):e12470. https://doi.org/10.1002/trc2.12470

Subject Index

A

Acridine orange 258–259, 262, 272–275, 277
Aggregation..........3, 13, 19, 24, 36, 46, 49–50, 54, 70, 87, 93, 103, 110, 149, 155, 160–162, 165–166, 169–171, 173–174
Aging 194, 199, 208–210, 212
Alzheimer's disease 13, 23–24, 35–36, 46, 69, 72, 87–88, 97, 110, 159, 162, 205, 215–216, 281, 289
Amplified brain derived tau oligomers.....110, 114–115, 125–126, 129–130, 132–135, 141–144
Amyloid-β plaque cores (APC).......................69, 72, 79
Autoinduction 15–16, 20

B

Bacterial expression..13
Baculovirus..3, 5–6, 9
Biomolecular condensates.......................................155
Biosensors 160, 173
Brain dissection.......................................227–229, 247
β-sheet.................................45–46, 109, 177, 205–206

C

Caenorhabditis elegans 103, 194–199, 203–206, 208–210, 212
Cation-exchange chromatography..................13, 15, 17
Cell-based assay 161, 178, 183, 186
Central nervous system (CNS)..................23, 252–253, 274, 276, 285, 287, 294
Chemical screen....................................46–47, 81, 183
Chromatography 13, 15, 88–89, 92, 98–99
Compounds.....................................38–43, 46, 54, 60, 63, 71–72, 81, 178, 183, 186, 253, 272–273, 276, 282–283, 285–289
Coomassie blue..............................9, 57–58, 60–61, 65

D

Dementia................................3, 69, 72, 110, 159, 194
Drosophila...... 145, 215–216, 218–220, 224–227, 229, 232–233, 241–242, 248–249

F

Fibril formation 13, 45–46, 48, 50
Flow cytometry..............9, 93–94, 167, 170, 172–173, 178–179, 181, 183, 186
Fluorescence resonance energy transfer (FRET)....88–90, 93–96, 160, 165–166, 170, 172–174, 178
Fluorescent amyloid multi emission spectra microscopy...............................74, 77, 124, 141
Fourier transform infrared spectroscopy 36–37, 43, 124, 140

G

GABAergic neurons.........................196, 199, 204, 209
GAL4-UAS 218, 254
Glutamate-responsive neurons 196

H

HEK293T.......89, 93, 95, 97, 103, 160, 164, 166–167
High performance liquid chromatography (HPLC)...............113, 115–116, 124, 129, 140
Hyperphosphorylation................3, 24, 36, 70–71, 281

I

Immuno-fluorescence 120, 122, 136, 138
Immunogold 41, 46–47, 51
Interactome 88–89

L

Liquid-liquid phase separation (LLPS) 150, 152, 154–156
Luminescence assay..........................217, 226, 245–247

M

Mechanosensory neurons.........................199, 209–210
Microtubule-associated protein tau (MAPT).......11, 23, 35, 127, 193, 230, 232–234, 252, 281
Movement assay...257, 267

Jessica S. Fortin (ed.), *Translational Research Methods for Tauopathies*, Neuromethods, vol. 233, https://doi.org/10.1007/978-1-0716-5234-3,

N

Native PAGE 208, 216–217, 226, 242–245, 248
Neurodegeneration....... 24, 36, 70, 162, 178, 194, 196, 199, 204, 208, 210, 215, 254, 256, 258, 274, 281–282, 286–289, 291–293
Neurofibrillary tangles (NFTs) 3, 11, 24, 35, 46, 69–72, 75–76, 160, 193–194, 251, 281, 286
Neurons 35, 69–71, 79–80, 110, 141–142, 150, 193–194, 208–209, 212, 252–253, 255–256, 274–275, 281

P

Paired helical filaments (PHFs) 11, 46, 70–71, 80
Photo-Induced Cross-linking of Unmodified Proteins (PICUP) 13, 19, 36–37, 40, 57–58, 60, 62, 65
Progressive supranuclear palsy 3, 87–88, 97, 110, 159, 251–254, 281
Protein aggregates 28, 38, 72, 76, 79–81, 155
Protein aggregation 48–49, 69–70, 155
Protein expression................ 6, 8, 16, 27, 127, 178, 219
Protein purification 8, 13, 25, 29, 31, 128
Proteomics 81–82, 90
Proximity Ligation Assay (PLA) 215–217, 229, 232–235, 238, 241
Ps19 mouse model.. 97, 282

R

Recombinant protein 13, 26, 103, 111, 128, 172
Recombinant tau oligomer............. 110, 113, 130–131, 134–135, 139, 141–143
RNA.. 19, 69, 149–155, 196, 253, 284

S

Sf9 insect cells.. 4–5
Small molecule........................ 258–259, 261–263, 271
Split-green fluorescent protein (GFP) complementation 178–179, 183, 186, 188, 199
Split-luciferase reporter 217, 249

T

Tau................ 3–4, 8–9, 11, 13, 19, 21, 23–25, 27, 29, 35–36, 46, 54, 71, 88, 97–98, 101, 149–150, 152, 154–156, 159–160, 165–166, 193–194, 196, 204–206, 208, 252–255, 264, 266, 269, 271, 275
Tau 0N4R....................... 13–14, 19, 21, 23, 151, 179, 183, 187, 224
Tau 2N4R.......................... 4–6, 23, 25, 46–48, 58–60, 62–64, 111, 224
Tau^{LUM}................... 216–218, 224, 232, 240–241, 245, 247–248
Tau oligomer bioactivity 88, 125, 133, 145
Tau oligomers......................... 24, 58, 60, 63, 129–130, 133, 135–136, 139–140, 142–143, 177–178
Tau oligomer specific antibodies.............. 144, 204, 217
Tauopathy 4, 24, 87–88, 91–92, 94, 159–160, 173, 195, 205, 215–216, 248, 251, 253–254, 282–283, 285–289, 291–292, 294
Tau-seed 87–89, 93–94, 96–98, 103
Tau-seeding 39, 87–89, 94, 96–98, 160, 162, 165
Thioflavin S (ThS) 36, 45–48, 73, 77–78
Thrashing assay.. 198
Transmission electron microscopy (TEM) 13, 15, 19, 36–37, 40–41, 45–47, 49, 51–52, 54, 80–81, 124

W

Western blot 4, 58, 60, 64–65, 97–98, 165, 169, 198, 216–217, 224–226, 242–244, 247–248, 282, 284–285

Z

Zebrafish 252–255, 257–262, 264–269, 271–273, 275–277

MIX
Papier aus verantwortungsvollen Quellen
Paper from responsible sources
FSC® C105338

If you have any concerns about our products,
you can contact us on
ProductSafety@springernature.com

In case Publisher is established outside the EU,
the EU authorized representative is:
Springer Nature Customer Service Center GmbH
Europaplatz 3, 69115 Heidelberg, Germany

Printed by Libri Plureos GmbH
in Hamburg, Germany